Geomorphological Processes and Human Impacts in River Basins

Some Other IAHS Publications

The Use of Suspended Sediment and Associated Trace Elements in Water Quality Studies
by Arthur J. Horowitz
The US Geological Survey provided financial sponsorship for this publication
Special Publication 4 (1995) price £14.50, 58 + vi pages
ISBN 0-947571-79-5

Human Impact on Erosion and Sedimentation
Proceedings of a symposium held at Rabat, Morocco, April–May 1997
edited by D. E. Walling & J.-L. Probst
Publ. 245 (1997) price £42.00; 312 + viii pages
ISBN 1-901502-30-9

Modelling Soil Erosion, Sediment Transport and Closely Related Hydrological Processes.
Proceedings of a symposium held at Vienna, July 1998
edited by W. Summer, E. Klaghofer & W. Zhang
Publ. 249 (1998) price £57.00, 454 + x pages
ISBN 1-901502-50-3

The Hydrology–Geomorphology Interface: Rainfall, Floods, Sedimentation, Land Use.
Proceedings of a conference held at Jerusalem, May 1999
edited by Marwan A. Hassan, Olav Slaymaker & Simon M. Berkowicz
The Hebrew University of Jerusalem provided financial sponsorship for this publication
Publ. 261 (2000) price £49.00, 326 + x pages
ISBN 1-901502-16-3

The Role of Erosion and Sediment Transport in Nutrient and Contaminant Transfer. Proceedings of a symposium held at Waterloo, Ontario, July 2000
edited by M. Stone
Publ. 263 (2000) price £48.00, 308 + xii pages
ISBN 1-901502-26-0

The Structure, Function and Management Implications of Fluvial Sedimentary Systems.
Proceedings of a symposium held at Alice Springs, September 2002
edited by Fiona J. Dyer, Martin C. Thoms & Jon M. Olley
Publ. 276 (2002) price £72.00, 484 + xii pages
ISBN 1-901502-96-1

Hydrology in Mediterranean and Semiarid Regions. Proceedings of a conference at Montpellier, France, 2003
edited by Eric Servat, Wajdi Najem, Christian Leduc & Ahmed Shakeel
Publ. 278 (2003) price £76.25, 498 + xiv pages
ISBN 1-901-502-12-0

Erosion Prediction in Ungauged Basins: Integrating Methods and Techniques
Proceedings of a symposium held at Sapporo, Japan, July 2003
edited by Dirk de Boer, Wojciech Froehlich, Takahisa Mizuyama & Alain Pietroniro
Publ. 279 (2003) price £46.50, 250 + xii pages
ISBN 1-901502-22-8

Erosion and Sediment Transport Measurement in Rivers: Technological and Methodological Advances. Proceedings of a workshop held in Oslo, Norway, June 2002
edited by Jim Bogen, Tharan Fergus & Des Walling
Publ. 283 (2003) price £45.50, 238 + viii pages
ISBN 1-901502-42-2

Sediment Transfer Through The Fluvial System
Proceedings a symposium held in Moscow, August 2004
edited by Valentin Golosov, Vladimir Belyaev & Des E. Walling
Publ. 288 (2004) price £45.50, 498 + x pages
ISBN 1-901502-67-8

Sediment Budgets 1
Part 1 of the proceedings of Symposium S1 held during the Seventh IAHS Scientific Assembly at Foz do Iguaço, Brazil, April 2005
edited by Des E. Walling & Arthur J. Horowitz
Publ. 291 (2005) price £63.00, 372 + xii pages
ISBN 1-901502-87-2

Sediment Budgets 2
Part 2 of the proceedings of Symposium S1 held during the Seventh IAHS Scientific Assembly at Foz do Iguaço, Brazil, April 2005
edited by Arthur J. Horowitz & Des E. Walling
Publ. 292 (2005) price £63.00, 358 + xii pages
ISBN 1-901502-92-9

The Ecohydrology of South American Rivers and Wetlands. International Symposium held in Manaus, Brazil, November 1999
edited by Michael McClain
The UNESCO Regional Office for Science and Technology for Europe (ROSTE) sponsored this publication
Special Publ. 6 (2002) price £39.50, 210 + xiv pages
ISBN 1-901502-02-3

Hydrology: A Question of Balance
by J. V. Sutcliffe
IWMI, the International Water Management Institute, Colombo, Sri Lanka, provided financial sponsorship for this publication
Special Publ. 7 (2004) price £30.00, 200 + xviii pages
ISBN 1-901502-77-5

*More details at **www.iahs.info** including abstracts of all papers since 2000*

Please send orders and enquiries about books to:
Mrs Jill Gash, IAHS Press, Centre for Ecology and Hydrology, Wallingford, Oxfordshire OX10 8BB, UK
tel: +44 1491 692442; fax: +44 1491 692448/692424; jilly@iahs.demon.co.uk

IAHS members benefit from discounts on book prices: 25% for all members, and 80% discount for members in financially disadvantaged countries on titles published before the current year; IAHS membership is free of charge.
All prices include delivery by surface mail. Payment may be made by credit card (VISA, MasterCard, EUROCARD).

***Hydrological Sciences Journal** is also published by IAHS. Please address enquiries to:*
Mrs Frances Watkins, IAHS Press, Centre for Ecology and Hydrology, Wallingford, Oxfordshire OX10 8BB, UK

Geomorphological Processes and Human Impacts in River Basins

Edited by

RAMON J. BATALLA
University of Lleida and Forestry Institute of Catalonia, Spain

CELSO GARCIA
University of the Balearic Islands, Mallorca, Spain

IAHS Publication 299
in the IAHS Series of Proceedings and Reports

Published by the International Association of Hydrological Sciences 2005

IAHS Publication 299
ISBN 1-901502-28-7

British Library Cataloguing-in-Publication Data.
A catalogue record for this book is available from the British Library.

The papers included in this volume have been reviewed and some were extensively revised by the editors, in collaboration with the authors, prior to publication.

IAHS is indebted to the employers of the Editors for the invaluable support and services provided that enabled them to carry out their task effectively and efficiently.

Publications in the series of Proceedings and Reports are available from:
IAHS Press, Centre for Ecology and Hydrology, Wallingford, Oxfordshire OX10 8BB, UK
tel.: +44 1491 692442; fax: +44 1491 692448; e-mail: jilly@iahs.demon.co.uk

Printed in The Netherlands by Krips BV, Meppel.

Preface

This volume contains a selection of papers presented at the International Conference on River and Catchment Dynamics: Natural Processes and Human Impacts, convened in May 2004 at the Forestry Institute of Catalonia, in Solsona (Catalonia, Spain). The conference was organized as a tribute to Maria Sala and it was held on the occasion of her retirement. The conference explored issues concerning river and catchment processes, with special reference to Mediterranean environments, topics that Maria Sala particularly cultivated during her extensive and pioneering scientific career. The conference focused on the linkages amongst basin and river dynamics and human impacts as a basis for environmental management. Sediment sources and transfer to fluvial systems, fluvial dynamics and sediment transport, erosion processes and land degradation in drainage basins, and water and sediment management, were all considered with the aim of progressing towards a present-day diagnosis of, and future strategies for, environmental management of rivers and catchments all over the world.

The papers forming this volume reflect a variety of geomorphological processes at both the basin and the river levels, as well as their interactions and their relations with human activities that can interfere with them. At the basin scale, the contributions cover aspects from runoff and erosion on the slopes, including the role of forest fires, sediment supply from mass movements, basin sediment yield, water quality in agricultural areas and desertification. At the river scale, papers deal with sediment transport, both in suspension and as bed load, and the relationship between river process and human actions such as channelization, impoundment, gravel-mining, disturbance of riparian forest and agriculture. A wide range of methodologies are presented to assess the various problems addressed, including modelling-approaches, sediment transport and reservoir sedimentation techniques, and lichenometry. The contributions have been divided in to five sections aiming at representing the main processes that occur at both the basin and river scales as well as the human activities interconnecting with them.

Slope Processes and Impacts The effects of anthropogenic disturbance or climate change on runoff and erosion rates in source areas are analysed by Kirkby & Bracken. The study is based on a 1-dimensional simulation model which explicitly represents spatial differences in runoff generation and the accumulation of runoff downslope across the frequency distribution of rainstorms. The model allows for dynamic interaction with slope-base streams, and computes the stability with respect to the formation of ephemeral gullies on hillsides.

Antronico *et al.* evaluate soil erosion in two basins of southern Calabria, southern Italy, with an integrated approach, using a qualitative method, based on aerial photograph analyses, and a quantitative method, based on the RUSLE model. The use of the combined approach allows assessment of the relative role and the spatial–temporal distribution of the parameters influencing the erosive processes in the area.

Coscarelli *et al.* review the most-used methodologies for the characterization of areas sensitive to desertification. They present an application of the methodology proposed in the MEDALUS project to an area of the Calabrian territory (southern Italy), emphasizing the physical aspects controlling desertification process.

Martínez-Fernández *et al.* study the runoff yield with respect to soil moisture in periods of maximum water deficit, in a small Mediterranean basin in the centre of Spain. They observe how the character of climatic conditions in the growing season and water interception and consumption by the forest affect the generation of runoff, despite the abundant precipitation in the area.

Úbeda & Bernia report on the changes in soil aggregate stability after a wildfire, focusing on the effects of fire intensity on soil structure in the experimental area of the Cadiretes Massif, northeast Spain. The fire did not disturb the stability of the aggregates immediately after burning and, remarkably, they were even more stable than those sampled in the control forest.

Peart *et al.* studied sediment delivery from a landslide to a stream in a Hong Kong basin. Data on suspended sediment concentration and observations on sediment colour and C and N content showed that suspended matter in streamflow was derived from the landslides, demonstrating the connectivity of hillslope and channel systems. Much debris from the landslides remained on the slope in an ephemeral drainage line.

Following a modelling approach, Wilkinson *et al.* describe how a sediment budget, developed using sediment transport modelling (SedNet), sediment tracing, and water quality monitoring, can be used to identify the primary sources of suspended sediment in a large basin in Australia. Targeted erosion control would provide more than a 3-fold reduction in suspended sediment export compared with random actions.

Catchment Hydrology and Water Quality Arterial drainage schemes (ADSs) were implemented on Irish rivers for the purposes of improving the land drainage and reducing the frequency and extent of overland flooding. The ADSs involved the deepening and widening of the channels in order to increase their discharge-carrying capacity. Bhattarai & O'Connor analyse 16 ADS and six non-ADS basins showing that the ADS (post-1945)-basins have experienced major changes in the peak discharge–return period relation over time.

Igouzal *et al.* study the hydraulic regime and water quality of Sebou River in the Gharb agricultural zone (Morocco). The work was carried out in a 90-km study reach upstream of a reservoir dam using the dynamic water quality model WASP5 and during the base flow period. Their results provide reasonable estimates of water quality variables in a short time period, useful for many environmental and health and safety considerations.

Hydrological characteristics and water quality were studied in a small agricultural basin in northwest Switzerland by Weisshaidinger *et al.* Substantial molybdate reactive phosphorus (MRP) input into the stream via subsurface drains was observed, while runoff exceeded eutrophication limits that are already at low recurrence MRP concentrations.

Suspended Sediment Transport and Yield The performance of a feed-forward back-propagation artificial neural network (ANN) and a multiple quadratic regression (MQR) model are tested by Schnnabel & Maneta using data from a wooded rangeland basin in southwest Spain. ANN and MQR performed similarly but better than rating curves of a single variable. In addition, both can reproduce the hysteretic sediment–discharge loops.

Ribes & Provansal describe how since 1994, the lower Durance River in France has experienced a notable increase in the hydro-sedimentary inputs, which results in a series of morphological impacts in the bankfull channel. The results highlight the difficulties of restoring sediment transit in a highly disturbed and vegetated river channel.

Farguell & Sala examine the temporal and spatial relations between suspended sediment concentrations and discharge in the Mediterranean basin of the Anoia River, northeast Spain. The scatter is mainly a consequence of the seasonality of events, with higher loads in autumn and spring, and also of the hysteretic loops in relation to either single and multiple peak events.

The first reported suspended sediment data for the Balearic Islands, Spain, are presented by Estrany & Garcia who describe the suspended sediment fluxes in the Na Borges basin, a lowland agricultural area in Mallorca. Results show different behaviour in the availability and depletion of the sediment according to the magnitude and extension of each event, and according to the mixing and routing of water from different sources.

Araújo & Knight present sediment yield assessments for one urban and six rural watersheds, located in the semiarid region of northeast Brazil, with a wide range of characteristics. Results show that reservoir storage capacity is being reduced at a 2%-per-decade rate. The Hidrosed model was applied and calibrated with field data. Hidrosed showed good agreement with field data.

Suspended sediment dynamics and trends in two Mediterranean mountainous basins in the southern Pyrenees, Spain, are analysed by Balasch *et al.* Following a nested basins approach, results show clockwise and anticlockwise hysteretic patterns and exhaustion effects that are explained in terms of temporal runoff generation, sediment sources distribution and stream channel sinks, demonstrating the different seasonal patterns of the processes involved.

Bed Load Transport and River-Channel Processes Bed load under low transport conditions in armoured gravel-bed channels is examined by Church & Hassan. Authors show that bed material transport remains at very low rates most of the time. Under such circumstances classical formulae overpredict transport since they are developed for fully mobile conditions. Consequently, the transport cannot be calculated merely by an adjustment of bed material parameterization to reflect the surface armour, though that may work for specific histories of bed development.

Teruggi *et al.* quantify the sediment transport in several geomorphological units of the Rio Quequén Grande basin, Argentina. Bed load discharge and suspended sediment load measured during bankfull floods are very low, given the relatively high stream power. The measured bed load discharge was up to five orders of magnitude smaller than that predicted by bed load equations, presumably due to low sediment availability.

Gob *et al.* report the application of lichenometry to the study of mountainous Mediterranean boulder-bed rivers, characterized by strong floods. They propose several diameter/critical specific streampower relations by associating the blocks' diameter, dated using lichenometry, to the specific streampower. The study highlights the uneven dissipation of energy due to the roughness of bed forms in the studied rivers.

Vericat & Batalla describe the bed load transport observed under low sediment transport conditions and constant discharge in the gravel-bedded lower Ebro River, northeast Spain. The low bed load transport rates are related to the development of an armour layer during small winter floods, which reduces sediment availability and, consequently, controls the subsequent bed-material dynamics (i.e. bed load and grain size).

Rovira presents a particle-size sediment budget for the Lower Tordera, also in northeast Spain, by measuring the bed load transport in two channel sections. Coarser fractions do not reach the outlet basin. This is due to a reduction of river competence and to the effect of the gravel extractions. The channel tends to accumulate sediment during dry periods when a hydrological bimodality (perennial to ephemeral) occurs.

River Dynamics and Impacts The impact of groynes along the Upper Odra channel, Poland, is analysed by Czajka. Constructions create a new environment for sediment deposition and the inter-groyne infill constitutes a significant sink for material transported at a mean annual rate of 5 cm during the last 40 years, deposition being especially active during floods.

Lu *et al.* describe the sediment discharge in the dry, hot valley of the Longchuanjiang basin, China. Sediment loads show a significant increase of up to 3.5% per year, despite numerous reservoir constructions and insignificant changes in water discharge. The phenomenon is attributed to the intensification of human activities, such as deforestation and afforestation, expansion of agriculture, and road construction.

Klein examines the flow regime, water quality and land use in the flood plain of the Lower Jordan River, Israel, for planning purposes. The average annual flow has severely decreased since early 20th century, although the remaining river flow is of good quality. In spite of changes, the environmental conditions allow further planning and development of the area, especially for agricultural purposes.

The effect of dams on the hydrological regime and the geomorphology of the Daliyyot River in Israel is reported by Shtober-Zisu & Inbar; the river studied has an impoundment ratio (IR) of 0.37. The average peak discharge and the annual runoff decreased by half and the recurrence interval increased for the highest peak flows. Small geomorphic changes are observed after major floods, including sedimentation in small pools downstream of dams.

Bailey *et al.* develop a theoretical riparian disturbance sensitivity index, which is tested using 15 before-and-after case studies of gravel-bed rivers, the majority of them subjected to direct riparian disturbance, in British Columbia, Canada. Results indicate that the riparian disturbance sensitivity index provides a good measure of expected channel sensitivity and change in width following riparian disturbance.

Wittenberg & Greenbaum examine river adjustment to an artificial disturbance in an ephemeral mountainous stream on Mount Carmel, Israel. A sedimentological and morphological comparison between two sections downstream of a pond was used to assess channel adjustment. The actual effect of the pond is limited to the immediate section downstream, and the intensity, duration and the long recession of the flows are sufficient to maintain the channel's stability.

Acknowledgements Several acknowledgements are necessary to the organizations and institutions that sponsored and supported the organization of the Conference. They are:
Forestry Institute of Catalonia (CTFC); Commission on Land Degradation (COMLAND); International Geographical Union (IGU); International Commission on Continental Erosion (ICCE); International Association of Hydrological Sciences (IAHS); University of Lleida (UdL); University of the Balearic Islands (UIB); University of Barcelona (UB); Spanish Ministry of Education and Science (MCYT); Department of Universities, Catalan Government (DURSI); European Society for Soil Conservation (ESSC).

We are also grateful to the members of the international scientific committee for their assistance and advice during the preparation of the conference, and to the reviewers that undertook thorough revisions of all the papers submitted for publication in this volume.

The Conference would not have been so fruitful without the endless and generous work undertaken by the members of the conference team at the Forestry Institute of Catalonia, the University of Lleida, the University of the Balearic Islands, and the University of Barcelona. Moshe Inbar was always there too, providing support, encouragement and wise personal and scientific advice. We are indebted to all of them.

Ramon J. Batalla
University of Lleida and Forestry Institute of Catalonia, Spain

Celso Garcia
University of the Balearic Islands, Mallorca, Spain

Contents

1 Slope Processes and Impacts

Modelling hillslope connectivity and channel interactions in semiarid areas: implications for hillslope restoration following land abandonment

MIKE KIRKBY[1] & LOUISE BRACKEN[2]

1 *School of Geography, University of Leeds, Leeds, UK*
m.j.kirkby@leeds.ac.uk

2 *Department of Geography, University of Durham, Durham, UK*

Abstract In semiarid areas, little locally generated overland flow and even less locally eroded sediment reaches channels during floods, due to poor connectivity in both natural and disturbed landscapes. Where anthropogenic disturbance or climate change alter runoff and erosion rates in source areas, there are short and long term adjustments to both hillslope and channel morphology. Abandonment of arable cultivation is used here as an example of such a change, explored though a one-dimensional simulation model which explicitly represents spatial differences in runoff generation and the accumulation of runoff downslope across the frequency distribution of rainstorms. Discharge and topography are used to forecast sediment transport and the consequent evolution of hillslope form. The model allows for dynamic interaction with slope-base streams, and computes the stability with respect to the formation of ephemeral gullies on the hillside. This model is run until the slope form is in balance with the processes, and a step-change is then applied to represent the short term impact of land abandonment. An initial sharp increase in sediment loss is followed by a slow recovery as the hillslope begins to evolve towards a new stable form. The effect of possible remedial measures is explored showing that the most effective treatments are those which increase infiltration, and that direct reductions in sediment transport generally lead to erosion downslope.

Key words connectivity; erosion mitigation; modelling; semiarid; soil erosion

INTRODUCTION

Recent developments in understanding semiarid hydrology have focused increasingly on the importance of differences in local runoff generation, and on the connectivity of runoff once generated. Runoff generation has been discussed in the context of Hydrological Response Units (HRUs; Flügel, 1995; Bull *et al.*, 2003) and Hydrologically Similar Surfaces (HYSS; Bull *et al.*, 2000; Kirkby *et al.*, 2002a), identifying areas of similar response, usually identified with combinations of land-use, soil type and topography. Natural connectivity follows a network of irregularities on the hillside (Prosser & Abernethy, 1996), but also depends on the duration of intense bursts of rainfall which generate runoff (Bracken & Croke, 2005). If these bursts are much briefer than the time for overland flow to traverse the length of the hillside, then points of runoff will be connected with other points both up- and downslope, but through connections will develop only in exceptionally intense and long-lasting storms.

The three-dimensional (3-D) form of the landscape at every point is determined by the mass balance between sediment transport rates and erosion or deposition (Dietrich & Dunne, 1993). Every change will lead to immediate and longer term responses in the landscape, both within a single hillslope and in coupling hillslopes and channels at their heads and along their banks. The main natural processes acting on semiarid hillslopes are rainsplash, water erosion (wash) and, locally, mass movements; with the addition of strong impacts due to tillage erosion in agricultural areas (Bryan, 2000). Channel processes are dominated by episodic fluvial transport and, in some steeplands, by debris flows.

These hillslope and channel processes are, in turn, driven by water flow and gradient, often estimated as power laws in gradient (exponent of 1–2) and distance or discharge (exponent 2–3; e.g. Musgrave, 1947; Yalin, 1977). The rates and balances between hillslope processes are significantly affected by small changes in surface characteristics, which immediately modify infiltration, erodibility and vegetation cover; and in the longer term affect surface grain sizes and regolith depth. Low connectivity means that discharge increases less than linearly down the length of a hillside, and approaches a stable maximum related to the intensity and duration of each storm (Kirkby *et al.*, 2005). Sediment transport therefore responds by showing a much weaker increase with slope length than for a humid landscape. Sediment transport in channels similarly increases only modestly with distance downstream or catchment areas except in the largest storms.

As channel networks extend or atrophy, there is also an important dynamic link between the hillslope and channel phases of the catchment. Periods of higher sediment transport by wash processes lead to extensions of the channel network (Smith & Bretherton, 1972; Dunne & Aubry, 1986), with consequent improvements in connectivity as overland flow then has a shorter travel path to a flow-line with low transmission losses, and this provides a positive feedback to further erosion. Similarly, periods of reduced wash erosion encourage vegetation growth, with a tendency towards accumulation of organic material and infilling of channels.

EFFECTS OF CHANGES IN DRIVERS OF EROSION

In many semiarid landscapes, climate and land-use are changing continually, making it difficult to assess the field relationships between changing drivers and changed landscapes. This difficulty is compounded by the long times required for each change to work through the landscape. In this paper we apply a simple 1-D deterministic model to simulate and assess the impact of sudden step-change on a system which has evolved under uniform conditions for a long time, as an initial stage in understanding the impact of more complex changes.

An initial sharp change gradually gives way to the development of a new stability, with rates of erosion in balance with the changed conditions, which generally revert towards conditions before the step-change, though this can take a very long time to evolve. Thus we need to distinguish the immediate effects of disturbing the pre-existing stability from the long term effects of the changed drivers.

These hillslope effects will also influence the physical relationship between hillslope and channel where there is dynamic coupling between them. Where, for example, hillslopes deliver increased coarse sediment, channels respond, commonly by aggradation or reduced incision (Harvey, 2001). In other examples, partial ponding of runoff from arable land can deposit sediment, so that overflowing clear water runoff is able to erode aggressively downstream, particularly where gradients are steep.

We also use the simulation model to investigate the impact of mitigation measures applied locally, to reduce the cost of remedial measures by applying them only to the parts of the hillside most at risk. Local mitigation may consist, for example, of treatments to protect the soil, through the use of geo-textiles, soil conditioners or enhanced re-vegetation, and these treatments can be simulated through their impact on local soil properties, such as infiltration or erodibility. These measures will, if successful, reduce local runoff generation and sediment transport within the treated area, and to a lesser extent in areas downslope. Over time, the accumulated effects will impact on a larger area, both up- and downslope of the initial treatment. Thus we can assess the extent to which remedial measures could be confined to a fraction of the slope, with consequent economic benefits.

The particular example described here is to examine the effects of land degradation following abandonment of agriculture, and its possible mitigation. This example is seen as one of the common scenarios for desertification in southern Europe, as prime land is cultivated more intensively, and marginal land is abandoned as no longer economic. Management by ploughing is one of the major controls of surface properties which removes vegetation, destroys crusting, and moves rock fragments to the surface (Wijdenes *et al.*, 1997), and so abandonment of previously farmed land has a complex relationship with runoff when ploughing stops (Lasanta *et al.*, 2000). Immediately after abandonment, erosion is at its most severe due to the loss of tillage roughness and vegetation cover, so that soil crusting develops strongly, leading to increased runoff and erosion (Bull *et al*, 2000; Lasanta *et al*, 2000). At this stage, the hillside commonly develops a network of rills which cut into the surface and improve connectivity for runoff and sediment. Over time, vegetation re-establishes itself, improving the soil organic matter and protecting the surface so that runoff and erosion decrease once more, and the channels may become ineffective as they slowly infill with debris. These changes are summarized in Fig. 1.

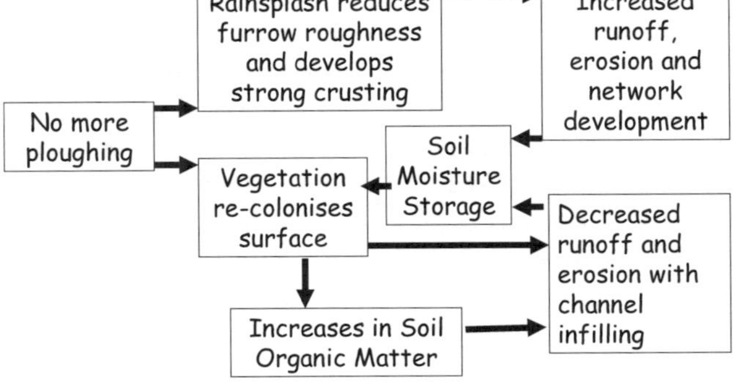

Fig. 1 Processes enhancing erosion and recovery of abandoned agricultural land.

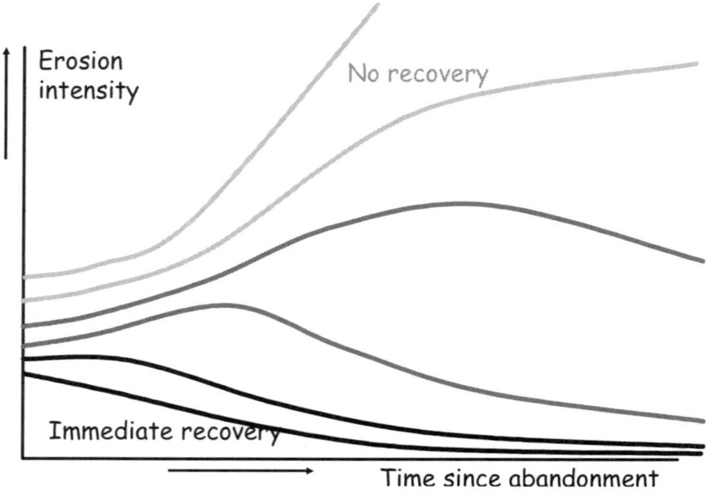

Fig. 2 Theoretical trajectories towards recovery or irreversible degradation.

The relative rates of these two processes, of erosion and re-vegetation, depends on local conditions which can favour one or the other, so that the trajectory of erosion may lie anywhere between the extremes of catastrophic gullying and prompt recovery of semi-natural vegetation as indicated schematically in Fig. 2. Erosion usually begins at once, although it may increase in rate as topsoil is removed, whereas re-vegetation usually takes at least a season to begin, is often partly offset by grazing (with or without fire management) and usually requires many years to reach a new equilibrium (Lasanta *et al.*, 2000; Bull *et al.*, 2003).

THE MODEL

The model used here has been adapted from a simple 1-D hillslope model, but with a number of additional features to adapt it to the purposes of this paper, using methods developed in the MEDALUS and MEDRUSH models (Kirkby *et al.*, 2002b). The basic model applies a mass balance approach to convert sediment transport rates to rates of hillslope erosion or aggradation. The deliberate focus is on long-term mean conditions, to distinguish the signal of land-use change and mitigation treatments from the noise of random variations in surface properties and the random incidence of intense storms. However, it is recognized that this does not answer the question of whether the effects described can, in practice, be observed in relation to the random scatter.

In Fig. 3, the basic slope model is represented by the bottom row of boxes. Sediment transport rates are specified in relation to topography (gradient and distance from divide). Transport rates are converted to local rates of erosion or deposition through the continuity equation, here for 1-D and with no external inputs of (e.g. wind blown) sediment:

$$\frac{\partial z}{\partial t}+\frac{\partial S}{\partial x}=U \tag{1}$$

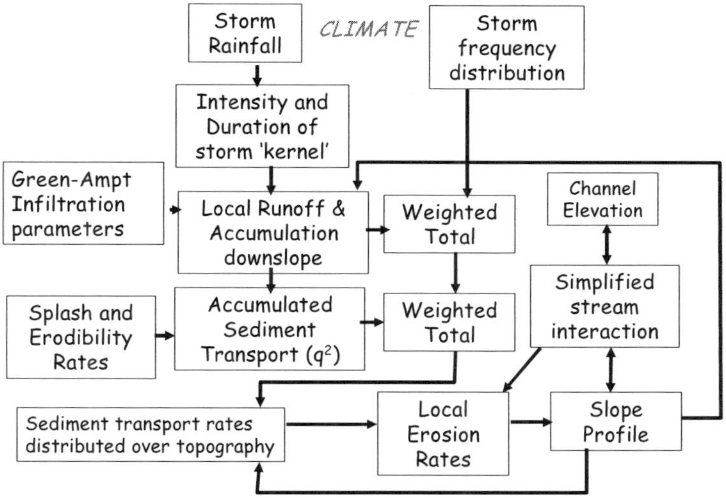

Fig. 3 Interactions incorporated in hillslope runoff and sediment model.

where z is elevation; S is sediment transport (in the downslope direction); U is the rate of tectonic uplift; x is distance from the divide; and t is elapsed time.

The basic model has been modified in two main ways. The first is to include explicit soil parameters which can be linked to surface condition and localized remedial treatments; and the second is to make the climatic response explicit through the frequency distribution of storms and the accumulation of local runoff contributions downslope. Together these changes allow a much more complete description of the hillslope overland flow runoff and its variations downslope and with storm frequency, providing an improved physical basis for estimating sediment transport. However, it has not been possible to include all relevant feedbacks, and it is recognized that the model presented here contains no seasonal variations in climate, and contains no explicit representation of grain sizes. The implications of these shortcomings are briefly discussed in the conclusions

Hydrology informs the model structure, and, for the semiarid conditions considered, is based on the assumption that storm events are random, drawn from a gamma distribution based on daily rainfall data (Kirkby *et al.*, 2000). Storms are assumed to be independent in the sense that the soil is considered to be dry at the beginning of each storm. A kernel is estimated for each storm size, associated with a kernel duration and kernel intensity. These kernel properties are used to estimate the infiltrated depth and the duration of overland flow associated with the storm. As storm size increases, the kernel intensity approaches an upper limit (150 mm h^{-1}), and the kernel total (the product of kernel intensity and kernel duration) is an increasingly large fraction of the storm total rainfall for storms of more than 100 mm.

The kernel intensity and duration are used, within a Green-Ampt infiltration equation, to estimate total local infiltration during the storm, and hence, by subtraction, local runoff generation. The parameters of the Green-Ampt equation are allowed to differ between treated and untreated sections of the slope, creating corresponding differences in local runoff generation, which are then accumulated downslope to estimate runoff discharge. The local runoff j can be calculated from the relationships:

$$j = pr_* \left[1 - \sqrt{\frac{2h}{r_*} - \left(\frac{h}{r_*} \right)^2} \right] \tag{2}$$

where the storage to ponding, $h = \dfrac{Bi_*}{(i_* - A)^2}$;
the final proportion of runoff, $p = 1/A/i_*$;
A and B are the parameters of the Green-Ampt equation $f = A + B/S$;
and r_*, i_* are the storm kernel effective total rainfall and intensity.

The kernel duration ($t_* = r_*/i_*$) is also used to estimate the overland flow travel distance $x_* = ut_*$, assuming a constant overland flow velocity (u = 2 cm s^{-1}), and the storm overland flow discharge volume q per unit width is accumulated as $\dfrac{dq}{dx} = j - \dfrac{q}{x_*}$. Under conditions of uniform flow generation, j, this leads to exponential saturation of discharge at $q = jx_*$, but for larger storms the distance to reach saturation increases (with x_*) so that the weighted average discharge increases steadily, although more and more slowly, downslope.

Sediment transport is then taken to combine a diffusive rainsplash term, proportional to gradient, and a wash term, proportional to discharge squared times gradient, scaled by an erodibility which reflects soil strength. Although the 1-D model cannot show any rill morphology, the Smith & Bretherton (1972) stability criterion can be calculated to show whether wash erosion processes are able to enlarge small depressions to initiate rilling. The criterion states that small depressions will grow unstably in this sense if and only if the ratio of $\partial S/\partial a : S/a > 1$, where S is the local sediment transport, expressed in terms of a and gradient a is the area drained per unit contour width and the partial differentiation is performed keeping gradient constant. When the stability ratio is less than 1.0 then small depressions tend to be filled in by diffusive processes, while sites with values substantially greater than 1.0 generally develop into ephemeral and permanent gullies.

With this formulation, changes in conditions can be explicitly incorporated into the model, both changes in rainfall amount and distribution, or surface properties via the infiltration parameters or through erodibility. The effect of treatments can also be simulated by allowing surface properties to vary along the length of the slope. The sediment transport calculated in this way can be incorporated directly into the simple slope model, and is now responding to the driving conditions. It is also possible to disaggregate the storm distribution to see the relative erosional and runoff impacts from storms of different sizes.

The last component of the model is the interaction with the basal stream. This has been dynamic, while retaining the 1-D structure, by assuming that the stream is able to transport an increment of sediment from the hillslope, and allowing this increment to increase (linearly at present) with basal elevation. This represents, in a simplified form, the relationship between stream elevation and stream gradient. This linkage provides a moderately realistic response, in that high rates of slope erosion can lead to channel aggradation, and slope stabilization to incision.

Vegetation has a complex relationship with variables included in the model and has a major influence on runoff and erosion at all scales. Vegetation has a positive effect on infiltration by increasing organic matter, lowering bulk density (Wilcox *et al.*, 1998) and increasing hydraulic conductivity (Nicolau *et al.*, 1996). At the plot scale, research has shown that infiltration rates beneath bushes can increase infiltration up to three times that of the inter-shrub area (Lyford & Qashu, 1969). Vegetation also reduces rainsplash and hence crust formation, but also increases surface roughness (hence hydraulic resistance) and can increase ponding. Many investigations have noted patterning of vegetation at the hillslope scale (Sanchez & Puigdefabregas, 1994) which can greatly increase infiltration compared to unvegetated areas. Encouraging the growth of vegetation is therefore one useful measure to encourage hillslope stability. However, vegetation is not explicitly included in this model because we are primarily looking at two conditions: first the gradual evolution of the hillslope under constant stable conditions (i.e. including uniform vegetation cover), and second the immediate response of the system to a step-change in conditions, with or without local treatment. Within each stage the properties of the vegetation cover (or lack of it) are represented through the infiltration and erodibility parameters, thus retaining a simple model structure. Where evolution over a longer period following a step-change is shown in the figures, it should be borne in mind that the model retains the parameters for conditions immediately following the step-change, and that there will be an additional dynamic interaction with vegetation growth that is not shown (and would require additional parameters). In a similar spirit, no allowance has been made for evolution of the surface grain size, again focusing attention on conditions immediately following the step-change. Most field and simulation evidence suggests that the processes of armouring (in stony soils) and vegetation regrowth will produce full or partial recovery as sketched in Fig. 2 above.

SIMULATION RESULTS

To illustrate the impact of external change on erosion rates, and to investigate the possibilities of mitigation, the simulation has proceeded in three stages. First an initially uniform slope has been allowed to evolve under uniform conditions to attain a stable state. Second the final form of this first stage has been subjected to a step-change of conditions, to observe the immediate and longer term changes in erosion and hillslope form. Thirdly the same sequence has been run with mitigation measures applied over a part of the slope length, to see how far such measures might be effective. In the first stage the profile changes from a uniform to a largely convex profile with little soil erosion, with no tendency to rilling on the average, as shown by the stability criterion, and adjusts to a balance with channel conditions. Figure 4 shows an increase and then a more gradual reduction in average rates of erosion accompanying this evolution, and these changes are accompanied by a more uniform distribution of erosion down the length of the hillside.

The step-change applied in the second stage of the simulation is to increase erodibility by a factor of 2.5× and to reduce the *B* parameter in the Green and Ampt infiltration equation from 50 to 10 mm^2 day^{-1} (over the whole slope except where there are treated areas). Further evolution with these conditions leads to the eventual development of a basal concavity. Figure 5 shows that there are immediate large increases in the rates of erosion and an eventual slow decrease to more uniform rates very similar to those before the step-change occurred, as the form of the slope profile as a whole adjusts to the changed conditions.

These changes are associated with an initial shift in the Smith & Bretherton (1972) stability criterion, towards an average tendency to rilling except near the top of the slope. This is associated with large storms (Fig. 6). Prior to the step-change, the entire profile was stable on average, although with the possibility of rilling in large events. Immediately after the step-change, instability occurs beyond 10 m for frequency-weighted average conditions. The stability criterion shows that there is the potential for rilling almost to the divides during large storms, but this is balanced by a tendency for these rills to fill in during smaller events, due to rainsplash.

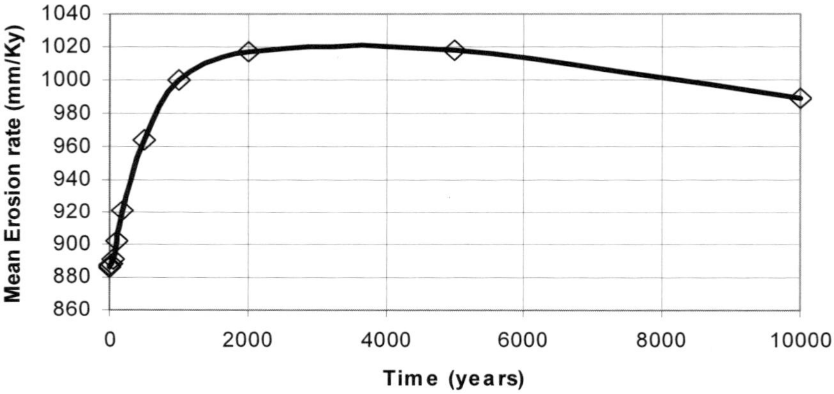

Fig. 4 Reduction in average erosion rate over time as profile stabilizes.

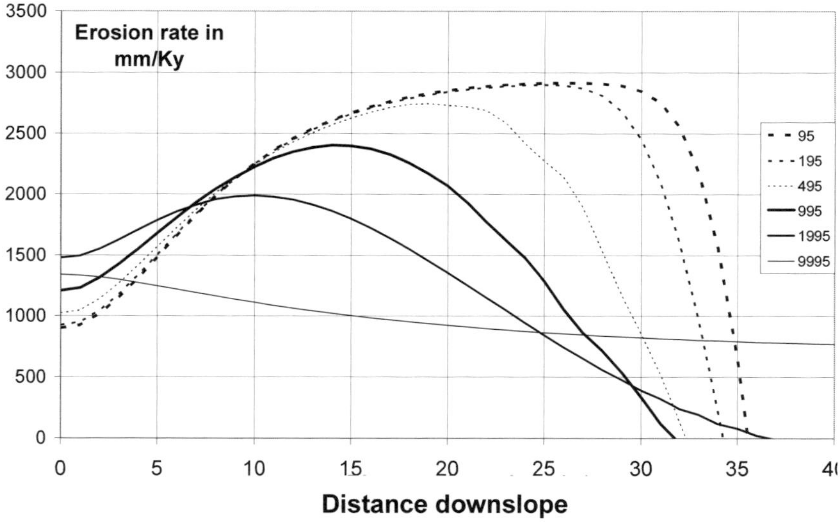

Fig. 5 Variation in erosion rate downslope after a step change towards higher erodibility and reduced infiltration. Key indicates evolution times in years.

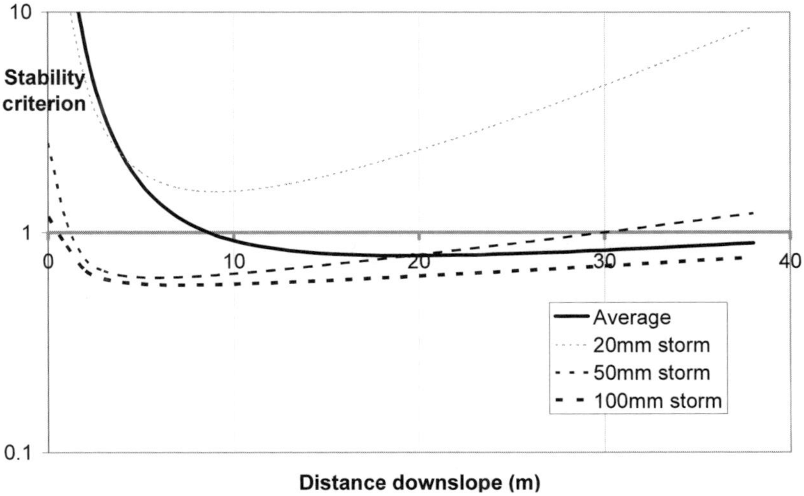

Fig. 6 Stability criterion shortly after step-change in conditions. Note that average conditions are then unstable at >10m from divide.

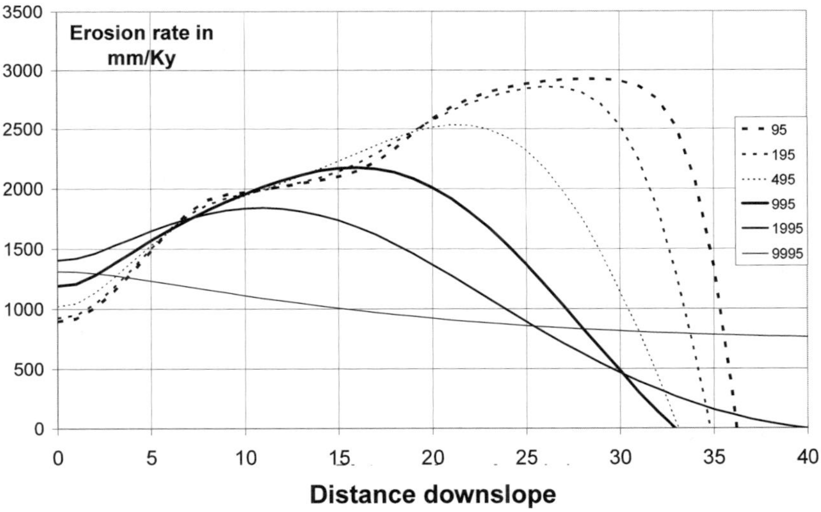

Fig. 7 Pattern of local erosion after step-change to more erosive conditions, as in Fig. 6, but with mitigation measures between 10 and 20 m. Key indicates evolution times in years.

The third stage is to apply remedial measures to a part of the slope at the same time as, or shortly after, the step-change in external conditions. These may consist of measures to promote infiltration through the application of geotextiles, soil conditioner or natural mulches, and they may also reduce erodibility and encourage plant establishment. Other measures promote re-vegetation directly through seeding and/or irrigation. Here we represent such treatments through the same descriptor for surface texture, namely erodibility and infiltration parameters.

In the example shown in Fig. 7, the same step-change has been applied as in Fig. 5, but mitigation has been applied between 10 and 20 m downslope, at the top of the area of potential instability and rilling, as indicated in Fig. 6. In comparing Figs 5 and 7, it may be seen that erosion, immediately after the step-change, is considerably reduced in the neighbourhood of the mitigation measures, but that the maximum of erosion at 30 m is little reduced. Over time, the impact of the mitigation is gradually reduced, even assuming, as here, that the treatment retains its full effectiveness in enhancing infiltration and reducing erodibility.

The effect of the mitigation may also be seen in the stability criterion, which shows a step increase within the treated section immediately after the change in conditions and application of treatment, but little impact outside the treated area. Indeed there is a slight reduction in stability immediately below the treatment, where erosion and runoff rates, although still low, show a particularly rapid local increase. However, application in this zone, at the top of the area of potential rilling, does appear to push the rill heads downslope, and this is seen as an important qualitative benefit by reducing the risk of gullying.

DISCUSSION

It is clear that there are advantages in applying some form of treatment to all or part of the hillside. While it is more effective to treat a larger fraction of the slope, it is also more costly, and economical to treat only the most critical areas. To apply treatments in practice, it is therefore necessary to consider: (a) how to select the most appropriate techniques; (b) the best areas to treat; and (c) how to assess the efficacy of treatments after application to refine the methodology.

Comparing simulated treatments that influence erodibility directly with those that increase infiltration rates (through the A and B parameters of the Green-Ampt equation), it is clear that those that reduce erodibility have a stronger impact on local erosion levels, but that the reduced erosion is offset by a zone of sharply increasing erosion immediately downslope of the treated area. The treated region tends to become steeper over time, and instability is increased immediately below the treated area to a level higher than without treatment. Thus the greater retention of sediment within the treated area produces a "clear water" effect which allows the flow to erode aggressively when it leaves the treated area, and the effect of the treatment is to move the channel incision downslope and, if anything, to make it more severe. This is similar to increased erosion noted downslope/downstream of control measures used to manage runoff and channel flow (Goldman *et al.*, 1986; Fifield, 2002).

Treatments which increase infiltration rates without any decrease in erodibility have a much smaller impact on local erosion. Figure 7 shows an example in which the steady rate of infiltration in the Green-Ampt equation (A in equation (2) above) has been doubled. In comparison with Fig. 5, there is some reduction in erosion and increased stability to accompany the lesser local runoff generation in the treated area between 10 and 20 m, but this is much less marked than where erodibility is reduced directly.

These results indicate some preferences in the type of physical treatment which is likely to be effective. It appears that silt fences and other methods which directly trap sediment, and methods like impermeable geotextiles which prevent erosion while enhancing runoff are likely to do more harm than good, except perhaps where an existing channel needs to be eliminated. Methods which seem to be closer to the ideal of promoting infiltration without reducing erodibility are the spreading of fine netting close above the surface, a surface dressing of stones resting on the soil surface and/or the incorporation of soil conditioners that improve near-surface soil structure. Clearly these provisional conclusions need to be confirmed in field experiments. Increasing infiltration should also increase soil moisture and so enhance vegetation regrowth in the treated areas, providing a dynamic positive feedback to further protect the soil.

The question of where to place protective treatment has focused above on the prevention of channel erosion, based on the long-term average. Thus the upper margin of treatment should be placed at the top of the area prone to rill incision, and extended as far downslope as economically practicable. Although the model presented here can be used to give guidance on this issue, the particular formulation of the model is not sufficiently proven to rely fully on the simulation results. It seems preferable, therefore, to make use of previous empirical results (e.g. Montgomery & Dietrich, 1988; Dietrich & Dunne, 1993; Vanderkerckhove *et al.*, 1998; Kirkby *et al.*, 2003) which have attempted to measure the critical area at the heads of rills or ephemeral gullies. These results have generally expressed the locus of critical points along an inverse relationship between catchment length, L, or area, A, and gradient, Λ, in the neighbourhood of the stream head. Stream heads are seen to form where the product $L\Lambda$ or $A^{0.5}\Lambda$ exceeds a critical value which depends on soils and climate. The advantage of this approach is that it can be applied from measurements of form in each area to be treated without the need for disproportionate research effort. Where a length–slope product is used, some qualitative allowance should be made for the 3-D form of the slope, to allow for the increased flow convergence within hollows.

The third issue raised above is of evaluating the effectiveness of treatment, and here no simple solution seems to be available. Figure 6 shows that the stability is very strongly dependent on storm size. Large events can therefore, even when the hillslope has been stabilized for the weighted average of long term conditions, erode ephemeral gullies above the treated areas, although the treatment should still show some benefit in diffusing the impact of this erosion. Furthermore the actual sequence of storm and low intensity rainfall events after treatment can either enhance or destroy the benefits of treatment. Most treated areas should benefit from multiplier effects following treatment, with local improvements in vegetation cover and soil organic matter, but these take time to develop. In terms of treatment, its impact should, where possible, be enhanced by local irrigation at the start of the rainy season to encourage vegetation regrowth. Monitoring the effectiveness of treatment should include measurements of rainfall as well as hillslope runoff and sediment movement. Comparisons can also be made with neighbouring untreated areas after a storm, to observe the differences in the critical area–slope or length–slope product at which ephemeral gullies are formed.

Throughout the conception of this study and this paper it has been assumed that the growth of vegetation benefits hillslope stability and that any treatment should encourage growth due to the positive effects outlined above. However, details about the type of vegetation, the density of plants, the patchiness of natural vegetation and the time taken for re-vegetation to occur are sparse in semiarid areas. For any treatment to be worthwhile, the benefits need to be greater and more immediate than those attributed to naturally occurring re-vegetation. Because of the random incidence of large storms, the main benefit of effective treatment is thought to be expressed through raising the threshold for the initiation of ephemeral gullies in the first storms after treatment, and so accelerating the process of natural recovery. This is seen as an important area for future research.

CONCLUSIONS

The model presented here has a number of novel features which improve our understanding of the interactions between hydrology and hillslope erosion. The present model however, also has a number of shortcomings which merit further development. These include the lack of feedback mechanisms to changed vegetation cover and surface grain size, the restriction to a 1-D slope profile and the lack of stochastic elements to assess model uncertainty. The processes included are qualitatively correct, but the mathematical formulation, although chosen to behave properly for extreme values, has not been validated against field data, so that detailed quantitative inferences should not be drawn.

Nevertheless, there are a number of conclusions which seem to transcend these limitations, and for which the general structure of this hydrologically responsive model is thought to be adequate. First, that the goal of remedial treatment on hillslopes should be to stabilize them against channel erosion which, once well developed, provides much better connectivity to remove water and sediment from the hillside, and so increases the risk that a hillside will degrade irreversibly following abandonment of agriculture. Second, that treatment should focus on methods which primarily enhance infiltration and vegetation recovery, and should only aim only to trap sediment or reduce erodibility to treat existing critical gullies.

Acknowledgements Authors thank Robert Millar whose comments greatly improved the first version of the manuscript.

REFERENCES

Bracken, L. J. & Croke, J. (2005) The concept of connectivity in geomorphology: definition and evaluation. *Hydrological Processes* (submitted).

Bryan, R. B. (2000) Soil erodibility and processes of water erosion on hillslope. *Geomorphology* **32**, 385–415.

Bull, L. J., Kirkby, M. J., Shannon, J. & Hooke, J. M. (2000) The variation in estimated discharge in relation to the location of storm cells in SE Spain. *Catena* **38**, 191–209.

Bull, L. J., Kirkby, M. J., Shannon, J. & Dunsford, H. (2003) Predicting Hydrological Similar Surfaces (HYSS) in semiarid environments. *Advances in Monitoring and Modelling* **1**(2), 1–26.

Dietrich, W. E. & Dunne, T. (1993) The channel head. In: *Channel Network Hydrology* (ed. by K. J. Beven & M. J. Kirkby), 175–220. John Wiley, Chichester, UK.

Dunne, T. & Aubry, B. F. (1986) Evaluation of Horton's theory of sheetwash and rill erosion on the basis of field experiments. In: *Hillslope Processes* (ed. by A. D. Abrahams), 31–53. Allen and Unwin, London, UK.

Fifield, J. S. (2002) *Field Manual on Sediment Control and Erosion Control; Best Management Practices for Contractors and Inspectors.* Forester Press, Santa Barbara, USA.

Flügel, W. A. (1995) Delineating hydrological response units by geographical information system analyses for regional hydrological modelling using PRMS/MMS in the drainage basin of the River Bröl, Germany. *Hydrol. Processes* **9**, 423–436.

Goldman, S. J., Jackson, K. & Bursztynsky, T. A. (1986) *Erosion and Sediment Control Handbook*. McGraw-Hill, New York, USA.

Harvey, A. M. (2001) Coupling between hillslope and channels in upland fluvial systems; implications for landscape sensitivity, illustrated from the Howgill Fells, NW England. *Catena* **42**, 225–250.

Kirkby, M. J., Le Bissonais, Y., Coulthard, T. J., Daroussin, J. & McMahon, M. L. (2000) The development of land quality indicators for soil degradation by water erosion. *Agriculture, Ecosystems and Environment* **81**, 125–136.

Kirkby, M. J., Bracken, L. J., & Reaney, S. (2002a). The influence of landuse, soils and topography on the delivery of hillslope runoff to channels in SE Spain. *Earth Surf. Processes Landf.* **27**, 1459–1473.

Kirkby, M. J., Abrahart, R. J., Bathurst, J. C., Kilsby, C. G., McMahon, M. L., Osborne, C. P., Thornes, J. B. & Woodward, F.I. (2002b) MEDRUSH, a basin scale physically-based model for forecasting runoff and sediment yield. In: *Mediterranean Desertification: a Mosaic of Processes and Responses* (ed. by N. A. Geeson, C. J. Brandt & J. B. Thornes), chap 16, 203–227. John Wiley, Chichester, UK.

Kirkby, M. J., Bracken, L. J., Poesen, J., Nachtergaele, J. & Vandekerckhove, L. (2003) Observed and modelled distributions of channel and gully heads—with examples from SE Spain and Belgium. *Catena* **50**, 415–434.

Kirkby, M. J., Bracken, L. J. & Shannon, J. (2005) The influence of rainfall distribution and morphological factors on runoff delivery from dryland catchments in SE Spain. *Catena* (in press).

Lasanta, T., Garcia-Ruiz, J. M., Perez-Rontome, C. & Sancho-Marcen, C. (2000) Runoff and sediment yield in a semiarid environment: the effect of land management after farmland abandonment. *Catena* **38**, 265–278.

Lyford, F. P. & Qashu, H. K. (1969) Infiltration rates as affected by desert vegetation. *Water Resour. Res.* **5**, 1373–1376.

Montgomery, D. R. & Dietrich, W. E. (1988) Where do channels begin? *Nature* **336**, 232–234.

Musgrave, G. W. (1947) Quantitative evaluation of factors in water erosion—a first approximation. *J. Soil Water Conserv.* **2**, 133–138.

Nicolau, J. M., Solé-Benet, A., Puigdefábregas, J. & Gutiérrez, L. (1996) Effects of soil and vegetation on runoff along a catena in semiarid Spain. *Geomorphology* **14**, 297–309.

Proser, I. & Abernethy, B. (1996) Predicting the topographic limits to a gully network using a digital terrain model and process thresholds. *Water Resour. Res.* **32**(7), 2289–2298.

Smith, T. R. & Bretherton, F. P. (1972) Stability and the conservation of mass in drainage basin evolution. *Water Resour. Res.* **8**, 1506–1529.

Vanderkerckhove, L., Poesen, J., Oostwoud Wijdenes, D. & Figueiredo, T. (1998) Topographic thresholds for ephemeral gully initiation in intensively cultivated areas of the Mediterranean. *Catena* **33**, 271–292.

Wijdenes, D. O., Poesen, J. & Vandekerckhove, L. (1997) Chiselling effects on the vertical distribution of rock fragments in the tilled layer of a Mediterranean soil. *Soil Till. Res.* **44**(1–2), 55–66.

Wilcox, B., Wood, M. & Tromble, J. (1998) Factors influencing infiltrability of semiarid mountain slopes. *J. Rangeland Manage.* **41**, 197–206.

Yalin, M. S. (1977) *Mechanics of Sediment Transport*, 2nd edn. Pergamon Press, London, UK.

Geomorphological Processes and Human Impacts in River Basins (Proceedings of the International Conference held at Solsona, Catalonia, Spain, May 2004). IAHS Publ. 299, 2005.

Surface erosion assessment in two Calabrian basins (southern Italy)

LOREDANA ANTRONICO, ROBERTO COSCARELLI & ORESTE TERRANOVA
Italian National Research Council, Institute of Research for Hydrogeological Protection (CNR-IRPI), Section of Cosenza, Via Cavour, 4–6, 87030 Rende (CS), Italy
antronico@irpi.cnr.it

Abstract The present study presents an assessment of the erosion for two basins located on the Ionian side of southern Calabria (southern Italy). Two different approaches are considered to evaluate soil erosion: a qualitative method, based on aerial photo interpretative analysis, and a quantitative method, based on the RUSLE model. The percentage of area affected overall by surface erosion proves to be greater using the quantitative approach (38.5%) compared to the qualitative (13.3%). The mean erosion estimated using RUSLE is about 3.5 mm year^{-1} and can be compared with the values for Calabria reported in the literature. The application of the two approaches has shown that the spatial–temporal distribution of the parameters influencing the erosive process plays a relevant role.

Key words aerial photographs; Calabrian basins, Italy; erosion; RUSLE

INTRODUCTION

Soil erosion is caused by the interaction of several factors, among which surface water runoff and human activities certainly play a significant role. This phenomenon is a serious problem since it determines the removal of the most useful portions of soil for the vegetation, of fine organic substances and materials, as well as diminishing the capacity of the soil to retain water and nutrients. The serious effects of erosion constitute a real environmental challenge that should not be underestimated. Also, in the last few decades, erosive processes have been appearing everywhere with greater frequency and importance. In Europe the worst situation can be found in the Mediterranean area, with about 50–70% of cultivated agricultural land affected by soil erosion phenomena. In the case of Italy it has been estimated that the area of potential high erosion risk is about 27% (Giordano, 1994). Several areas of the Calabria region (southern Italy) are subject to widespread and intense erosion phenomena. Reported erosion rates can not be directly compared since they use differing spatial and temporal scales. Sorriso-Valvo (1996) estimated that 6 mm year^{-1} from the slopes in some Calabrian areas are lost (considering overall mass movements and surface erosion). The same author estimated an erosion rate of 12 mm year^{-1} in a badland zone of the southern Ionian coast with a small reservoir based on a temporal scale of 15 years (Sorriso-Valvo *et al.*, 1992). In addition, Raglione *et al.* (1980) experimentally observing the slopes of a strip of the Calabrian Ionian coast for about a year, estimated erosion at a rate of 10 mm year^{-1}.

The methodologies proposed in the literature for the study of erosion processes can be divided into the following categories: quantitative, semi-quantitative and qualitative. The quantitative approach includes those models (empirical or concentrated) for which the evaluation of the erosion is based on the application of a single empirically obtained equation owing to the superimposition of effects. Among the said procedures, one of the most used is the Universal Soil Loss Equation (USLE), (Wischmeier & Smith, 1978). The semi-quantitative models are simple models applicable in particularly complex climatic cases and when the basic information about the factors that determine an erosion phenomenon is insufficient (PSIAC, 1968). The qualitative approach includes all those methodologies based on direct observation of soil degradation phenomena using such techniques as remote sensing and photo interpretation, and on the creation of geomorphological maps, enabling the analysis of erosion patterns and of the deposition produced by specific morphogenetic processes. This paper reports an erosion assessment for a Calabrian Southern Ionian area to which both a qualitative type methodology, based on the observation of aerial photographs, and a quantitative methodology, based on the Revised version of the USLE (RUSLE) proposed by Moore & Burch (1986), have been applied.

STUDY AREA

Most of Calabria consists of a sector of the Alpine chain formed of nappes derived from oceanic and continental crust (Fig. 1). The whole region is undergoing tectonic uplift which is the main cause of its high relief. Geological trends in association with the Mediterranean climate (e.g. heavy seasonal rainfall) and the intense seismic and human activity play an important role in producing highly unstable slopes. In particular, mass movement and badland erosion are very common throughout the region, except on alluvial plains and flat areas at the top of mountain ranges.

The study area is located on the eastern slopes of the Serre Massif and corresponds to the basins of the Assi and Guardavalle *fiumare*. These two basins are representative of the geomorphology and hydrology of the Ionian side of the Serre Massif. The geological basement of the study area is represented by crystalline and metamorphic rocks that outcrop in the medium-high part of the basins. Continental and marine sedimentary terrains outcropping in the medium-low part of the basins lie transgressively on crystalline-metamorphic unit. The Assi and Guardavalle *fiumare* basins have an overall surface area of 95 km^2 (Assi 66.4 km^2 and Guardavalle 28.6 km^2). The main water courses are 28.8 km and 18.2 km long, respectively, on a west–east axis debouching to the Ionian Sea with a mean slope around 5%. The hydrographic networks of the two basins are similar in appearance, but different in drainage density (7.7 km^{-1} for the Assi and 5.5 km^{-1} for the Guardavalle) and channel frequency (46.3 km^{-2} and 20.3 km^{-2}, respectively). The altimetric characteristics are described in Fig. 2. Calabria presents a rather heterogeneous rainfall distribution both on a spatial and a time scale. In the western slopes (Tyrrhenian coastal zone) rainfall often exceeds 2000 mm $year^{-1}$ while on the Ionian coastal zone annual rainfall varies between 600 and 1000 mm $year^{-1}$. Eight pluviometric stations located in the surroundings of the basins were used in the analysis: Badolato, Ferdinandea, Pietracupa, Mammone, Stilo, Punta Stilo, Riace and Serra San Bruno (see Fig. 3). Average annual rainfall, weighted by the number of years of observation, is 1265 mm with 122 days of rainfall on average. The

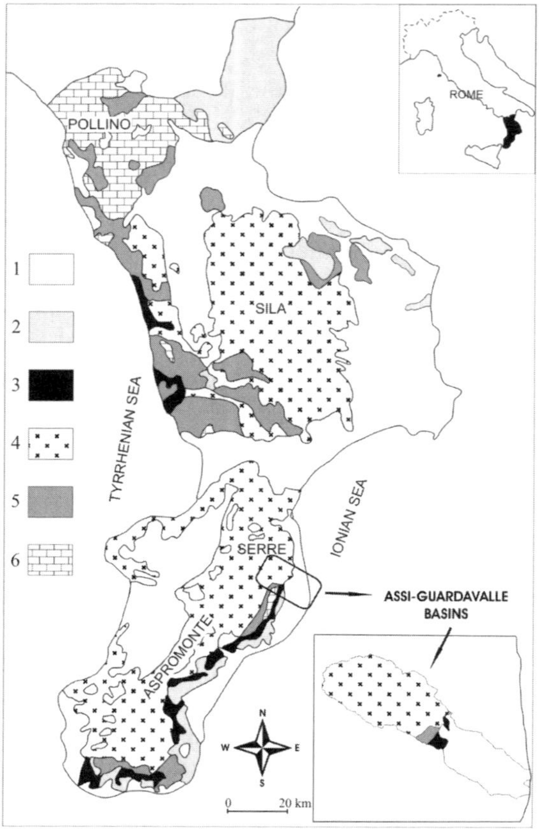

Fig. 1 Geological sketch of Calabria. Units are grouped according to their mechanical behaviour. Legend: (1) Miocene to Holocene marine and continental sedimentary units; (2) Mesozoic and Tertiary flysh-type sequence; (3) Neogene hard sedimentary units; (4) Paleozoic, medium to high-grade metamorphic and intrusive rocks of the Alpine units; (5) Paleozoic and Mesozoic very low to low grade metamorphic rocks; (6) carbonate units of the Appennine range and limestone cover of Alpine units. From Sorriso-Valvo (1993), modified.

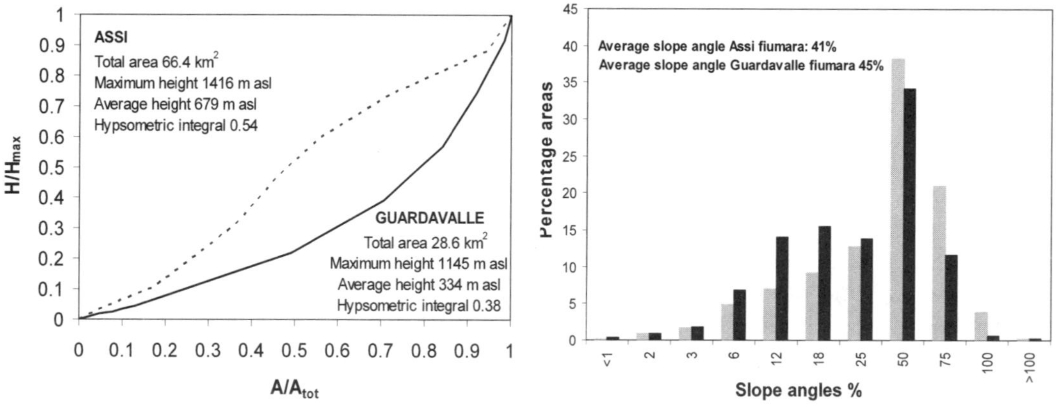

Fig. 2 Percentage hypsometric curves of Assi (dotted line) and Guardavalle (full line) basins; valleyside slope angle distributions of Assi (grey) and Guardavalle (black).

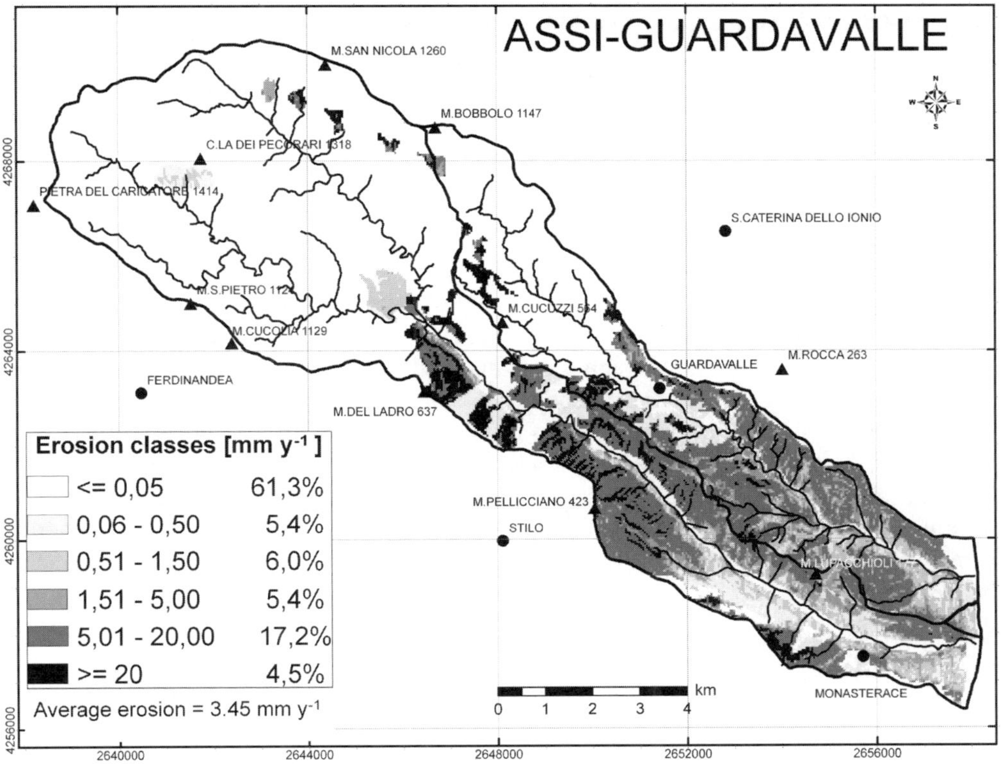

Fig. 3 Map of estimated average soil loss for the Assi-Guardavalle basins.

most intense were those that occurred during October 1951, October 1953 and September 2000. In particular, during the event of 14 October 1951, around 1046 mm of rain fell in 5 days in Badolato, with a maximum intensity of 371 mm day^{-1}. Considerable rain (357 mm in the previous 10 days) preceded the event. During the event of 1953, 675 mm were recorded at the Ferdinandea station (19 and 28 October). Finally, 634 mm were recorded in Ferdinandea between 8 and 11 September 2000, with maximum intensity of 271 mm day^{-1}, and 463 mm were recorded in Punta Stilo between 8 and 10 September with maximum intensity of 301 mm day^{-1}.

METHODOLOGY

In order to estimate the erosion processes by a qualitative method, an aerial photography investigation was carried out. Black and white aerial photographs of the years 1954–1955 and 2000, respectively, at 1:33 000 and 1:25 000 scale, were used. Aerial photographs of 1954–1955 were taken after the events occurred in central-southern Calabria in October 1951 and October 1953. Therefore the use of these aerial photographs provided an indication as to the sum of the effects produced in terms of landslides and erosion by the intense rainfall events of 1951 and 1953. Aerial photographs of 2000 were taken after the intense rainfall event that affected a sector

Table 1 Results of the application of qualitative method (aerial photo interpretation).

Degree of erosion	% of erosion relative to the total basin areas (1954 aerial photos)	% of erosion relative to the total basin areas (2000 aerial photos)
Weak	3.32	2.97
Moderate	3.06	1.05
Moderate with landsliding	5.86	1.77
Severe	0.40	0.00
Severe with landsliding	0.66	0.30

along the southern Ionian side of Calabria on 8–10 September 2000. Through the interpretation of aerial photographs, two inventory maps at 1:25 000 scale related to the two periods were prepared. These maps charted erosion processes and landslides. Regarding erosion, three main classes have been identified: (a) areas with a high degree of erosion; (b) areas with a moderate degree of erosion; (c) areas with a low degree of erosion. Two other classes were added, representing those areas where simultaneous landslide and erosion process have been described: areas with a moderate degree of erosion plus landslides and areas with a high degree of erosion plus landslides (Table 1).

For the quantitative approach, a specific software (Ero.Cal) was developed. Ero.Cal allows the management of a geo-database, furnishes digital terrain maps in GIS format and evaluates hydric erosion using mathematical modelling. The RUSLE model, applied in this paper, is expressed as follows:

$$A = R\ K\ L\ S\ C\ P \qquad (1)$$

where A (t ha^{-1} year^{-1}) is the estimated average soil loss, R (MJ mm ha^{-1} h^{-1} year^{-1}) is the rainfall–runoff erosivity factor, K (t h MJ^{-1} mm^{-1}) is the soil erodibility factor, L (nondimensional) is the slope length factor, S (nondimensional) is the slope steepness factor, C (nondimensional) is the cover-management factor, P (nondimensional) is the support practise factor. Regarding the evaluation of the rainfall aggression factor (R), the "isoerodents" have been calculated using the following relations proposed by Ferro *et al.* (1991) for Sicily on the basis of the empirical relation elaborated by Renard *et al.* (1997), and considered valid for the present study context:

$$R/100 = 10.80 + 0.082\ (I_{1,2}\ I_{6,2}\ I_{24,2})\ (\text{MJ mm ha}^{-1}\ \text{year}^{-1}) \qquad (2)$$

$$R/100 = -36.46 + 1.84\ IF - 0.158\ P\ (\text{MJ mm ha}^{-1}\ \text{year}^{-1}) \qquad (3)$$

Equation (2) is valid for the pluviographic stations, (3) for the pluviometric stations. $I_{1,2}$, $I_{6,2}$ and $I_{24,2}$ represent the values regarding a return period of 2 years of the annual maximums of rainfall intensity of the respective durations of 1, 6 and 24 h; IF is the modified Fournier index, referring to the mean annual values (Arnoldus, 1980). The areal distribution of the factor K has been evaluated on the basis of the grain-size characteristics of the surface material, taken from the soil map drafted by the Regional Agency for Development and for Agricultural Services (ARSSA, 2003). In this way 19 types of soil and the corresponding K values have been identified. The values varied from null erodibility R of the bottom of the valley, to a maximum of 1.27 t ha^{-1} year^{-1} in the slopes surrounding the high part of the basins where altered and degraded

Table 2 Cover-management factor values and land use (from ARSSA, 2003).

Description	*C* values
Annual agricultural crops (irrigated or not, tree-cover or not)	0.4
Permanent grasslands; vines; orchards; orange (lemon) grove; olives	0.3
Mixed forest	0.00015
Degraded forest	0.006
Shrubs; natural pasture	0.05
Mediterranean Macchia; riparian vegetation	0.003
Bare lands	1
Rare vegetation	0.3

crystalline ground outcrops. The *S* and *L* factors have been calculated, using a 40-m grid DTM with the help of SURFER (1999) software. The *C* factor has been estimated on the basis of the soil use map resulting from photointerpretation, on-site tests and historical surveys (Table 2). Finally, regarding factor *P*, a unitary value has been adopted, as usually it is the case for natural slopes with no conservation practices.

RESULTS AND DISCUSSION

The results of the erosion evaluation carried out using photointerpretation are shown in Table 1. The reduction in the areas affected by observable erosion in 2000 compared to 1954–1955 is to be put in relation to the different pluviometrical characteristics of the rainfall events. In general it has been observed that all the lithologic types outcropping in the area are affected, even though to differing degrees, by the various erosion classes. Figure 3 shows the average soil loss estimated by RUSLE model for the Assi-Guardavalle basins. The results are analysed with regard to three homogeneous zones. In the mountain zone, with protection by vegetation cover, a few limited areas present surface erosion and the average erosion rate is about 0.9 mm $year^{-1}$. In the middle zone, the erosion is more widespread (about 28% of the area falls in a class with an erosion rate of 5.0–20.0 mm $year^{-1}$) and the average erosion rate is about 7.0 mm $year^{-1}$. Finally, in the valley zone, the average estimated erosion value (5.0 mm $year^{-1}$) is inferior to the middle zone average value even if the erosion is very widespread (38% of the area has an erosion rate of 5.0–20.0 mm $year^{-1}$). The mean erosion estimated using RUSLE is about 3.5 mm $year^{-1}$ and it is comparable with values given in the literature for Calabria.

The comparisons among the estimates obtained using the two different approaches permit important observations to be carried out. The percentage of area affected overall by surface erosion proves to be greater using the quantitative approach (38.5%) compared to the qualitative (13.3 %). Photointerpretation allows the areas affected by erosion to be recognized objectively, in a qualitative way, due to a particular sequence of rainfall events, land uses, etc. On the contrary, the quantitative approach permits soil erosion risk to be evaluated on the basis of various factors like soil use, agriculture practices, soil management, etc. For instance, badlands prove the most subjected to erosion on the basis of photointerpretation (1954 and 2000 years) but, being unaffected by agriculture, RUSLE simulations furnish middle values. On the contrary, the RUSLE model is not able to simulate a snapshot. For this reason aerial photography constitutes an indispensable tool to integrate knowledge about the physical factors affecting the erosive process.

CONCLUSIONS

An estimate of erosion in a Calabrian Southern Ionian area has been carried out in this paper. The application of the quantitative and qualitative approaches has shown the need to pay special attention to the spatial–temporal distribution of the parameters influencing the erosive process. In fact, only the evaluations carried out with aerial photographs of 2000 (qualitative method) prove to be in correspondence with the quantitative approach values which, even if referring to mean rainfall data, take the soil use and vegetation conditions into account, approximately relative to the same temporal period. Continuation of the study will enable the evaluations to be extended over the whole of Calabria and an improvement in understanding regarding the estimate of the parameters based on the application of different approaches.

Acknowledgements The authors are grateful to J. Carles Balasch for the relevant remarks and comments which led to improve the initial version of this paper.

REFERENCES

Arnoldus, H. M. (1980) An approximation of the rainfall factor in the Universal Soil Loss Equation. In: *Assessment of Erosion* (ed. by M. De Boodt & D. Gabriels). Wiley, Chichester, UK.

ARSSA (2003) I suoli della Calabria. Carta dei suoli della Calabria 1:250.000 della Regione Calabria. Servizio Agropedologia, Monografia divulgativa.

Ferro, V., Giordano, G. & Iovino, M. (1991) La carta delle isoerodenti e del rischio erosivo del territorio siciliano come strumento per la pianificazione degli interventi di conservazione del suolo. In: *I Piani di bacino per la difesa del suolo, la gestione delle acque e la tutela dell'ambiente* (Atti Convegno Nazionale Taormina, 23–24 Novembre 1990).

Giordano, A. (1994) Valutazione del rischio di erosione del suolo nell'ambiente mediterraneo. Programma CORINE della CEE. In: *Difesa e Valorizzazione del Suolo e degli Acquiferi* (Atti del IV Convegno Internazionale di Geoingegneria, Torino), 151–160.

Moore, I. D. & Burch, G. J. (1986) Modelling erosion and deposition: topographic effects. *Trans. ASAE* **29**, 1624–1630.

PSIAC (Pacific Southwest Inter Agency Committee) (1968) Report of the water management subcommittee on factors affecting sediment yield in the pacific southwest area and selection and evaluation of measures for reduction of erosion and sediment yield. Pacific Southwest Inter Agency Committee.

Renard, K. G., Foster, G. R., Weesies, G. A., McCool, D. K. & Yoder, D. C. (1997) Predicting soil erosion by water: a guide to conservation planning with the Revised Universal Soil Loss Equation. *USDA Agric. Handbook 703*, Washington, USA.

Raglione, M., Sfalanga, M. & Torri, D. (1980) Misura dell'erosione in un ambiente argilloso della Calabria. *Annali Ist. Sper. Studio e Difesa Suolo* **XI**, 159–181.

Sorriso-Valvo, M. (1993) The geomorphology of Calabria—a sketch. *Geogr. Fis. Dinam. Quat.* **16**, 75–80.

Sorriso-Valvo, M. (1996) Relationship between mass-movement, bedload budget and beach dynamics with special regards to streams with coarse-grained riverbeds. In: *The Coast: Geological/Geomorphological Processes and Resources Management* (Perth Seminar., 4 June, Perth, Australia).

Sorriso-Valvo, M., Antronico, L. & Borelli, A. (1992) Recent evolution of badland-type erosion in Southern Calabria (Italy). *Geoöko plus* **3**, 69–82.

SURFER (1999) User's guide. Contouring and 3D. Surface mapping for scientist and engineers. Golden software, Inc., Golden, Colorado, USA.

Wischmeier, W. H. & Smith, D. D. (1978) Predicting rainfall erosion losses. *USDA Agriculture Handbook 537*, Washington, USA.

Methods for the characterization of areas sensitive to desertification: an application to the Calabrian territory (Italy)

ROBERTO COSCARELLI, IVANA MINERVINO & MARINO SORRISO-VALVO

Italian National Research Council – Institute of Research for Hydrogeological Protection (CNR-IRPI), Section of Cosenza, Via Cavour, 4–6, 87030 Rende (CS), Italy

coscarelli@irpi.cnr.it

Abstract Desertification, as it has been defined by the United Nations Convention to Combat Desertification (UNCCD), represents the whole phenomena of land degradation, resulting from many causes. Southern European countries are considered as areas of new expansion of desertification, caused by inadequate agricultural practices, overgrazing, and deforestation, which increase anthrophic pressure. During the last few years, in the context of several international research projects, methodologies were proposed, based on various parameters. These are grouped in uniform classes and weighted by the attribution of scores in order to assess the so-called "quality indexes". By computation of the previous indexes, "desertification indicators" are defined. This work reviews the most-used methodologies and presents an application of the methodology proposed in the European MEDALUS project to an area of the Calabrian territory (southern Italy), relative to the physical aspects.

Key words Calabrian territory, Italy; desertification; ESAI method

INTRODUCTION

The term "desertification" was defined by the United Nations Convention to Combat Desertification (UNCCD) in 1994 as "land degradation in arid, semiarid and dry subhumid areas, resulting from various factors, including climatic variations and human activities". In the last few years several studies have shown that a quarter of the Earth's land is threatened by desertification; more than one billion people and more than 100 nations are at risk of desertification as a consequence of lost soil productivity (UNEP, 1992). For instance, in Africa 73% of drylands is degraded; in America this value rises up to 76%. The destruction of the land's productive capacity costs (1990 estimate from UNEP, 1992) more than 42 billion US$ per year in the world (Africa: nine billion US$; Asia: 21 billion US$; North America: five billion US$; South America: three billion US$; Australia: three billion US$; Europe: one billion US$).

The causes of desertification can be divided into natural causes (drought, increase of average storm precipitation intensity with a consequent increase of rainfall erosivity, positive trends of the air temperature, etc.) and human causes (inappropriate land-use, uncontrolled deforestation, overgrazing, uncontrolled use of chemical fertilizers and pesticides, excessive coastal urbanization, etc.). As a consequence of the phenomenon, soils become particularly vulnerable to runoff, drought, floods and fire; moreover, salinization can threaten soil and groundwater, in this way decreasing soil productivity and water quality.

All these factors cause changes in the distribution of food and water, obviously with consequences on social and economic aspects: migration of populations affected by average changes in the distribution of resources might increase the risks of political instabilities and conflicts.

Several projects to study and evaluate desertification were developed all over the world. Among the projects funded by the European Commission, the most famous is the MEDALUS Project—Mediterranean Desertification and Land-use (Brandt & Thornes, 1996). It was developed in three phases during the period 1991–1999, involving about 50 groups from universities, research institutes and industry. The "sister" project of MEDALUS was the EFEDA Project—European Field Experiment in Desertification threatened Areas—developed in 1991–1993 (Bolle *et al.*, 1993). Among the projects aimed at analysing the phenomenon by remote-sensing data, the RESMEDES (Remote Sensing of Mediterranean Desertification and Environmental Changes) project (1998) can be mentioned. The DESERTLINKS project (Combating Desertification in Mediterranean Europe Linking Science with Stakeholders) is a recent project, with the goal of linking science with the stakeholders (UNCCD, 2002).

METHODOLOGY

Methodologies used to evaluate the desertification phenomenon are based on indicators. The Organization for Economic Co-operation and Development (OECD, 1993) has defined three basic criteria for indicator evaluation: policy relevance and utility for users, analytical soundness and measurability.

The European Environment Agency (EEA) approach is based on the above criteria, with special care for the target audience. EEA introduced the DIPSIR model, developing the Pressure–State–Response model adopted by OECD and adding references to the causes (driving forces) and impacts on the environment. The DIPSIR framework shows a complex chain of causes-effects ranging from Driving forces (activities) to Pressure, to changes on the State of the environment, to Impacts and Response. It is based on the assumption that economic activities and society's behaviour affect environmental quality. This framework applies well to desertification.

The most accredited method on the topic of desertification is that introduced in the MEDALUS project. The most important object of the third phase of the MEDALUS project was the development and the application of a methodology for the identification of Environmental Sensitivity Areas (ESAs) to desertification (European Commission, 1999). It was developed and validated in the Agri basin (Italy) and also first applied to the following target areas: the Isle of Lesvos (Greece), the Alentejo region (Portugal), and the Gualentin basin (Spain).

The model takes into consideration a broad system of key indicators of desertification (European Commission, 1999), each of which gives information on topic characteristics able to describe the capability of the land to withstand further degradation, or the suitability of the land for supporting specific types of land use. These indicators are parameters relative to soil, vegetation, climate and management.

In the original version of the procedure, a score, ranging from 1 to 4, is assigned to each indicator to evaluate the minor or maximum sensitivity to degradation processes. Indicators are grouped and combined into four categories representing the four Quality Indexes of Soil (SQI), Climate (CQI), Vegetation (VQI) and Management (MQI).

All the data are introduced into a regional Geographical Information System (GIS). Every quality index is structured into layers, one for each indicator; the index value is calculated for each elementary unit as a geometric average of the scores attributed to the indicators.

Finally, the Index of Environmental Sensitive Areas (ESAI) is calculated as a geometric average of the four quality indexes:

$$ESAI = (SQI*CQI*VQI*MQI)^{1/4} \quad (1)$$

Three main classes of sensitivity to desertification, with three subclasses in each type, are considered (European Commission, 1999). “Critical ESAs” are already highly degraded areas in which affecting phenomena can threaten the environment of the surrounding areas. “Fragile ESAs” are areas with a delicate balance between natural and human activity, in which any climate or land-use change could cause desertification. “Potential ESAs” are areas threatened by desertification under significant climate change if a correct land-use plan is not applied. The areas with the best soil characteristics and under semiarid or wetter climatic conditions, independently of vegetation cover, are not threatened by desertification.

Modifications of the original approach for the characterization of ESAs have been introduced by many authors. Montanarella (2002) brought modifications in relation to the regional scale at which the model has been applied and by the introduction of an Erosion Quality Index (EQI); the ESA Index in this case is calculated as a geometric average of five parameters:

$$ESAI = (SQI*CQI*VQI*MQI*EQI)^{1/5} \quad (2)$$

Much more detailed modifications were proposed by other authors (Ladisa *et al.*, 2002) consisting of supplementary indicators: an integrated Land-use and Management Quality Index (LU_MQI) and a new Human Pressure Index (HPI). On the basis of these modifications, the ESAI can be obtained from the following formula:

$$ESAI = (SQI*CQI*VQI*LU\_MQI*HPI)^{1/5} \quad (3)$$

APPLICATIONS

The ESAI approach was partially adopted in the present study to evaluate the environmental “physical” sensitivity to desertification in the Cirò district (Calabria, southern Italy). It is the first application to the Calabrian territory at the basin scale and it is different from other partial or complete applications of the procedure to the Calabrian territory (Piro *et al.*, 2004) at a regional scale, like those that the experts of the Calabrian Regional Agency for Environmental Protection (ARPACAL) are developing. The Cirò district is an area well-known for wine production, facing the Ionian sea, about 20 000 ha wide, entirely characterized by Tertiary and Quaternary deposits. A morphogenetic, climatic and vegetational overview of the Ionian coast in

Calabria shows that this strip of land, including the foothills of the immediate hinterland, could be threatened by desertification. It is an area exposed both to pressure from unsustainable human activities and to the stress of climatic conditions. In fact, hydrological studies (Coscarelli *et al.*, 2004; Cotecchia *et al.*, 2004) carried out for the territory of southern Italy show trends towards aridity conditions.

Soil and vegetation data were provided by the Regional Agency for Agriculture (ARSSA) which carried out a pedological investigation of the Cirò district (ARSSA, 2002), obtaining soil data from more than 450 soil profiles. Vegetation data are related to the CORINE Land Cover project (ARSSA, 2002). Hydrological records were collected from five meteorological stations (Crucoli, Cirò Marina, Montagna cc, Scala Coeli and Strongoli) of the Italian Hydrographic Service, two of which are located in the immediate surroundings of the study area. A DTM with grid cell size of 20 m was used for the elaborations and the acquisition of the local morphometric terrain parameters: slope gradient and aspect (slope direction). The processes of data building, modelling analysis and map display (GIS elaboration) were obtained by ArcInfo software.

In the present study the management layers were excluded to analyse only the physical components, as also indicated by Ferrara & Kosmas (2004). In the opinion of the authors the evaluation of the Management Quality Index through the parameters proposed in literature would not be greatly indicative given the scale of this application. For this reason the Index of Environmental Sensitivity Areas was calculated as a geometric mean of only the three quality indexes SQI, VQI, CQI: the ESAI evaluated in this way can be defined as a "Physical Index" of Environmental Sensitivity Areas. Table 1 shows the classification of indicators with the relative scores for each "physical" quality index as proposed in literature (European Commission, 1999).

Regarding the range of the ESA Physical Index (ESAPI), the minimum value is equal to 1; the maximum one was calculated by the geometric mean of the three maximum values of the SQI, VQI, CQI:

$$ESAPI_{max} = (2.24*2.52*2)^{1/3} = 2.24 \quad (4)$$

The whole range of ESAPI was divided in to the same classes as the ESAI (European Commission, 1999). For this reason, the scores of each class of ESAI were re-ranged from 0 to 100%; the same percentages were applied to the whole range of ESAPI (1–2.24) obtaining the range of each class as reported in Table 3.

RESULTS

Four maps were prepared by the application of the ESAs approach: the ones relative to soil, vegetation and climate quality and the one relative to the ESA "Physical" Index. Table 2 shows the results relative to the Soil, Vegetation and Climate Quality Indexes.

Regarding CQI, this index varied only with the "aspect": the rainfall and aridity parameters had scores equal, respectively, to 1 and 1.1 for the whole study area. These last two parameters, in fact, are not so variable at a basin scale approach: as Basso *et al.* (2000) showed, they are representative of a smaller scale (1:500 000), instead of the "aspect" that has a representative scale 1:25 000. For VQI and SQI there is no difference regarding the scales of the layers used for the evaluation of these indexes

Table 1 Classification of indicators with the relative scores for each quality index.

Soil quality

Soil texture classes	Scores	Parent material classes	Scores	Rock fragments cover classes (%)	Scores
L, SCL, SL, LS, CL	1.0	Shale, schist, basic and ultrabasic rocks, conglomerates, unconsolidated deposits.	1.0	>60	1.0
SC, SiL, SiCL	1.2	Limestone, marble, granite, rhyolite, ignibrite, gneiss, siltstone, sandstone.	1.7	20–60	1.3
Si, C, SiC	1.6				
S	2	Marl, pyroclastics.	2	<20	2.0

Soil depth classes (cm)	Scores	Drainage classes	Scores	Slope classes (%)	Scores
>75	1.0	Well drained	1.0	< 6	1.0
30–75	2.0			6–18	1.2
15–30	3.0	Imperfectly drained	1.2	18–35	1.5
< 15	4.0	Poorly drained	2.0	>35	2.0

Vegetation quality

Fire risk classes	Scores	Erosion protection classes	Scores	Drought resistance classes	Scores
Bare land, perennial agricultural crops, annual agricultural crops (*maize, tobacco, grasslands*)	1.0	Mixed Mediterranean macchia/evergreen forests, rocks	1.0	Mediterranean macchia, mixed Mediterranean macchia/evergreen forests	1.0
Annual agricultural crops (*cereals, sunflower*), deciduous oak, (mixed), mixed Mediterranean, macchia/evergreen forests	1.3	Mediterranean macchia, pine forests, permanent grassland, evergreen perennial crops, olives, shrubs	1.3	Conifer, deciduous, olives	1.2
Mediterranean macchia	1.6	Deciduous forests	1.6	Perennial agriculture trees (*almonds, vines, orchard*)	1.4
Pine forests	2.0	Deciduous perennial agricultural crops (*almonds, orchard*)	1.8	Perennial grassland	1.7
		Annual agricultural crops (*grasslands, cereals, maize, tobacco, sunflower…*), vines, horticulture, barren, riverbed	2.0	Annual agricultural crops, (*grasslands, cereals, maize, tobacco, sunflower…*), horticulture, barren, riverbed	2.0

Plant cover classes (%)	Scores
>40	1.0
10–40	1.8
<10	2.0

Climate quality

Aspect classes	Scores	Rainfall classes (mm year^{-1})	Scores	Aridity index (Bagnouls-Gaussen Index) classes	Scores
North	1.0	>650	1.0	<50	1.0
South	2.0	280–650	2.0	50–75	1.1
		<280	4.0	75–100	1.2
				100–125	1.4
				125–150	1.8
				>150	2.0

Table 2 Classes and extensions of Quality Indexes related to the study area.

Classes	Description	Range	Extension (km^2)	Percentage (%)
Soil Quality Index				
1	High quality	<1.13	6.98	3.68
2	Moderate quality	1.13–1.45	154.74	81.47
3	Low quality	>1.45	28.22	14.86
Vegetation Quality Index				
1	High quality	<1.13	6.22	2.96
2	Moderate quality	1.13–1.38	85.40	40.62
3	Low quality	>1.38	118.61	56.42
Climate Quality Index				
1	High quality	<1.15	105.36	52.48
2	Moderate quality	1.15–1.81	95.39	47.52
3	Low quality	>1.81	0.00	0.00

Table 3 Classes and distribution of the Physical Index of ESAs related to the study area.

Type	Subtype	Range	Extension (km^2)	Percentage (%)
Critical	C3	>1.56	0.21	0.11
Critical	C2	1.43–1.56	8.18	4.34
Critical	C1	1.39–1.43	11.68	6.20
Fragile	F3	1.33–1.39	40.59	21.55
Fragile	F2	1.27–1.33	47.08	24.99
Fragile	F1	1.23–1.27	37.07	19.67
Potential	P	1.18–1.23	25.78	13.68
Non-affected	N	<1.18	17.83	9.46

(Basso *et al.*, 2000). These scale differences can appear clearly in applications of ESAI procedure to a basin scale, even if they do not influence the validity of the whole procedure, largely validated and tested.

As regards the results of the ESA Physical Index (Fig. 1), the largest part of the Cirò district (about 66% of the area) is classified as "Fragile" (mainly F2); about 11% of the area is classified as "Critical" (only 0.11 as C3, the worst subtype) and about 23% as "Potential" or "Not Threatened". More detailed results are shown in Table 3.

It is necessary to stress that the different sensitivity values of the areas to desertification should be validated comparing the "actual situation" with a future situation after a certain number of years; in this way it should also be possible to evaluate the influence of the various Quality Indexes in calculating the Index of Environmental Sensitivity Areas.

CONCLUSIONS

Southern European countries are considered as areas of new expansion of desertification, often caused by an increasing anthropic pressure. The case study presented in this paper relates to an area located along the Ionian coast of Calabria(southern Italy). The application of the ESAI procedure to the aforementioned

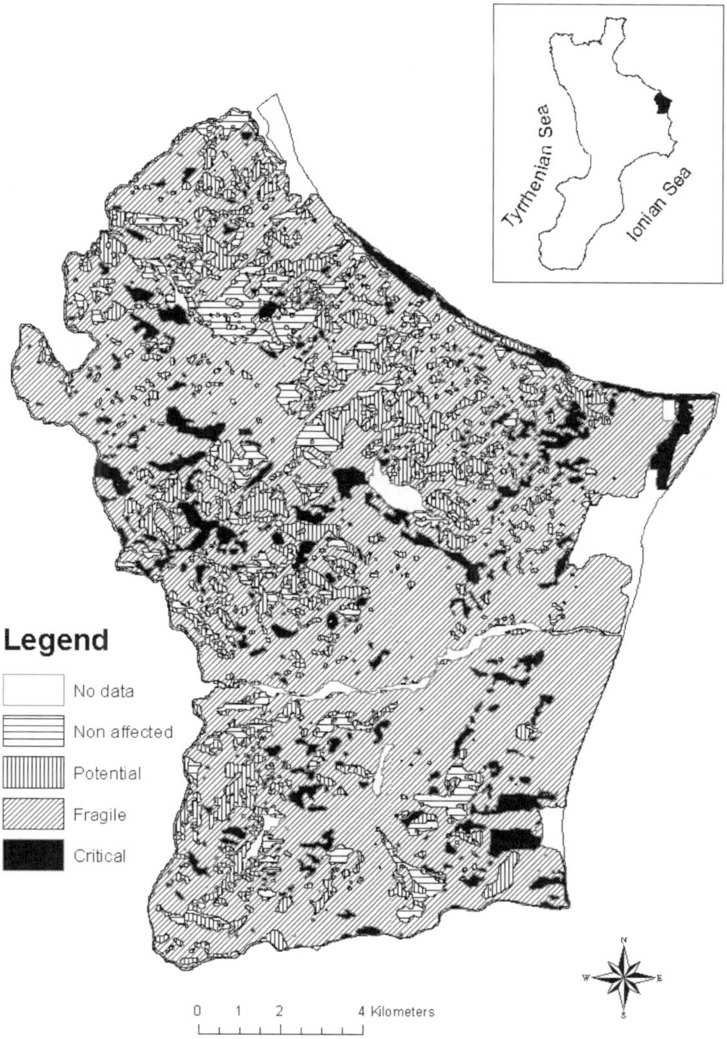

Fig. 1 Distribution of the Physical Index of ESAs in the study area.

area, limited to the physical aspects, showed that about 11% of the territory can be classified as "critical". Appling the methodology at the basin scale to the Calabrian territory some considerations were carried out, regarding the different significant geographic scales of the Quality Indexes, the procedure validation and the data availability. Some of these considerations are strictly linked to the peculiarity and the characteristics of the Calabrian territory.

The authors are now working on a research project called "ISPARIDE" (Identification and estimate of parameters for evaluating desertification risk). The project is partially funded by the Italian Ministry of Education, University and Research; the duration is two years, ending in April 2006. It is composed of three Work Packages and its aim is to develop, validate and test other methodologies using both on-site measurements and remote sensing data, comparing also the results obtained with these new methodologies and the results of ESAI procedure, partial and complete. The authors hope to have important results by developing this research project.

Acknowledgements The authors thank Emilio Catalano of CNR-IRPI—Section of Cosenza, Italy, for implementation of the computing procedure used and for figure elaborations. Helpful suggestions made by an anonymous reviewer are gratefully acknowledged.

REFERENCES

Agenzia Regionale per lo Sviluppo e per i Servizi in Agricoltura (ARSSA) (2002) *Carta dei Suoli e Zonazione Viticola del Cirò DOC (scala 1: 25.000)*. Regione Calabria. Monografia Divulgativa.

Basso, F., Bove, E., Dumontet, S., Ferrara, A., Pisante, M., Quaranta, G. & Taberner, M. (2000) Evaluating environmental sensitivity at the basin scale through the use of geographic information systems and remotely sensed data: an example covering the Agri basin (southern Italy). *Catena* **40**, 19–35.

Bolle, H. J., André, J. C. & Arrue, J. L. (1993) EFEDA. European Field Experiment in a Desertification threatened Area. *Ann. Geophys.* **11**, 173–189.

Brandt, C. J. & Thornes, J. T. (eds) (1996) *Mediterranean Desertification and Land-use*. John Wiley, Chichester, UK.

Coscarelli, R., Gaudio, R. & Caloiero, T. (2004) Climatic trends: an investigation for a Calabrian basin (southern Italy). In: *The Basis of Civilization—Water Science?* (ed. by J. C. Rodda & L. Ubertini), (Rome, December 2003), 255–266. IAHS Publ. 286. IAHS Press, Wallingford, UK.

Cotecchia, V., Casarano, D. & Polemio, M. (2004) Characterisation of rainfall trend and drought periods in southern Italy from 1821 to 2001. In: *New Trends in Hydrology* (ed. by R. Gaudio) (Proc. first Italian-Russian Workshop), CNR-GNDCI Publ. 2823.

European Commission (1999) The Medalus project Mediterranean desertification and land-use. Manual on key indicators of desertification and mapping environmentally sensitive areas to desertification (ed. by C. Kosmas, M. Kirkby & N. Geeson). EUR 18882.

Ferrara, A. & Kosmas, C. (2004) Expert system for evaluating the Environmental Sensitivity Index (ESI) for a local area: Methodology. In: www.unibas.it/desertnet/esi/method.htm.

Ladisa, G., Todorovic, M. & Trisorio-Liuzzi, G. (2002) Characterization of areas sensitive to desertification in southern Italy. In: *New Trends in Water and Environmental Engineering for Safety and Life: Eco-compatible Solutions for Aquatic Environments* (Proc. Capri 2nd Int. Conf., Capri, Italy).

Montanarella, L. (2002) Indicazione delle aree vulnerabili alla desertificazione in Puglia (2002). In: *Il Sistema Informativo Territoriale della Regione Puglia*. Documenti del territorio, XIII, Centro Interregionale di Coordinamento e Documentazione per le Informazioni Territoriali, Roma.

Organization for Economic Co-operation and Development (OECD) (1993) OECD core set of indicators for environmental performance reviews. OECD Environmental Directorate Monographs 83.

Piro, P., Carbone, M. & Frega, F. (2004) Analisi della inclinazione alla desertificazione del territorio assistita da un sistema informativo geografico (GIS). Applicazione alla regione Calabria. In: *Tecniche per la difesa dall'inquinamento* (Preprints of 25th Corso di Aggiornamento).

RESMEDES (1998) Final Report of EC Contract. Contract No.: ENV4-CT95-0094, European Commission, Directorate-General, Science, Research and Development.

United Nations Environment Programme (UNEP) (1992) *World Atlas of Desertification*. Edward Arnold, London, UK.

United Nations Convention to Combat Desertification (UNCCD) (2002) Second annex subregional report for the implementation of the UNCCD. Athens, Greece.

Runoff and soil moisture relationships in a small forested basin in the Sistema Central Ranges (Spain)

JOSÉ MARTÍNEZ-FERNÁNDEZ, ANTONIO CEBALLOS, SEGISMUNDO CASADO, CARLOS MORÁN & VIRGINIA HERNÁNDEZ

Department of Geography, University of Salamanca, Cervantes 3, 37002 Salamanca, Spain
jmf@usal.es

Abstract In the present work we analyse runoff yield in a small Mediterranean basin with respect to soil moisture dynamics. The aim was to observe the behaviour of the basin in terms of the generation of runoff in periods in which maximum water deficit occurs in the soil. Despite the abundant precipitation in the mountainous regions of the Mediterranean area, the joint action of the climatic conditions prevailing in the growing season and water interception and consumption by the forest may seriously affect the generation of runoff over a considerable part of the year.

Key words evapotranspiration; interception; Mediterranean climate; runoff yield; soil moisture; water deficit

INTRODUCTION

The reduction in runoff as a result of factors such as the interception or consumption of water by forests is well known (Bosch & Hewlett, 1982; Andreassian, 2004). In areas where forest has been recovered, the runoff yield has been severely affected (Beschta *et al.*, 2000). Thus, the basins of mountainous zones where changes in land use have led to a recovery of the forest have undergone an appreciable decrease in runoff (Beguería *et al.*, 2003). In Mediterranean environments it has been found that soils under tree cover are subject to a considerable reduction in surface runoff (Joffre & Rambal, 1993). In regions with strong seasonal contrasts in precipitation, the forest itself may exert a very powerful influence on runoff during the dry season. Here, we studied the relationship between soil moisture dynamics and runoff yield in a forest basin under subhumid Mediterranean conditions.

FIELD SITE AND METHODOLOGY

Study site

The experimental basin at Rinconada has been in operation since 1999. It occupies an area of 62.3 ha and is located 70 km south of the city of Salamanca (Fig. 1), forming part of the western sector of the Sistema Central range (Spain). Altitude ranges between 1140 and 1450 m a.m.s.l. with a SW–NE orientation. The basin overlies

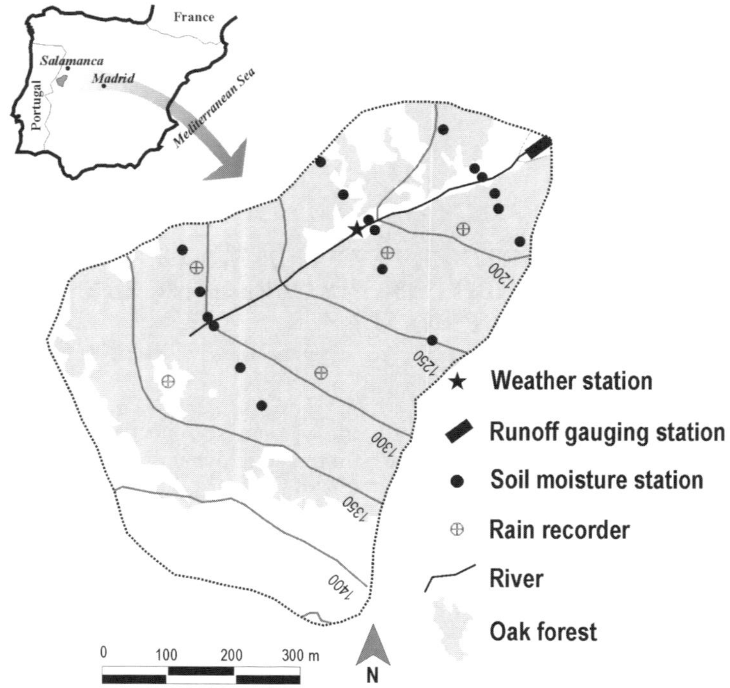

Fig. 1 Location map of the Rinconada Experimental Basin (Spain) and experimental layout.

a diverse geological substrate composed mainly of sandstones, limestones and quartzites. The main soils are Regosols, Leptosols and Cambisols and the predominant drainage class corresponds to a well-drained soil. The average of available soil water is 77 mm (0.094 cm^3 cm^{-3}). The climate of the zone is subhumid Mediterranean. Mean annual temperature is 10°C; December is the coldest month, with 1.8°C, and July the hottest, with 19.9°C. Mean annual precipitation ranges around 1000 mm, November being the wettest month, with 129 mm, and August the driest, with 15 mm. Mean annual potential evapotranspiration (ET) is close to 850 mm. Almost 70% of the surface of the basin is occupied by oak (*Quercus pyrenaica*) forest and the basin is used for extensive livestock raising and forestry.

MATERIALS AND METHODS

The Rinconada experimental basin has an automatic weather station that measures the main climatic variables at intervals of 10 min. Runoff is measured with an H-flume gauge with an automatic water level recorder that records data at 10-min intervals. Under the forest there are four automatic rain recorders and 68 raingauges for estimating rainfall interception.

Soil moisture is measured at fortnightly intervals at 18 measurement stations, applying the TDR (Time Domain Reflectometry) technique. The stations are located on three transects perpendicular to the valley bottom, these transects being representative of the different exposures and stretches of the slopes of the basin. Each station has five double-wire TDR probes in a horizontal position at depths of 5, 15, 25, 50 and 100 cm, respectively. The database reported here covered a period of 45 months, from April 2000 to December 2003.

RESULTS AND DISCUSSION

The evolution of mean soil moisture reflects a very prominent seasonal contrast, as is characteristic of Mediterranean areas. The high precipitation recorded between October and May (1264 mm in 2000–2001; 633 mm in 2001–2002, and 1093 mm in 2002–2003) did not prevent soil moisture from falling as low as wilting point values during the growing season. When the soil hydrological behaviour is referred to that of the vegetation, it is appropriate to express soil moisture values in terms relative to plant available water. Accordingly, in the analysis we introduced the concept of Water Deficit Index (WDI) with a view to gaining a better understanding of the analysis of water availability in the soil and its evolution. The Water Deficit Index was defined as follows:

$$WDI = \left(\frac{\theta - \theta_{fc}}{\theta_d} \right) \times 100$$

where θ is the soil moisture content, and θ_d is the difference between soil moisture at field capacity, θ_{fc} and the wilting point, θ_{wp}. When the WDI is below zero, the soil is subject to a water deficit and at a value of –100% it is absolute and passes to the wilting point. Positive WDI values represent conditions of excess water.

Analysis of mean monthly WDI values likewise revealed the great seasonal contrast, and, at the same time, allowed clear identification of the periods with excess water and those of water deficit. Along the nearly 4-year period of follow-up, it was observed that as the cycle of vegetative activity of the oaks began (second fortnight in May) the soil began to reach a situation of water deficit (WDI $<0\%$) and this stage persisted until October (Fig. 2). The most striking feature of this type of behaviour was that during the summer months the soil became completely depleted (WDI $\leq -100\%$) in terms of water availability and this level of water stress persisted over one or two months (August–September). This situation of maximum water deficit during the summer has been detected previously at comparable sites in nearby forested mountainous areas (Moreno *et al.*, 1996a,b), although to a lower extent, probably because of the use of different thresholds to define soil moisture availability. The reasons for this situation of marked and prolonged water deficit must be sought in both the reduction in precipitation and the increase in evapotranspiration (maximum ET in June, close to 160 mm), and in water uptake by the forest.

It was observed that marked episodes of precipitation during the period of water deficit (for example, 162 mm from 5 to 22 October 2001, and 183 mm from 15 to 23 September 2002) had little effect on the generation of runoff. The fact that soil water is depleted over a prolonged period, coinciding with the moment of greatest biological activity, means that in a great number of events, all the rainfall is absorbed and consumed and this does not show up either in the hydrographs or in the evolution of

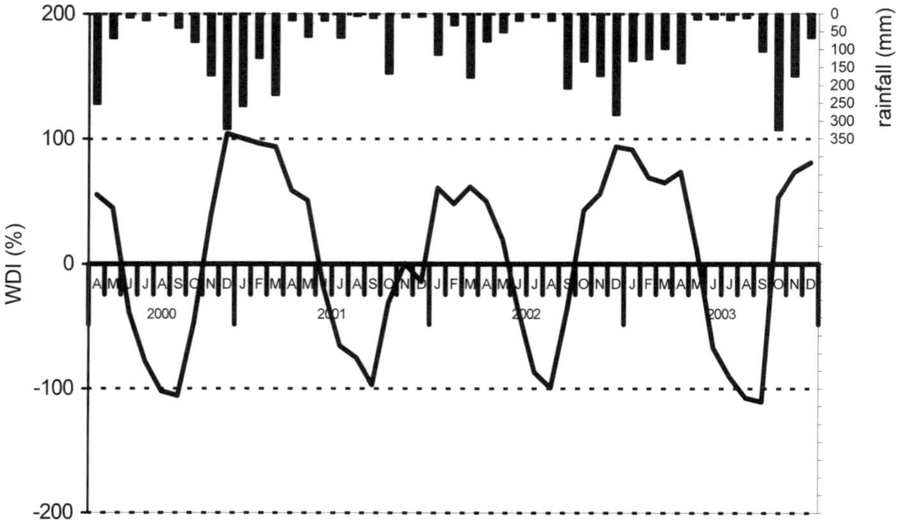

Fig. 2 Evolution of mean monthly rainfall and Water Deficit Index (WDI) values from April 2000 to December 2003.

soil moisture. Therefore, soil water recharge is incomplete during the dry season and conditions to generate runoff are inadequate because saturated flow was observed as the main discharge component on an annual scale basis. Saturated flow operates only when soil moisture content is high and such conditions do not occur during the dry season.

As a result of the foregoing, it was seen that during the period of soil water deficit the basin had a low capacity to generate runoff during the rainy periods (sometimes with considerable amounts of precipitation). In the four years of observations, the mean precipitation of the dry period (May–September) reached almost 200 mm. When the water in the soil-forest system had become totally depleted, the rainfall, however intense, from each episode was absorbed and the capacity to generate runoff was very low or absent. The interception of rainfall by the forest was around 17% during the wet season and 32% during the vegetative period, when the forest canopy provides maximum cover. Rainfall interception, the recharging of a soil avid for water and the intense ET occurring in the summer, consume the water from any rainy event during this period of the year.

Figures 3 and 4 show the relations between maximum rainfall intensity in 10-min intervals (I_{10}) and maximum discharge in 10 min and between daily rainfall and daily mean discharge at the Rinconada experimental basin for the period from January 2000 to October 2003. The behaviour of those variables during the wet season (October–April) and the dry season (May–September) is compared in both figures. In terms of both the maximum discharge generated and the mean discharge, the difference between the wet period and the dry one is very marked: not in absolute values (which is logical and obvious) but in relative terms; that is, in the capacity to generate runoff. The differences in the slope of the fitted lines show the different types of hydrological behaviour between both seasons. Intensity and peak discharge are better correlated in the dry season than in the wet season, whereas daily precipitation and discharge are not

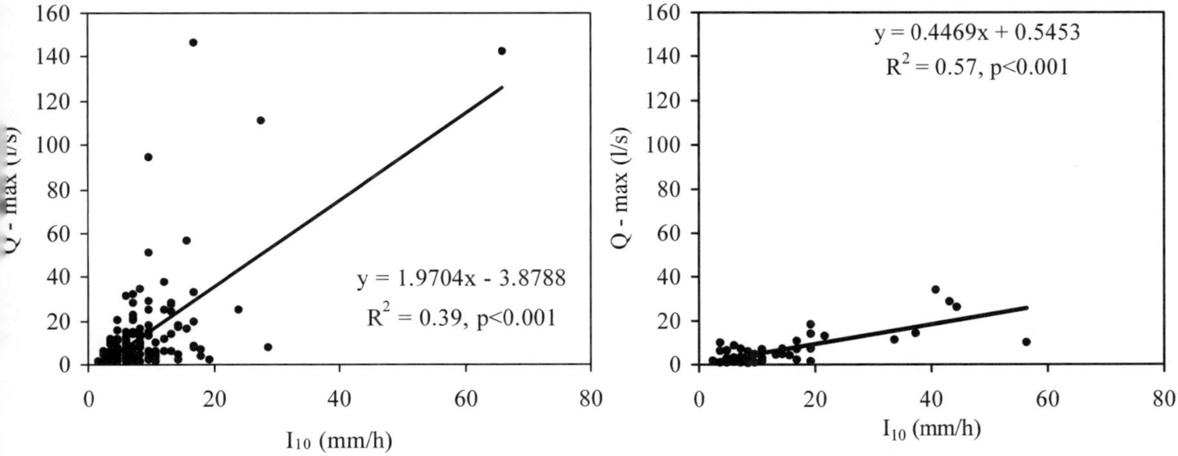

Fig. 3 Relationship between 10 min maximum rainfall intensity and 10 min maximum discharge for all events from October to April, 157 runoff events (left) and from May to September, 57 runoff events (right).

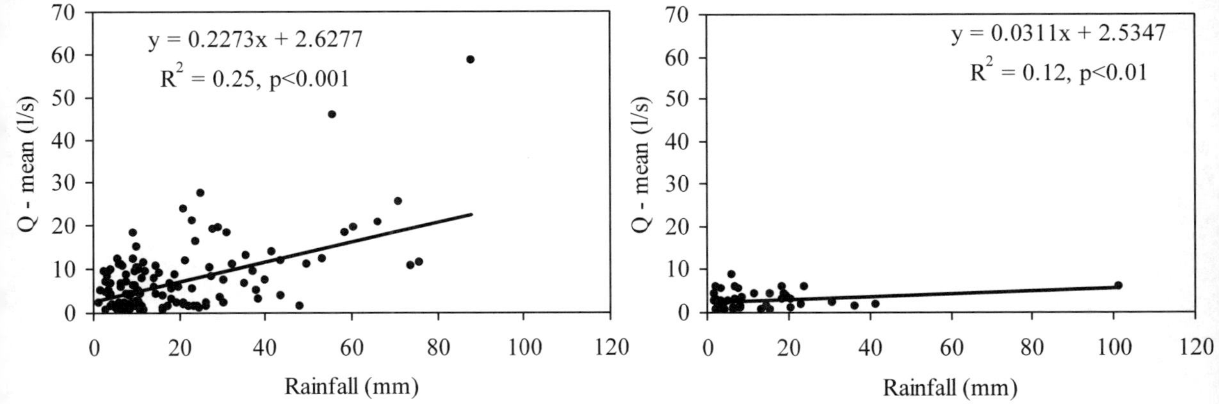

Fig. 4 Relationship between daily rainfall and daily mean discharge from October to April, 157 runoff events (left) and from May to September, 57 runoff events (right).

correlated in the dry season. During the wet season, runoff is fully controlled by saturated flow mechanisms. During the period in which the forest depleted the available water, the runoff yield from the basin was dramatically reduced, both as regards extreme values and in terms of mean values. The scarce discharge generated during the dry season only depends on Hortonian overland flow episodes linked to small deforested surface areas next to the outlet.

Acknowledgements This study was fully supported by the Spanish Ministry of Science and Technology (Project REN2003-00381) and the Regional Government of

Castilla-León (Project SA016/03). Authors thank Rosa M. Poch for helpful comments on the paper during the revision process.

REFERENCES

Andreassian, V. (2004) Waters and forests: from historical controversy to scientific debate. *J. Hydrol.* **291**, 1–27.

Beguería, S., López-Moreno, J. I., Lorente, A., Seeger, M. & García-Ruiz, J. M. (2003) Assessing the effect of climate oscillations and land-use changes on streamflow in the Central Spanish Pyrenees. *Ambio* **32**, 283–286.

Beschta, R. L., Pyles, M. R., Skaugset, A. E. & Surfleet, C. G. (2000) Peakflow responses to forest practices in the western cascades of Oregon, USA. *J. Hydrol.* **233**, 102–120.

Bosch, J. M. & Hewlett, J. D. (1982) A review of catchment experiments to determine the effect of vegetation on water yield and evapotranspiration. *J. Hydrol.* **55**, 3–23.

Joffre, R. & Rambal, S. (1993) How tree cover influences the water balance of Mediterranean rangelands. *Ecology* **74**, 570–582.

Moreno, G., Gallardo, J. F., Ingelmo, F., Cuadrado, S. & Hernández, J. (1996a) Soil water budget in four *Quercus pyrenaica* forests across a rainfall gradient. *Arid Soil Res. and Rehab.* **10**, 65–84.

Moreno, G., Gallardo, J. F., Schneider, K. & Ingelmo, F. (1996b) Water and bioelement fluxes in four *Quercus pyrenaica* forests along a pluviometric gradient. *Ann. Sci. Forest.* **53**, 625–639.

The effect of wildfire intensity on soil aggregate stability in the Cadiretes Massif, NE Spain

XAVIER ÚBEDA & SARA BERNIA
Departament de Geografia Física i AGR. Facultat de Geografia i Història, Universitat de Barcelona, C/ Baldiri Reixac s/n, 08028 Barcelona, Spain
xubeda@ub.edu

Abstract This paper reports on the changes in soil aggregate stability after a wildfire, with the focus on the effects of fire intensity on soil structure. The aggregate stability of soils burned at different fire intensities was tested using the TDI method. The samples were taken from three areas burned at three fire intensities and the results of the aggregates were compared with those from an unburned forest. The study area is located in the Cadiretes massif, northeast Spain. The analyses were undertaken just after the fire in 1994, and again in 1995 and 1997. The most stable aggregates compared to those from the control area were those sampled just after the fire. The disaggregation was highest in 1995 and lowest in 1997. The fire did not disturb the stability of the aggregates immediately after burning. After the fire the aggregates did not become unstable and, in contrast, they were even more stable than those sampled in the control forest.

Key words fire intensity; forest fire; soil aggregate stability; TDI test

INTRODUCTION

Fires are frequent in the Mediterranean environment and have been a significant environmental agent over the last few decades in Spain (Vélez, 1997). The variation in the intensity of these forest fires has an important effect on the changes in the soil after burning: on occasions these changes may be almost imperceptible, whereas elsewhere they may be particularly pronounced (Úbeda & Sala, 2001; Dragovich & Morris, 2002). Fire intensity also determines the losses of solute and suspended soil as Gimeno *et al.* (2000) reported in burnt shrublands in eastern Spain, although this is mainly found for the months or few years after fire.

During a forest fire, changes in the physical, chemical and biological properties of a soil are rapid, while in the period following a forest fire such changes can be either locally confined (erosion processes due to storm events) or widespread (regrowth of roots and its consequent effects). Ternan & Neller (1999), in an analysis of soil properties after forest fires, concluded that there are complex interactions and feedback between a soil's physical and chemical properties, aggregate stability and pore size characteristics, which vary spatially with time as the soil and vegetation recover. They claim that there is a need for more studies to shed more light on these interactions.

Soil aggregate stability indicates the resistance of its aggregates to the impact of rainfall, its consistency and the ease with which it disaggregates (Imeson & Vis, 1984). It is widely assumed that aggregate stability is related to soil erodibility. This paper focuses on the Mediterranean area where environmental conditions clearly favour soil degradation. Furthermore, given that in this environment most fires occur during the summer, the abandoned lands are exposed to frequent, heavy autumn storms and so the soils are extremely vulnerable (Martínez & Diaz, 1994).

Following the breakdown of soil aggregates, a crust of fine particles is formed, sealing the soil surface and further reducing the rates at which water can be absorbed by the soil (Farres, 1978). As a result overland flow increases, leading to more erosion (Ellison, 1947). This suggests that aggregate stability is a key determinant of erosion behaviour, particularly if we bear in mind that environmental hazards, such as forest fires, can modify soil properties.

The relationship between aggregate stability and soil properties has been widely studied (Tisdall & Oates, 1982; Rengasamy *et al.*, 1984; Ternan *et al.*, 1996). Organic matter content, the quantity of fine particles (principally clay), micro-organisms, and iron and aluminium oxides, are thought to be essential characteristics for ensuring good aggregate stability. Furthermore, all these soil characteristics can be modified by soil management practices (Imeson & Jungerius, 1976; Ternan *et al.*, 1996). In Mediterranean climatic conditions the relationship between soil structure and soil properties has been studied by Boix (2000) and the results of this study confirm the positive influence of clay, organic matter and earthworm activity on soil aggregate stability. The aim of this paper is to determine the effects of fire and fire intensity on aggregate stability during the recovery period.

STUDY AREA

The results reported here refer to soils located in the Cadiretes Massif, in the northernmost zone of the Catalan Coastal Ranges, northeast Spain, at an altitude of around 190–250 m a.m.s.l. The Cadiretes Massif is predominantly granite, although soils on Palaeozoic metamorphic rocks such as schist and slates can also be found. Erosion processes have largely shaped the granitic section of the Cadiretes Massif, though in some areas metamorphic features underlie this relief. The massif is covered by dense Mediterranean vegetation, e.g. *Quercus suber*, *Arbutus unedo*, *Erica arborea*, and in some places *Pinus pinaster* plantations are found. The area receives 700–800 mm of annual rainfall, with a fairly marked seasonal variability. The maximum is registered in autumn (36% of the total), while winter (26%) and spring (26%) are also wet seasons. Summer (12%) is the driest. Autumn is also the season with greatest rainfall intensity, and values exceeding 15 mm h^{-1} are frequent. Summer temperatures often surpass 25°C, while in winter temperatures are generally mild and frost is very rare. Potential evapotranspiration exceeds precipitation from June to August, which results in a very low soil moisture content.

The predominant soil type in Cadiretes is classified as a Lithic Xerochrept (Úbeda, 1998), with a 15-cm deep sandy-loam A horizon. In the control forest area, this horizon is protected by a 3-cm deep O horizon of moder humus. However, in the most intensely burned areas, the humus has disappeared, initially because of the combustion episode, and subsequently because of erosion. These top centimetres of the soil contain large quantities (35–40%) of coarse particles (>2 mm) and there are also high quantities of organic matter content (10%). The organic matter content in the burnt areas decreased over time reaching values below 5%. The pH of the soil ranges from 6.0 to 6.5.

DETERMINATION OF THE BURNED AREAS

A field survey was undertaken immediately after the forest fire. Three different zones were identified according to the tree remains, the state and number of leaves and branches, the soil surface and the quantity and colour of ashes (black-low, grey-medium and white-high combustion) (Moreno & Oechel, 1989). After identifying these three zones, three sampling areas—one from each zone—were chosen. The areas had the same orientation (S) and slope (5%). The same characteristics were recognized in an adjacent unburned area, which served as a control for the forest sites. The samples used in determining aggregate stability were collected just after the fire (July 1994), eight months after the fire (March 1995), when the main erosion events had already occurred, and 32 months after the fire (March 1997), at the end of the study period. Fourteen samples were collected in each of the four forest areas during each sampling, and 20 aggregates were tested in each sample. A total of 3360 aggregates were analysed. All samples were collected using rings (8 cm wide and 5 cm high) from the top of the soil surface. A total of 496.5 mm of rain fell between the first and second samplings and 949 mm between the second and third samplings. This is a normal volume in this zone although a particularly intensive rainfall event was recorded on 20 October 2004 with a peak intensity of 55 mm h^{-1}.

LABORATORY AND STATISTICAL METHODS

The TDI (Ten Drop Impact) simulates rainfall impact on aggregates and is described in Low (1954). Aggregates of 4–4.8 mm were selected by sieving. Ten aggregates (4–4.8 mm air-dried) were subjected to the impact of 10 drops from a burette fixed at a height of 1 m. The aggregates were placed on a 2.8 mm sieve to allow the disaggregated sample to flow away. The drops of distilled water weighed 0.1 ± 0.001 g and had a diameter of 5.8 mm. The aggregates were weighed before the test and the disaggregated material, which passed through the sieve during the test, was also weighed after drying at 105°C for 24 h. The results are expressed as a percentage of the total aggregate weight.

It is widely assumed that burned soil aggregates are selected by the surface wash. Thus, the weaker aggregates disappear and only the stronger aggregates are sampled. To verify this assumption, the first aggregates to be analysed in this study were sampled just after the fire, before any rainfall was recorded.

The statistical tests were applied using SPSS 11.0 and were initiated with a summary analysis. Afterwards, a contrast analysis was performed on the data to determine whether there were any differences between the variable values recorded at the three sites of varying fire intensity and the control area during the study period. Before this test could be applied, we first need to determine whether the data were parametric or non-parametric. In order to detect this characteristic, three conditions must first be met (Mateos-Aparicio & Martín, 2003):

(a) Normality, which was verified by performing the Kolmogorov-Smirnov test.
(b) Homoscedasticity or the homogeneity of the population's variance, which was checked by applying Levene's variance equality test.

(c) Independence of the observations, which was verified by examining the scatter areas of the residuals compared with the values observed for each variable and parameter (observations are considered independent when no underlying trends are observed in the scatter areas).

After ensuring that these requirements were met, we performed Kendall's *W* non-parametric contrast test for related samples which provided us with the Kendall's *W* and the Chi-squared indexes. The hypotheses applied for this test are:

(a) H_0: (null hypothesis, $P_{\text{value}} > 0.05$) the significance values in the comparison of the three different zones of fire intensity and the control area for each year studied are not statistically representative and they can be explained by chance.
(b) H_A: (alternative hypothesis, $P_{\text{value}} < 0.05$) there are statistically significant differences between the different zones of fire intensity and the control area for each studied year and these differences cannot be explained by chance.

RESULTS

A summary of the statistical results of the four areas (high intensity, medium intensity, low intensity and control area) in 1994, 1995 and 1997 are shown in Table 1. The most stable aggregates were those sampled just after the fire (1994), presenting 1.89, 2.09 and 2.26% of disaggregation as opposed to 5.40% in the control forest. The disaggregation was higher in 1995 (6.15, 7.78 and 8.70%) and 1997 (3.95, 4.50 and 3.05%). If we examine data variability, we find that the results from the area of low fire intensity are the most homogeneous, presenting the lowest standard deviation and variance. In the case of the area of medium fire intensity, the variability of the data is quite similar to that of the control plot for 1995 and 1997. In the case of the area of high fire intensity, the period studied here is not long enough for the values of standard deviation and variance to have reached those of the control plot. Figure 1 clearly shows that aggregate stability at the control plot is the most constant over time, whereas the highest stability in the burned areas was recorded immediately after the fire, although one year later aggregate stability was at its lowest point during the period of study.

Table 1 Statistics of the eroded aggregate data from the four areas in each studied year.

Year	Fire intensity	N	Average	Standard deviation	Variance
1994	high	14	1.89	0.74	0.55
	medium	14	2.26	0.86	0.74
	low	14	2.09	0.38	0.15
	control	14	5.40	1.85	3.41
1995	high	14	9.13	7.75	60.07
	medium	14	7.78	5.22	27.21
	low	14	8.70	4.38	19.21
	control	14	6.15	5.02	25.17
1997	high	14	7.06	7.75	60.01
	medium	14	4.51	4.16	17.30
	low	14	3.96	1.99	3.95
	control	14	7.05	4.23	17.85

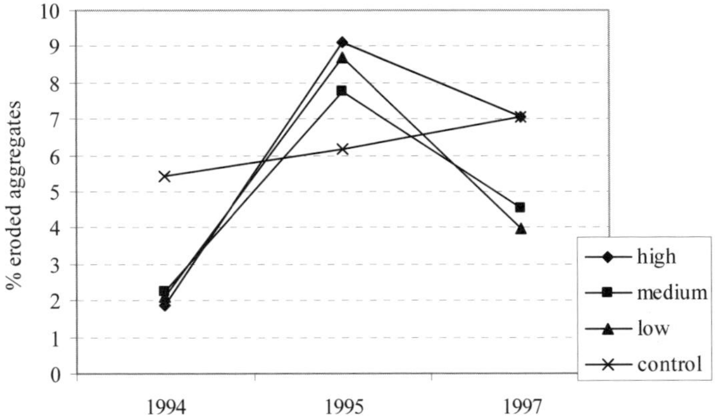

Fig. 1 Aggregate stability throughout time from the four studied areas.

In order to conduct the contrast analysis, we first verified the condition of normality by applying the Kolmogorov-Smirnov test. This test was run for each fire intensity and for each sample year. Results are shown in Tables 2 and 3. In the case of fire intensity (Table 2), the analysed data does not fit the normal distribution for the areas of high and medium fire intensity—the probability of the *Z* statistic is smaller than 0.05. Only the data from 1995 fits the normal distribution, as here the significance is greater than 0.05 (Table 3).

Table 2 Kolmogorov-Smirnov Test for each fire intensity.

		High	Medium	Low	Control
N		42	42	42	42
Normal parameters	Average	6.03	4.85	4.92	6.20
	Standard deviation	6.91	4.43	3.92	3.90
Differences	Absolute	0.27	0.20	0.18	0.18
	Positive	0.27	0.20	0.18	0.18
	Negative	–0.21	–0.17	–0.15	–0.09
Kolmogorov-Smirnov *Z*		1.72	1.32	1.18	1.20
Significance		0.01	0.06	0.12	0.11

Table 3 Kolmogorov-Smirnov Test for each study year.

		1994	1995	1997
N		56	56	56
Normal parameters	Average	2.91	7.94	5.64
	Standard deviation	1.81	5.70	5.05
Differences	Absolute	0.23	0.13	0.22
	Positive	0.23	0.13	0.22
	Negative	–0.14	–0.09	–0.15
Kolmogorov-Smirnov *Z*		1.72	0.98	1.65
Significance		0.01	0.29	0.01

To verify the second condition, we studied the homocedasticity of the data. This was done by applying Levene's variance equality test. The results for each fire intensity area and year are presented in Tables 4 and 5, respectively. Levene's index (Table 4) presented the lowest variability in the control area (Fig. 1). In each of the areas studied the differences between the variances are statistically significant. Only in the case of the sample year 1997 is the variance of the data homogeneous, as here the significance value is higher than 0.05 (Table 5). This is due to the fact that Levene's index is the lowest for this year. This finding was previously noted in Table 1 on observing the variance, and is here demonstrated numerically. The variability of the data in the burnt areas over time is close to that in the data for the control plot. Given the results of these last two tests, we did not verify the last condition—the independence of the observations.

Table 4 Test of variance homogeneity between fire intensities.

	Levene index	df1	df2	Sig.
High	9.66	2	39	0.00
Medium	5.09	2	39	0.01
Low	12.44	2	39	0.00
Control	3.31	2	39	0.05

Table 5 Test of variance homogeneity between study years.

	Levene index	df1	df2	Sig.
1994	7.75	3	52	0.00
1995	3.21	3	52	0.03
1997	2.61	3	52	0.06

Table 6 Kendall's *W* test for fire intensities.

High	N	14
	Kendall W	0.24
	Chi-squared	6.86
	df	2
	Significance	0.03
Medium	N	14
	Kendall W	0.43
	Chi-squared	12.00
	dfl	2
	Significance	0.00
Low	N	14
	Kendall W	0.68
	Chi-squared	19.00
	dfl	2
	Significance	0.00
Control	N	14
	Kendall W	0.01
	Chi-squared	0.14
	df	2
	Significance	0.93

Finally, we performed Kendall's *W* non-parametric contrast test for related samples. The Kendall's *W* and the Chi-Squared indexes were used to analyse the contrast between fire intensities and years. Table 6 shows the results for the different fire intensities in each year. Only the control area showed no statistical differences between the results in each year, since the significance recorded is higher than 0.05. All the burned areas, however, presented significant statistical differences between the values for the years 1994, 1995 and 1997 ($P_{value} < 0.05$), which means that these differences cannot be explained by chance. In the range of values, sample year 1994 was most markedly different, as this was when the effects of the forest fire were most significant.

Table 7 shows whether there were significant differences between the results in each area and between each year (1994, 1995 and 1997) because of the varying degrees of fire intensity. As mentioned above, 1994 was found to be the only year with statistically significant differences between fire intensities. This is shown in Table 6 where the significance value is lower than 0.05. The year with the highest disaggregation values was 1995 (7.9%). In 1997, 5.6% disaggregation was recorded while the lowest value was recorded in 1994, just after the fire, with 2.9% disaggregation, which can be considered a particularly low value.

Table 7 Kendall's *W* test by year between studied areas.

1994	N	14
	Kendall *W*	0.40
	Chi-squared	16.89
	df	3
	Significance	0.00
1995	N	14
	Kendall *W*	0.08
	Chi-squared	3.17
	df	3
	Significance	0.37
1997	N	14
	Kendall *W*	0.14
	Chi-squared	5.89
	df	3
	Significance	0.12

DISCUSSION

De Meester & Jungerius (1978) found that in soils with a high percentage of sand and silt the aggregate stability is closely related to the K-factor in the Universal Soil Loss Equation. However, Bergsma & Valenzuela (1981) did not find a significant correlation between drop test results and the K-value.

The differences in aggregate stability in soil samples at different periods of the year may be due to differences in their moisture content. Rasiah *et al.*, (1992) found that aggregate stability decreased linearly with increasing moisture content, although this does not constitute a standard correlation for all soil types. However, certain soil characteristics, such as clay content, ensure this correlation.

Mataix-Solera *et al.* (2002), in a study conducted in southeast Spain, found a decrease in the number of stable aggregates in the summer. This can be accounted for by the fact that the number of carbohydrates in the soil, which favour aggregation, decreases because of an increase in microbial activity (Lax & García-Orenes, 1993). Mataix-Solera *et al.* (2002) reported a recuperation of aggregate stability after a forest fire coinciding with the recovery in organic matter content.

The samples taken here from the burned areas presented a high aggregate stability. Those collected in the first sampling period—just after the fire—showed the highest stability, which is probably due to the fact that no rain fell in the study area in the period between the fire and the sampling episode. The aggregates were clearly more stable in the three burned areas than those sampled in the unburned area. Mataix-Solera & Doerr (2004), who reported similar findings, explain that aggregate stability increases with hydrophobicity. The increased stability after fire is due, it seems, to increase hydrophobicity of soil aggregates, which prevents slaking. However, these soils may be more susceptible to erosion due to increased runoff.

A few months (1995) and years (1997) after the fire, the aggregates had become increasingly more unstable. This might have been due to the repeated impact of rainfall on the soil surface and the various changes recorded at the soil surface (raindrop impact, crusting, soil erosion) (Úbeda, 1998). Yet despite this, the aggregates remained quite stable, if we compare them with the aggregates sampled in the control forest.

Other studies report the fact that exposure to high temperatures make the aggregates stronger. Giovannini (1994), for example, describes changes in aggregate stability according to thermal reactions that make the aggregates more stable, suggesting that the most pronounced increase in aggregate stability occurs in a temperature range of 220–460°C. This seems to be typical in dry soils, but is less frequent in sandy-loam soils.

The differences in stability recorded here in the control forest during the three sampling episodes may be due to the initial soil moisture content when sampling was performed. Cerdà (1998), in a study involving different Mediterranean vegetation types, found that aggregates sampled at the same site show a marked predisposition to disaggregation by rainfall impact after dry periods. He observed that a higher runoff sediment concentration is to be expected after a dry period when aggregates are more breakable.

CONCLUSIONS

Aggregate stability was high in the soils of the Cadiretes Massif and forest fires were found not to disturb the high stability of their soil structure. A significant increase in aggregate stability was recorded immediately after the fire, which decreased after one year. Fire intensity seems to have a marked impact on soil aggregate stability.

Acknowledgements This study was supported by the European Community Project EV5V-CT91-0043, "Desertification Risk Assessment and Land Use Planning in a Mediterranean Coastal Area" and the project "RESEL" from the Spanish Ministry of

Environment. We are also grateful for the useful suggestions on the paper by Artemi Cerdà and by an anonymous reviewer.

REFERENCES

Bergsma, E. & Valenzuela, C. R. (1981) Drop testing aggregate stability of some soils near Mérida, Spain. *Earth Surf. Processes Landf.* **6**, 309–318.

Boix, C. (2000) Procesos geomórficos en diferentes condiciones ambientales mediterráneas: el estudio de la agregación y la hidrología de los suelos. PhD Thesis, Faculty of Geography and History, University of Valencia, Valencia, Spain.

Cerdà, A. (1998) Soil aggregate stability under different Mediterranean vegetation types. *Catena* **32**, 73–86.

De Meester, T. & Jungerius, P. D. (1978) The relationship between the soil erodibility factor K, aggregate stability and micromorphological properties of soils in the Hornos Area, S. Spain. *Earth Surf. Processes Landf.* **3**, 379–391.

Dragovich, D. & Morris, R. (2002) Fire intensity, slopewash and bio-transfer of sediment in Eucalyptus forest, Australia. *Earth Surf. Processes Landf.* **27**, 1309–1319.

Ellison, W. D. (1947) Soil erosion studies. Parts I-VII. *Agricultural Engineering* **28**, 145, 197, 245, 297, 349, 402 & 442.

Farres, P. (1978) The role of time and aggregate size in the crusting process. *Earth Surf. Processes Landf.* **3**, 243–254.

Gimeno, E., Andreu, V. & Rubio, J. C. (2000) Changes in organic matter, nitrogen phosphorous and cations in soil as a result of fire and water erosion in a Mediterranean landscape. *European J. Soil Sci.* **51**, 201–210.

Giovannini, G. (1994) The effect of fire on soil quality. In: *Soil Erosion and Degradation as a Consequence of Forest Fires* (ed. by M. Sala & J. L. Rubio), 15–27. Geoforma Ediciones.

Imeson, A. C. & Jungerius, P. D. (1976) Aggregate stability and colluviation in the Luxembourg Ardennes as exemplified by hydrological and micropedological study. *Earth Surf. Processes Landf.* **1**, 259–271.

Imeson, A. C. & Vis, M. (1984) Assessing soil aggregate stability by water-drop impact and ultrasonic dispersion. *Geoderma* **34**, 185–200.

Lax, A & García-Orenes, F. (1993) Carbohydrates of municipal solid wastes as aggregation factor of soils. *Soil Technol.* **6**, 157–162.

Low, A. J. (1954) The study of soil structure in the field and in the laboratory. *J. Soil Sci.* **5**, 57–74.

Martínez, J. & Díaz, E. (1994) Changes of the physical and chemical properties in a soil affected by forest fire in Sierra Larga (Murcia, Spain). In: *Soil Erosion and Degradation as a Consequence of Forest Fires* (ed. by M. Sala & J. L. Rubio), 67–77. Geoforma Ediciones.

Mataix-Solera, J., Gómez, I., Navarro-Pedreño, C., Guerrero, C. & Moral, R. (2002) Soil organic matter and aggregation affected by wildfire in a *Pinus halepensis* forest in a Mediterranean environment. *Int. J. Wildland Fire* **11**, 107–114.

Mataix-Solera, J. & Doerr, S. H. (2004) Hydrophobicity and aggregate stability in calcareous topsoils from fire-affected pine forest in southeastern Spain. *Geoderma* **118**(1–2), 77–88.

Mateos-Aparicio, G. & Martín Dávila, M. (2003) Análisis de la varianza y de la covarianza. In: *Análisis Multivariable para las Ciencias Sociales* (ed. by J. P. Levy & J. Varela), 145–213. Pearson, Prentice Hall, Madrid, Spain.

Moreno, M. & Oechel, W. C. (1989) A simple method for estimating fire intensity after a burn in California chaparral. *Soil Sci. Soc. Am. J.* **55**, 235–240.

Rasiah, V., Kay, B. D. & Martin, T. (1992) Variation of structural stability with water content: influence of selected soil properties. *Soil Sci. Soc. Am. J.* **56**, 1604–1609.

Rengasamy, P., Greene, R. S. B. & Ford, G. W. (1984) The role of clay fraction in the particles arrangement and stability of soil aggregates—a review. *Clay Res.* **32**, 53–67.

Ternan, J. L. & Neller, R. (1999) The erodibility of soils beneath wildfire prone grasslands in the humid tropics, Hong Kong. *Catena* **36**, 49–64.

Ternan, J. L., Williams, A. G., Elmes, A. & Hartley, R. (1996) Aggregate stability of soils in central Spain and the role of land management. *Earth Surf. Processes Landf.* **21**, 181–193.

Tisdall, J. M. & Oates, J. M. (1982) Organic matter and water-stable aggregates in soils. *J. Soil Sci.* **33**, 141–163.

Úbeda, X. (1998) Efectes de les diferents intensitats de foc, durant els incendis forestals, en els paràmetres físico-químics del sòl i en l'increment de l'escolament i de l'erosió. PhD Thesis, Universitat de Barcelona, Barcelona, Spain.

Úbeda, X. & Sala, M. (2001) Chemical concentrations in overland flow from different forested areas in a Mediterranean Environment: burned forest at different fire intensity and unpaved road. *Z. Geomorphol.* **45**(2), 225–238.

Vélez, R. (1997) Recent history of forest fires in Mediterranean area. In: *Proceedings of the European school of Climatology and Natural Hazards Course*, 15–26. European Commission, Luxembourg.

Geomorphological Processes and Human Impacts in River Basins (Proceedings of the International Conference held at Solsona, Catalonia, Spain, May 2004). IAHS Publ. 299, 2005.

Sediment delivery from a landslide to a stream in a drainage basin in Hong Kong

MERVYN R. PEART[1], MARK E. RUSE[2] & LINCOLN FOK[1]

1 *Department of Geography, The University of Hong Kong, Hong Kong, China*
mrpeart@hkucc.hku.hk

2 *Fugro (Hong Kong) Ltd, Hong Kong, China*

Abstract Landslides commonly occur on undeveloped hillslopes in Hong Kong, China. Heavy rainfall in June 2001 and May 2003 resulted in small landslides in a headwater drainage basin. Measurements of suspended sediment concentration made during the storm events, along with observations on sediment colour and C and N content, indicate that the suspended matter in streamflow was derived from the landslides, evidencing the connectivity of hillslope and channel systems. Estimates of the volume of failed material when compared to those of deposition revealed that much of the debris from the landslides remained on the slope in an ephemeral drainage line. At-a-point erosion from the landslides is much greater than that recorded by erosion pins in the same catchment.

Key words hysteresis; landslide; sediment colour; suspended sediment

INTRODUCTION

In Hong Kong, landslides commonly occur both within the developed urban area, often in association with cut and fill slopes (cf. Brand, 1985), and on undeveloped hillslopes, termed "natural terrain" (Ng *et al.*, 2002). The Natural Terrain Landslide Study identified some 8804 recent landslides for hillslopes in Hong Kong (Evans *et al.*, 1999) from approximately 50 years of aerial photographic records. Research on landslides in "natural terrain" has focused upon identifying controlling factors in their location and generation (e.g. Franks, 1999; Dai *et al.*, 2001; Ruse *et al.*, 2002) and assessing susceptibility and hazard (e.g. Franks, 1999; Dai & Lee, 2002; Parry *et al.*, 2002). However, comparatively little attention has been given in Hong Kong to other geomorphological aspects of landslides, and in particular the linkage, or coupling, between hillslope and channel systems. Harvey (2002) suggests that understanding coupling within the fluvial system is important because it is influential in determining the geomorphic response to environmental change. He also observes that within headwater fluvial systems the most significant aspect of the linkage is hillslope channel coupling. Knowledge of hillslope channel connectivity in headwater streams is also important because, as Haigh *et al.* (1998) and Gomi *et al.* (2002) indicate, headwater catchments exert a control upon downstream fluvial systems. This study documents the linkage between hillslope and channel systems consequent upon landslides in terms of sediment production for a small headwater drainage basin in Hong Kong.

STUDY AREA

Monitoring of suspended sediment concentration began in 1993 in a small (0.052 km^2) drainage basin located near Wong Chuk Hang in the New Territories of Hong Kong, China. The basin, termed RDH, is a headwater stream of the Kam Tin River. The average slope gradient is 30°, with some areas being over 50° and the highest point of the drainage basin being around 380 m. The upper parts of the basin consist of granodiorite and volcanic tuffs with siltstone and sandstone interbeds, all of which have been subjected to weathering. Colluvium deposits occur on the slopes whilst soils are shallow, moderately acid Acrorthoxes. The hillslopes are covered in a mix of *Dicranopteris* fern, grassland, shrubland and a small area of *Acacia* trees. Hill fires have occasionally affected the area.

Climatically the area is subtropical monsoonal with about 80% of the rain falling between May and September. The mean annual rainfall of 2214 mm for the period 1961–1990, measured at the Hong Kong Observatory, hides a considerable range. The extremes, since 1884, are 901.1 mm in 1963 and over 3330 mm in 1997. At the Kadoorie Agricultural Research Centre (KARC), located close to the study basin, the mean annual rainfall for the period 1991–2001 is 2747 mm, with an annual maximum of over 3600 mm in 2001. May to August are hot and humid, with a mean daily temperature of 28.8°C in August (1961–1990), whilst the winters are more temperate with an average daily temperature of 15.8°C in January over the same period.

METHODS

Sampling to monitor suspended sediment concentrations at the outlet of the small headwater catchment began in 1993. The samples (a water-sediment mixture) have been collected over a range of flow conditions and have been collected manually and by means of an ISCO 2700 automatic sampler. Typical sample volumes are around 400 to 500 ml for the automatic sampler with larger volumes sometimes being collected by manual sampling in order to ensure sufficient sediment for analysis of C and N. Upon return to the laboratory the suspended sediment was separated using pre-weighed GF/C filter papers. Total C and N was measured on a Perkin Elmer 2400 elemental analyser after the sediment was disaggregated and passed through a 250 micron mesh sieve. Sediment colour was determined using standard soil colour charts. Water level was manually recorded at the time of sampling by means of a stage board at a v-notch weir located at the catchment outlet.

A number of erosion plots (6 × 20 m) with 30 erosion pins, regularly spaced in three rows of 10 (to aid relocation), have been maintained to measure the change in ground surface. The pins were inserted to a minimum depth of 40 cm in the soil to provide a stable datum. Pins were measured at least annually in January of each year since observations began on six plots in 1992. Average slope angles for plots 1 to 6 are 15°, 15°, 26°, 27°, 25° and 26°, respectively.

The landslides that activated during rainfall in June 2001 and May 2003 were accessible and field surveyed to gather data. The information recorded included data relating to both the landslide source (e.g. slope angle above the main scarp, rupture surface angle, main scarp depth, width) and the associated debris trail (e.g. depth, width, length). Volume calculations of both source and debris trail were obtained by multiplying measured area by their average measured depth. These were based upon a tape/chain field survey of the failure. These calculations permitted the determination of a mass balance for the landslide in terms of volume of material produced from the source and the volume of debris remaining in storage in the debris trail.

LANDSLIDES

Two rainfall events have generated landslides in the basin since observations began in 1992. The first occurred on 9 June 2001, in association with a low pressure trough near the South China coast. Nearly 100 mm of rainfall was recorded at KARC from 09:00 h on 8 June to 09:00 h on 9 June, with subsequent hourly rainfall totals of 68 and 52 mm. The monthly rainfall of 1083.6 mm at the Hong Kong Observatory was roughly three times the normal. In the second rainfall event, a trough of low pressure arising from an unstable maritime airstream produced rainfall sufficient for the issuance of a landslide warning on 5 May 2003. Around 190 mm of rainfall were recorded at KARC between 09:00 h on 4 May to 09:00 h on 5 May, with an hourly maximum of around 40 mm. Based upon the record of landslides in Hong Kong from the Natural Terrain Landslide Inventory for the period 1985 to 1994 inclusive, Evans *et al.* (1999) report the possible existence of rainfall thresholds. They suggest that for areas receiving an annual rainfall of between 2000 and 2400 mm landsliding will start with a 24-h rainfall total of 60–70 mm whilst in areas receiving an annual rainfall in excess of 2400 mm the threshold for landslide initiation is 70–110 mm. Both the landslide events observed in the study basin support these thresholds. Evans *et al.* (1999) state that their thresholds are average values and that site-specific rainfall thresholds will depend on a complex combination of factors and may vary widely.

Both landslides occurred at the head of an ephemeral drainage line between two minor spurs, at an altitude of around 336 m and in an area of low shrub and grass. The studies of Franks (1999), Dai & Lee (2002) and Ruse *et al.* (2002) in Hong Kong reveal that many landslides occur within or adjacent to significant drainage lines or hollows. Relict failures are evident on the western spur, with minor retrogression of the main scarp observed in 2000. The June 2001 failure occurred immediately upslope of the relict scarp and the landslide of May 2003 was a further scarp retrogression. Franks (1998) and Ruse *et al.* (2002) report that landslides in Hong Kong may be associated with areas of previous instability. The failures occurred in colluvium containing considerable amounts of sub-angular cobbles and some boulders. Both were relatively shallow; just over 1 m deep in 2001 and around 0.8 m deep in 2003. Pre-failure slope angles and rupture surface angles were around 44° and 46°, respectively, in 2001, 37° and 42° in 2003. The 2001 event was around 9 m wide, about 7 m long and had an estimated source area volume of around 50 m^3. The rather smaller 2003 mass movement had a maximum width and length of 4 and 2 m, respectively, and an estimated volume of 7 m^3.

The main debris trails extended around 20 m on the 2001 failure and around 26 m for the 2003 event and are within the range of values reported by Franks (1999) for a study based on landslides that occurred on hillslopes in Lantau Island, Hong Kong. Based upon the area of failed material deposition and the depth of the materials, deposition in the two events was estimated at 52 and 14 m^3 for 2001 and 2003, respectively. These simplistic estimates of deposition illustrate that much of the failed material remained in storage on the slope and in the ephemeral drainage channel. No pronounced scouring of the channel downstream of the failure was observed, suggesting that for these two events additional erosion caused by the landslide itself scouring the hillslope and channel was not important. This may reflect the fact the two failures did not develop into debris flows, in which scouring can lead to the incorporation of material into the landslide, a process referred to as bulking by Scott *et al.* (2001). They provide a number of examples of bulking whilst the work of Franks (1999), Ruse *et al.* (2002) and King & Williamson (2002) show that the process can occur in association with landslides in Hong Kong. Evidence that some of the failed material left the slope is provided by observations of suspended sediment transport made in a small stream downstream of the failures.

SUPSPENDED SEDIMENT TRANSPORT

For both the landslide affected storm events of June 2001 and May 2003 pre-storm values of suspended sediment were low, around 10 mg l^{-1}, levels to which suspended sediment concentrations returned at the end of the two events. However, exceptionally high suspended sediment concentrations were recorded during the June 2001 failure, with four samples exceeding 2600 mg l^{-1}. The much smaller landslide event of May 2003 produced, in comparison to the June 2001 storm, rather lower levels of suspended sediment, with a maximum of 227 mg l^{-1}. This may reflect the considerably smaller size of the 2003 failure. During 1993–2000, only 10% of all storm period samples (n = 566) exceeded 200 mg l^{-1}. This indicates that even the lower May 2003 suspended sediment maximum value of 227 mg l^{-1} is unusually high for the basin, while the concentrations recorded in June 2001 are exceptional. Prior to this event, the recorded maximum for this stream was 1419 mg l^{-1}, and only two samples had sediment concentrations in excess of 1000 mg l^{-1}.

The time series plots in Fig. 1(a) reveal that peak suspended sediment concentrations for both events lag behind the maximum water level, although this trend is less clearly developed for the June 2001 event with its two maxima of suspended sediment. Figure 1(b) reveals the hysteresis, a feature of storm period response reported by a number of studies (e.g. Wood, 1977; Williams, 1989), is present for the two events. Figure 1(b) reveals that for both landslide events an anti-clockwise loop is present, which results when sediment concentration peaks on the recession limb of the hydrograph. As Williams (1989) observes, one possible cause of such a feature is the relative travel time of the flood wave and the sediment flux with the former travelling faster than the latter and the lag time between the two increasing with distance downstream. He reports a second cause of anti-clockwise loops as being the combination of highly erodible soil and prolonged erosion during the flood, whilst a third cause relates to annual hydrographs on large rivers and is not applicable to this small basin. In this well-vegetated basin highly erodible conditions do not exist and this is supported by the generally low concentrations of suspended sediment that have been observed in the basin with, as noted previously, only 10% of all storm period samples exceeding 200 mg l^{-1}. Given the small size of the basin, differences in travel time between sediment and water peaks may be discounted as causing anti-clockwise hysteresis. For these particular storms anti-clockwise hysteresis is likely to reflect sediment release/production consequent upon the mass movement event and the fact that the failures occurred in the upper reaches of the basin, thereby maximizing travel distance. It is also noteworthy, given the small size of the study basin, that Williams (1989) indicates that in small streams the maximum sediment concentration usually occurs prior to the peak water discharge.

SEDIMENT PROPERTIES

Evidence for the contribution from the landslides to the sediment being transported in the stream is provided by sediment colour. Dry soil colour of bulk samples in the soil horizons is dull yellowish brown or yellowish brown for the organic topsoil horizon and dull yellow/yellowish and light yellow orange for the underlying colluvial substrate. In the storms preceding both landslides, sediment colour is associated with

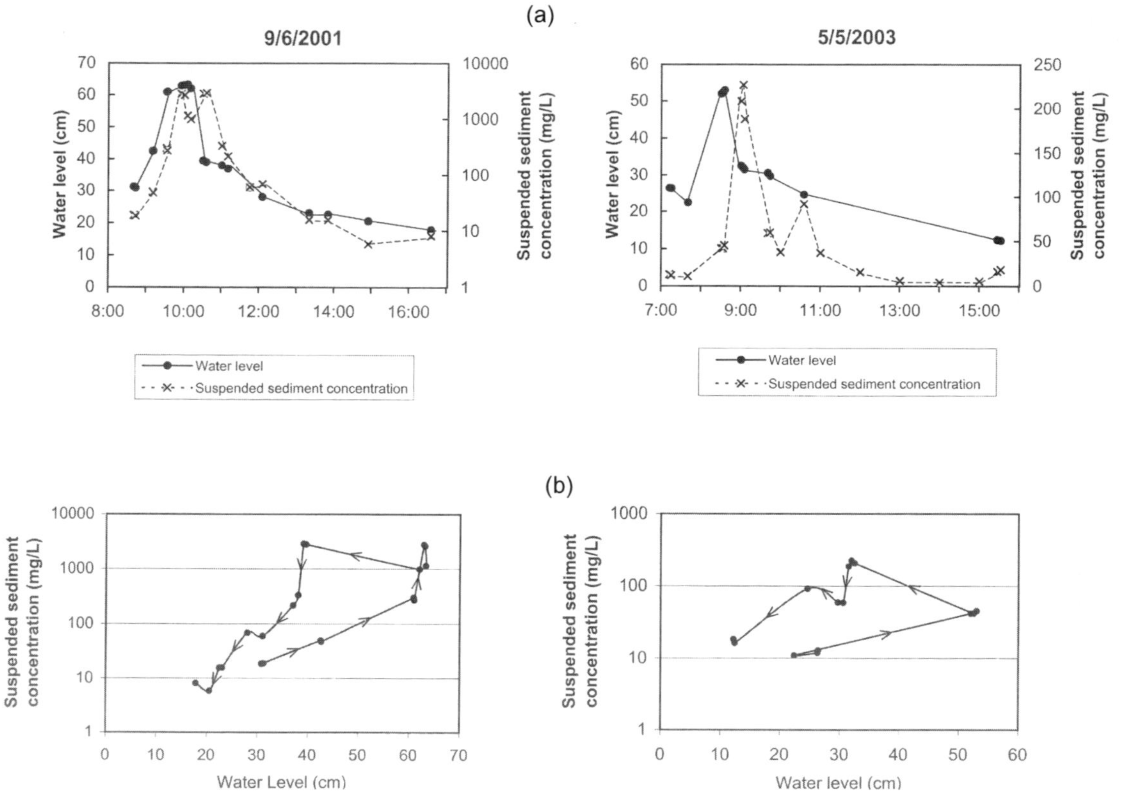

Fig. 1 Temporal variation of water level and suspended sediment for the two events (a), and hysteresis plots (b).

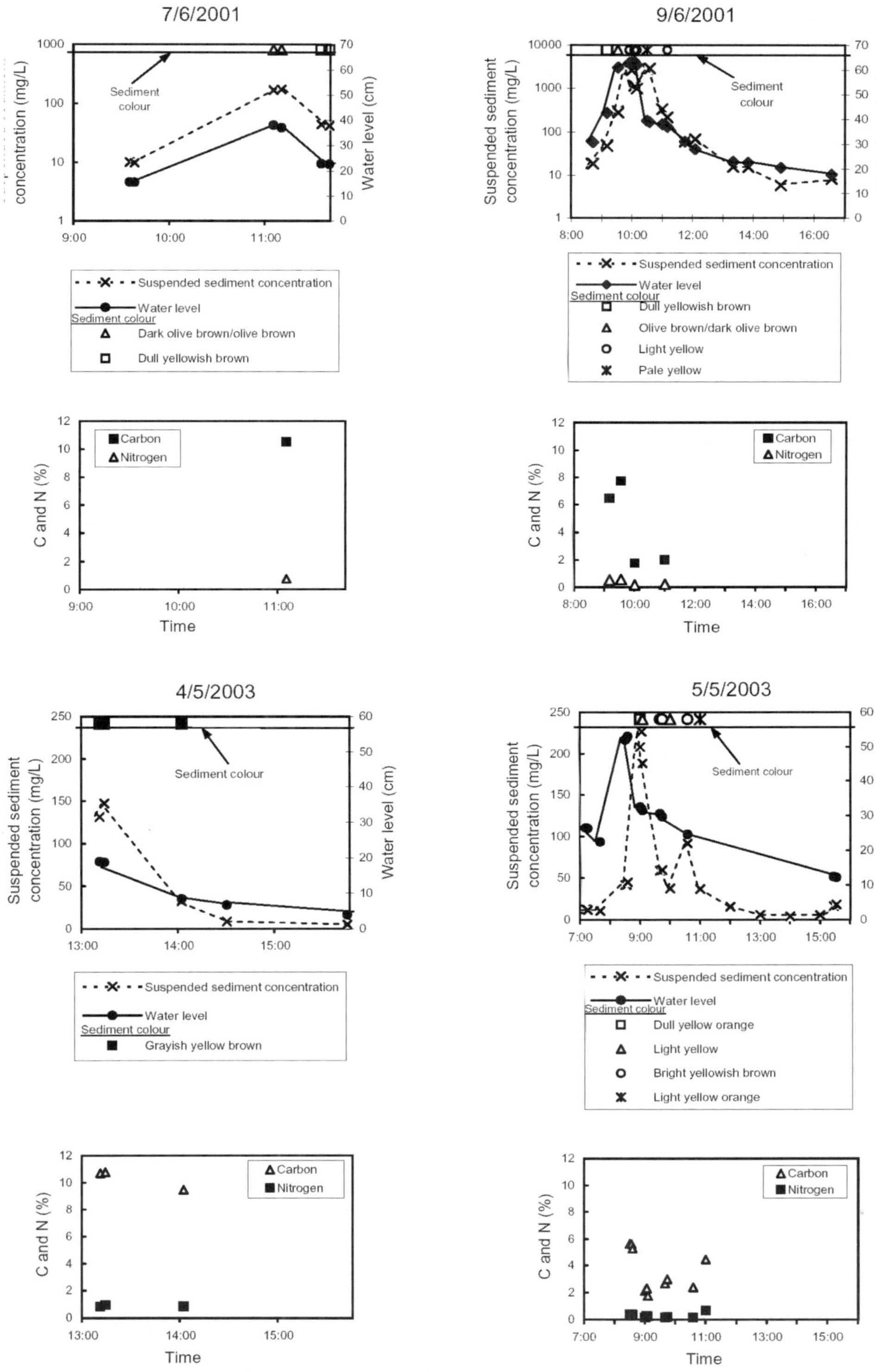

Fig. 2 Sediment property variation.

the organic topsoil colour descriptions, with, for example, dark olive brown or dull yellowish brown recorded in the 7 June 2001 storm preceding the first landslide event (Fig. 2). During the landslide event of 9 June 2001, sediment colour changes, with yellow colours predominating from 09:55 h, coinciding with the marked increase in suspended sediment concentration. The event of May 2003 also exhibits the bright yellow/brown and orange sediment colours that are associated with the description of soil colour for the deeper layers of the landslide scar.

The C and N content of the suspended matter provide further support for a change in source materials. Figure 2 reveals that the storm of 7 June 2001 and the early part of the 9 June 2001 landslide generating event are characterized by C and N contents that are considerably higher than those observed later in the 9 June event from around 1000 h, with, for example, C declining from 6–8% to 2% or less. A similar trend can be observed in May 2003, with the preceding storm event having suspended matter C and N contents that are considerably higher than those in the 5 May landslide event. This decline in C and N content is coincident with the large increase in suspended sediment concentration.

Sediment colour changes have been reported by Grimshaw & Lewin (1980) during individual storm events on the River Ystwyth in Wales. The colour change was interpreted by Grimshaw & Lewin (1980) as reflecting a change from "channel" sources to "non-channel" hillside erosion sites. A change in sediment colour in association with the spatial change of sediment source has been reported by Weaver (1967) and Schreier *et al.* (1998). Moreover, the changes observed in the small headwater basin of this study of sediment colour, in addition to C and N content, indicate that the occurrence of a landslide can influence the physical and chemical properties of the material being transported from the basin. The work of Schreier *et al.* (1998) on P further supports the influence of headwater areas upon nutrient content in sediment of downstream areas.

SEDIMENT DELIVERY FROM HILLSLOPE TO CHANNEL SYSTEMS

Data presented above indicates that the two landslide events delivered sediment to the drainage system of a small headwater catchment which was reflected in the suspended sediment transport. With the notable exception of the study by Butler & Malanson (1996) on the Flathead River in Montana, few direct observations have been made reporting changes in suspended sediment concentration consequent upon landslides. This study illustrates the linkage between hillslope and channel systems in the headwater basin, a point recognized as being important by Harvey (2002). It also indicates that in Hong Kong, as elsewhere, an understanding of sediment transport and production will need to consider processes in headwater catchments (cf. Haigh *et al.*, 1998; Gomi *et al.*, 2002).

The studies of Franks (1999) and Ruse *et al.* (2002) in Hong Kong permit this study to be placed in perspective. Ruse *et al.* (2002) investigated 121 landslides developed on the Tsing Shan Range. They classified 70 of the landslides as being open hillside failures and imply that failed debris remained on the hillslope. In contrast 51 landslides were classified as channelized debris flows suggesting that the debris reached drainage channels on the hillslope. The estimated total source volume of the

121 landslides according to Ruse *et al.* (2002) was 8600 m^3, with an additional 3100 m^3 being entrained by the landslide. Deposition within the landslide scars was estimated at 9700 m^3 or 83% of the sediment produced by mass movement. As for the two landslides in this study, much of the landslide debris remained in storage. Franks (1999) also reports examples of sediment delivery to drainage systems from landslides in Hong Kong. He notes that for a small source volume of less than 400 m^3 deposition predominated along the landslide debris trail: an observation supported by the two small landslides of this study.

Investigations outside of Hong Kong have also documented the role of mass movement and sediment production to the fluvial system. Researchers such as Becht (1995) and Cannon *et al.* (2001) have reported only small amounts of sediment reaching streams from mass movement, and are similar to the observations of this study. In contrast the work of Temple & Rapp (1972), Dietrich & Dunne (1978), Newson (1980), Page *et al.* (1994) and Harden *et al.* (1995) report an important contribution from mass movement to the fluvial system which affords a contrast to the sediment budgets for the landslides of this study.

EROSION

The existence of erosion pin data for the catchment permits some comparison to be made of at-a-point erosion. Surface change for the six erosion plots in 2001 ranged from a small gain of material (2.3 mm) on plot 4 to a loss of 3.2 mm on plot 6. In 2003 the values ranged from a small gain of 0.3 mm (plot 5) to a loss of 2.9 mm (plot 2). The two landslide events reveal far greater at-a-point changes than the erosion pin data, with an average scar depth of a little over 1.1 m in 2001 and that of the 2003 landslide being 0.8 m. Relatively shallow failures of this kind are considered common for Hong Kong in the general literature (c.f. Franks, 1999; Pinches *et al.*, 2002). The study of Lam (1977) also suggests that erosion by flowing water may produce much smaller rates of surface change, with an average surface lowering of 2.13 cm over a 15 month observation period on "badland" slopes, that is to say slopes with no protective cover of vegetation. Whilst the surface change is far greater than recorded by the erosion pins in the present study area, it is far below the at-a-point changes consequent upon a landslide.

Where erosion pins recorded a surface lowering in 2001, the estimated total volumes range from 41.6 to 166.4 m^3 if the rates are extrapolated over the whole catchment area. For 2003, the estimated volumes associated with surface decline range from 10.4 to 150.8 m^3 of material. In both years, the lower range is approximately equivalent to the volume of debris produced by the mass movement. The higher rates of surface decline recorded by the erosion pins produce volumes of debris that are around 3.3 and 21.5 times those produced by mass movement for 2001 and 2003, respectively.

CONCLUSIONS

Heavy rainfall resulted in two small landslides in a headwater drainage basin. Observations made during the storm events of June 2001 and May 2003 of suspended sediment concentrations and timing, along with changes in sediment properties that included colour in addition to C and N, indicate that the suspended matter in streamflow was derived from the landslides. Few direct observations of this linkage between hillslope and channel exist in the literature. Given the prevalence of landsliding in Hong Kong this connectivity has important implications for the management of the sedimentation problem in downstream areas. Further investigation needs to be made of the frequency of this connectivity and the factors influencing the linkage. Moreover, the data on N content of suspended sediment indicate that consideration should be given not only to the volume of sediment delivered but also to the nature and type of materials. The nature of the materials provided from upstream sources has implications in terms of downstream ecosystems. However, the study also reveals that much of the debris from the landslides remained on the slope in an ephemeral drainage line. Further work on the factors controlling the amount of material that remains in storage is important as this also controls the amount of sediment delivered downstream. This study has revealed that simple observations of sediment colour can provide useful information concerning erosion processes and sediment sources.

Acknowledgements CRCG and Dr Stephen S. F. Hui Trust Fund financial support is gratefully acknowledged. Prof. R. D. Hill kindly provided the erosion pin data. Helpful suggestions made by an anonymous reviewer are gratefully acknowledged.

REFERENCES

Becht, M. (1995) Slope erosion processes in the Alps. In: *Steepland Geomorphology* (ed. by O. Slaymaker), 207–222. John Wiley, Chichester, UK.

Brand, E. W. (1985) Landslides in Hong Kong. In: *Proceedings of 8th Southeast Asian Geotechnical Conference*, Kuala Lumpur, Malaysia. V2., pp. 1–15.

Butler, D. R. & Malanson, G. P. (1996) A major sediment pulse in a Subalpine River caused by debris flows in Montana, USA. *Z. Geomorphol.* **40**(4), 525–535.

Cannon, S. H., Kirkham, R. M. & Parise, M. (2001) Wildfire-related debris-flow initiation processes, Storm King Mountain, Colorado. *Geomorphology* **39**, 171–181.

Dai, F. C. & Lee, C. F. (2002) Landslide characteristics and slope instability modelling using GIS, Lantau Island, Hong Kong. *Geomorphology* **42**, 213–228.

Dai, F. C., Lee, C. F., Li, J. & Xu, Z. W. (2001) Assessment of landslide susceptibility on the natural terrain of Lantau Island, Hong Kong. *Environ. Geol.* **40**(3), 381–391.

Dietrich, W. E. & Dunne, T. (1978) Sediment Budget for a Small Catchment in Mountainous Terrain. *Z. Geomorphol. Supplementband* **29**, 191–206.

Evans, N. C., Huang, S. W. & King, J. P. (1999) *The Natural Terrain Landslide Study: Phases I and II.* GEO Report No. 73. Geotechnical Engineering Office, Civil Engineering Department, The Government of the Hong Kong SAR, Hong Kong.

Franks, C. A. M. (1998) Study of rainfall induced landslides on natural slopes in the vicinity of Tung Chung New Town, Lantau Island. *GEO Report no. 57. Geotechnical Engineering Office, Civil Engineering Department, the Government of the Hong Kong SAR, Hong Kong, China.*

Franks, C. A. M. (1999) Characteristics of Some Rainfall-induced Landslides on Natural Slopes, Lantau Island, Hong Kong. *Quart. J. Engng. Geol.* **32**, 247–259.

Gomi, T., Sidle, R. C. & Richardson, J. S. (2002) Understanding processes and downstream linkages of headwater systems. *Bio Science* **52**(10), 905–916.

Grimshaw, D. L. & Lewin, J. (1980) Source Identification for Suspend Sediment. *J. Hydrol.* **47**, 151–162.

Haigh, M. J., Singh, R. B. & Krecek, J. (1998) Headwater Control: Matters Arising. In: *Headwaters: Water Resources and Soil Conservation* (ed. by M. J. Haigh, J. Krecek; G. S. Rajwar & M. P. Kilmartin), 3–24. Balkema, Rotterdam, The Netherlands.

Harden, D., Colman, S. M. & Nolan, K. M. (1995) *Mass Movement in Redwood Creek Basin, Northwestern California.* US Geol. Survey Prof. Paper 1454-G, G1–G11.

Harvey, A. M. (2002) Effective Timescales of Coupling within Fluvial Systems. *Geomorphology* **44**(3–4), 175–201.

King, J. P. & Williamson, S. J. (2002) Erosion along debris avalanche trails. In:, *Natural Terrain—A Constraint to Development* (ed. by C. A. M. Franks, J. Hall, D. Martin, S. Ng, S. Parry, G. Pinches, & M. Thorn), 197–205. Institute of Mining and Metallurgy, Hong Kong Branch, Hong Kong, China.

Lam, K. C. (1977) Patterns and Rates of Slopewash on the Badlands of Hong Kong. *Earth Surf. Processes* **2**, 319–332.

Newson, M. (1980).The geomorphological effectiveness of floods – a contribution stimulated by two recent events in mid-Wales. *Earth Surf. Processes* **5**, 1–16.

Ng, K. C., Parry, S., King, J. P. Franks, C. A. M., & Shaw, R. (2002) Guidelines for Natural Terrain Hazard Studies. *Special Project Report SPR 1/2002. Geotechnical Engineering Office, The Government of the Hong Kong SAR, Hong Kong, China.*

Page, M. J., Trustrum, N. A. & Dymond, J. R. (1994) Sediment budget to assess the geomorphic effect of a cyclonic storm, New Zealand. *Geomorphology* **9**, 169–188.

Parry, S., Massey, C. I. & Williamson, S. J. (2002) Landslide susceptibility analysis for natural terrain hazard studies—Tsing Shan foothills area. In: *Natural Terrain—a Constraint to Development?* (ed. by C. A. M. Franks, J. Hall, D. Martin, S. Ng, S. Parry, G. Pinches, & M. Thorn), 113–123. Institute of Mining and Metallurgy, Hong Kong Branch, Hong Kong, China.

Pinches, G. M., Smallwood, A. R. H. & Hardingham, A. D. (2002) The study of the natural terrain hazard of Yam O Lantau. In: *Natural Terrain—a Constraint to Development?* (ed. by C. A. M. Franks, J. Hall, D. Martin, S. Ng, S. Parry, G. Pinches, & M. Thorn), 207–221. Institute of Mining and Metallurgy, Hong Kong Branch, Hong Kong, China.

Ruse, M. E., Waring, D. P., Kaldy, A., Chan, K. S. & Ng, K. C. (2002) Initiation and runout characteristics of a swarm of 121 landslides in the Tsing Shan Foothills, Hong Kong. In: *Natural Terrain—A Constraint to Development?* (ed. by C. A. M. Franks, J. Hall, D. Martin, S. Ng, S. Parry, G. Pinches, & M. Thorn), 77–87. Institute of Mining and Metallurgy, Hong Kong Branch, Hong Kong, China.

Schreier, H., Brown, S., Carver, M. & Shah, P. B. (1998) Linking land degradation to nutrient and sediment transport in a middle mountain watershed in Nepal. In: *Headwaters: Water Resources and Soil Conservation* (ed. by M. J. Haigh, J. Krecek, G. S. Rajwar & M. P. Kilmartin), 315–328. Balkema, Rotterdam, The Netherlands.

Scott, K. M., Macias, J. L., Naranjo, J. A., Rodriguez, S., McGeehin, J. P. (2001) catastrophic debris flows transformed from landslides in volcanic terrains: mobility, hazard assessment and mitigation strategies. *US Geol. Survey Prof. Paper 1630.*

Temple, P. H. & Rapp, A. (1972) Landslides in the Mgeta Area, Western Uluguru Mountains, Tanzania. *Geogr. Ann. A* **54**(3–4), 157–193.

Weaver, C. E. (1967) Variability of a river clay suite. *J. Sed. Petrol.* **37**, 971–974.

Williams, G. P. (1989) Sediment concentration versus water discharge during single hydrologic events in rivers. *J. Hydrol.* **111**, 89–106.

Wood, P. A. (1977) Controls of variation in suspended sediment concentration in the River Rother, West Sussex, England. *Sedimentology* **24**, 437–445.

Geomorphological Processes and Human Impacts in River Basins (Proceedings of the International Conference held at Solsona, Catalonia, Spain, May 2004). IAHS Publ. 299, 2005.

Targeting erosion control in large river systems using spatially distributed sediment budgets

SCOTT N. WILKINSON, JON M. OLLEY, IAN P. PROSSER & ARTHUR M. READ
CSIRO Land and Water, PO Box 1666, ACT 2601, Australia
scott.wilkinson@csiro.au

Abstract Targeting management actions aimed at improving water quality and habitat in large rivers is difficult. Resources for such actions are often limited; consequently, catchment works to reduce sediment supply need to focus on areas that will have the largest in-river effect. We describe how a sediment budget, developed using sediment transport modelling (SedNet), sediment tracing, and water quality monitoring, can be used to identify the primary sources of suspended sediment in the 29 000 km^2 upper-middle Murrumbidgee Basin, Australia. It is shown that targeted erosion control will provide more than three times the reduction in suspended sediment export that would occur if the actions were implemented randomly. The results show that 70% of the sediment exported comes from 20% of the catchment area, and that channel bank erosion is the dominant source of sediment. The method has application in all large river systems where water quality and habitat sedimentation are issues.

Key words catchment management; channel erosion; sediment budget; sediment tracing; SedNet; suspended sediment load

INTRODUCTION

The structure and health of riverine ecosystems depends on diversity and extent of the physical habitat. Physical habitat includes the size and shape of the channel, the form of the bed load sediments, the quality of the water and the supply of nutrients, all of which are influenced by the supply and characteristics of the sediments (Prosser *et al.*, 2001a). For example, the proportion of phosphorus transported in particulate form has a mean of 75% across 69 Australian water quality monitoring sites (Young *et al.*, 2001). Settlement of European immigrants 180 years ago brought widespread changes in land use across the southeast of Australia that delivered massive volumes of sediment to the rivers, and changed the form of rivers dramatically (e.g. Olley & Wasson, 2003). Increased sediment supply continues to affect the health of Australian rivers. Unfortunately, the magnitude of work required to reduce sediment loads to "natural", pre-European, rates exceeds the resources available. Erosion control works must therefore be targeted to where they will have the greatest benefit to river health, for example, reducing suspended sediment loads, or habitat sedimentation downstream.

In this paper, a combination of spatial modelling, sediment tracing techniques and analysis of water-quality and flow data are used to determine the major source of suspended sediments in the Murrumbidgee River basin. We demonstrate how spatial modelling can be used to target erosion control to areas contributing most to export of suspended sediment, to reduce impacts on downstream or estuarine reaches.

BASIN DESCRIPTION

The Murrumbidgee River drains approximately 84 000 km^2 of the Murray–Darling Basin (Fig. 1). Its catchment can be divided into three distinct geomorphic regions: the upper, mid, and lower Murrumbidgee. The upper region is mountainous and hilly, and is separated from the mid-region by two large storage reservoirs, Burrinjuck and Blowering (Fig. 1). These dams trap most of the sediment delivered from the upper basin (Wasson *et al.*, 1987). The mid-basin has undulating terrain dissected by incised stream and gully networks. These networks have been shown to be the primary source of sediment delivered to the lower Murrumbidgee River (Olive *et al.*, 1996; Wallbrink *et al.*, 1998). The main land use in the upper and middle basin is grazing, with some cropping in the middle basin. There is extensive grazing on the plains that make up the lower basin, and cropping and horticulture in the irrigation areas. A strong rainfall gradient exists across the basin; average annual totals vary from 1600 mm in alpine areas in the east to 300 mm at the catchment outlet. The river is now a major source of irrigation water and flow is heavily regulated, by the two major dams located in the upper catchment, and also by a series of weirs below Wagga Wagga (Fig. 1).

METHODOLOGY

Constructing the suspended sediment budget

The SedNet model developed for the National Land and Water Resources Audit (NLWRA) (Prosser, *et al.*, 2001b) is a physically-based process model. It constructs

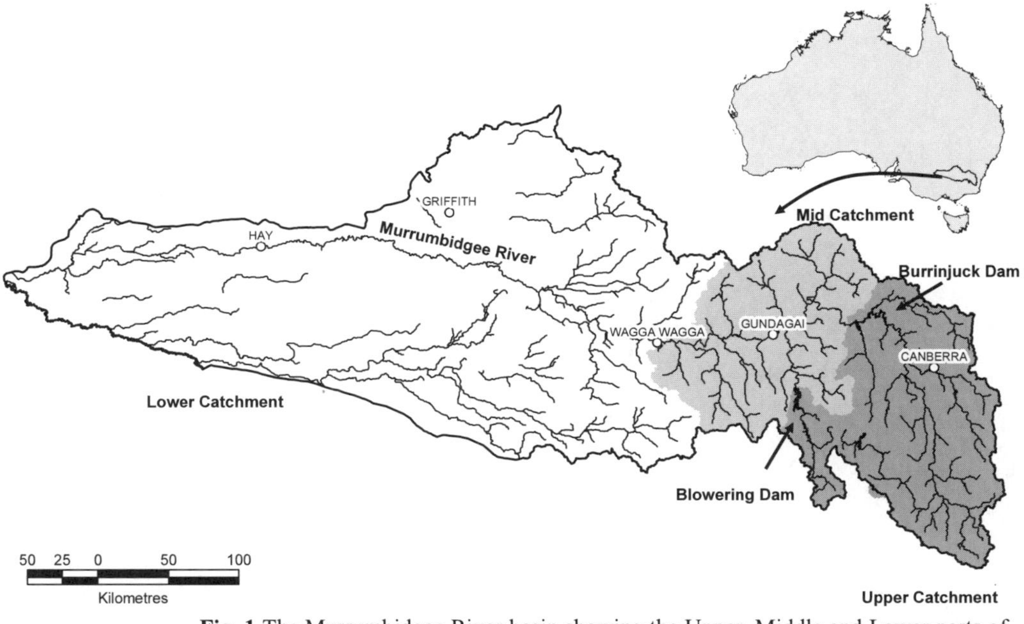

Fig. 1 The Murrumbidgee River basin showing the Upper, Middle and Lower parts of the basin.

sediment budgets of a river network and identifies the major sources, sinks and loads of material. In the model the river network is divided into a series of links which are the basic unit of calculation for the sediment budget. Each link extends between adjacent stream junctions or nodes and has a sub-catchment, which drains into the link between its upper and lower nodes (Fig. 2). ArcInfo™ AML scripts are used to define the river network and sub-catchments from a 25 m Digital Elevation Model (DEM). Using this environment, separate budgets for bed load and suspended sediment are then computed for each link; here we focus on the suspended sediment budget. The suspended sediment yield (Y_x) from a link is given by:

$$Y_x = H_x + G_x + B_x + T_x - F_x - R_x \qquad (1)$$

Sediment inputs to each link come from hillslope (H_x), gully (G_x) and riverbank (B_x) erosion, and from upstream tributaries (T_x) (Fig. 3). Sediment is either deposited within the link on flood plains (F_x), in reservoirs (R_x), or is transported downstream and delivered to the next link (Y_x). This process is carried out in each river link, from first to highest Shreve order (Shreve, 1966), so that Y_x for the outlet link represents mean annual export. Sediment budgets for both pre-European and present-day conditions were computed, to indicate relative changes.

Hillslope erosion The input from hillslope erosion is estimated using the Revised Universal Soil Loss Equation (RUSLE) (Renard *et al.*, 1997), as applied to the Australian continent by Lu *et al.* (2003). Most of the sediment mobilised on hillslopes is trapped on the hillslope and so the sediment delivered to streams (H_x) is modified

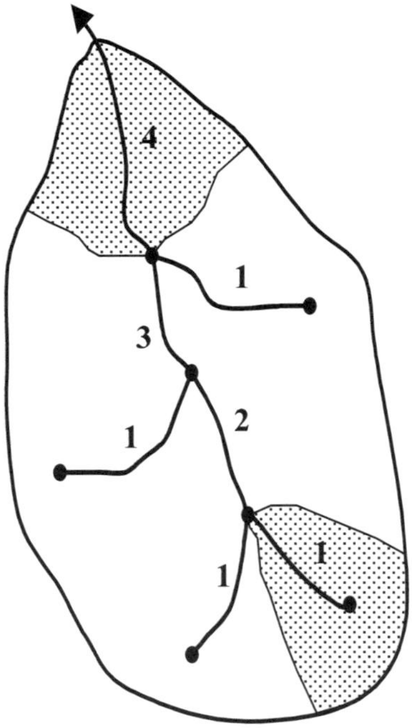

Fig. 2 A river network showing links, nodes, Shreve magnitude of each link (Shreve, 1966) and the internal sub-catchment areas of an order one and an order four link.

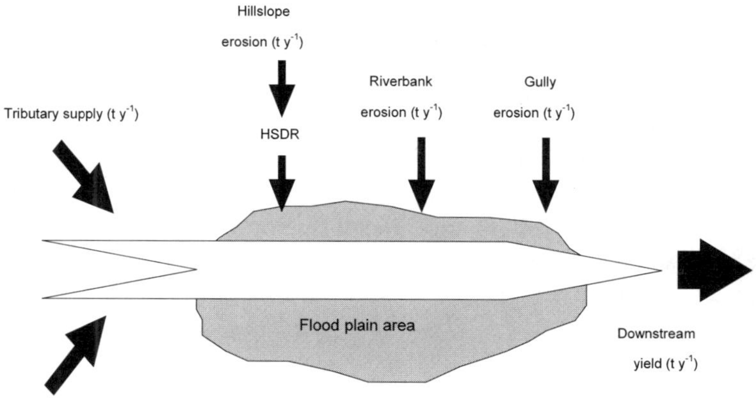

Fig. 3 Conceptual diagram of the SedNet river sediment budget for one river link.

by the hillslope sediment delivery ratio (*HSDR*). A uniform *HSDR* of 0.05 has been found to provide good agreement between hillslope erosion and yield in regions dominated by hillslope processes (Prosser *et al.*, 2001b). It is assumed that all hillslope sediment contributes to the suspended sediment budget. Cover factors for natural vegetation cover were used to calculate pre-European hillslope erosion.

Gully erosion Erosion from gully and stream banks can generate up to 90% of the total sediment yield from a catchment (Olley *et al.*, 1993; Prosser & Winchester, 1996; Wallbrink *et al.*, 1998; Wasson *et al.*, 1998). The linear extent of gully erosion in the Murrumbidgee Basin has been mapped by the New South Wales Department of Infrastructure, Planning and Natural Resources. Gully erosion is assumed to be negligible prior to the arrival of Europeans. The average rate of suspended sediment supply since, from gullies in each sub-catchment (G_x), is the product of gully length (L_x), estimated average cross-sectional area of 12 m^2 (a), the proportion of fine sediment, set at 0.5 (p_f), and average dry bulk soil density, estimated at 1.5 t m^{-3} (ρ_s), divided by the average time over which gullies have developed (τ = 120 years). A historical survey in the Murrumbidgee (Wasson *et al.*, 1998) and recent measurements (Caitcheon, 2004) indicate that current rates of gully suspended sediment generation have declined from their peak; and are of the order of 50% of the long term average. Consequently, suspended sediment supply from the gullies in each sub-catchment (G_x) is estimated as:

$$G_x = 0.5 \times \frac{p_f \rho_s a}{\tau} \times L_x \quad (2)$$

Riverbank erosion Riverbank erosion is the most uncertain of the sediment source terms in the modelling. Bank erosion rate, BE_x (m year^{-1}), is determined as proportional to stream power $\rho g Q_{bf} S_x$, where ρ is the density of water, g the acceleration due to gravity, Q_{bf} is bankfull discharge in m^3 s^{-1} and S_x is the energy slope (dimensionless) approximated by the channel gradient (Rutherfurd, 2000). Bankfull discharge was estimated as the flow with recurrence interval 2.27 years on the annual maximum series (Page, 1988). Using daily records from 17 river gauges, bankfull flow was regionalized as a power function of mean annual flow (*MAF*; Ml year^{-1}), which was also regionalized (Wilkinson *et al.*, 2005). Downstream of reservoirs, regionalized flow was adjusted to account for flow regulation.

It is known that degradation of riparian vegetation has resulted in widening of Australian river channels. We assume negligible bank erosion in the proportion of the link length that has fully intact riparian vegetation (PR_x), as determined from LANDSAT imagery with 30 m pixels (Barson *et al.*, 2000). We also reduce the bank erosion rate in narrow valleys having exposure of rock and other unerodible materials; limited measurements suggest an exponential relationship between rock exposure and flood plain width F_w, (m) (Hughes *et al.*, 2003). Thus:

$$BE_x = 0.0001\rho g Q_{bf} S_x (1 - PR_x)[1 - \exp(-0.008 F_w)] \tag{3}$$

The coefficient 0.0001 was calibrated to achieve a maximum bank erosion rate of 0.5 m year^{-1}; the rate required to achieve observed present channel width in some non-vegetated and historically eroding foot-hill streams (Rustomji, 2001). The amount (t year^{-1}) of suspended sediment supplied from bank erosion, $B_x = hp_f\rho_s BE_x l_x$ where h is bank height (3 m), and p_f, ρ_s have the same values defined for gully erosion, and l_x is link length (m).

Flood plain deposition Flood plain deposition, F_x (t year^{-1}) was determined as a function of flood plain area and the median value of daily overbank flows (Prosser *et al.*, 2001a). Flood plain mapping for each river link utilised hydrological and hydraulic modelling, (Pickup & Marks, 2001). Median overbank flow was calculated using a regionalization of the same form as for Q_{bf}.

Deposition in reservoirs Sediment deposition in reservoirs R_x is predicted using an empirical function of the mean annual reservoir inflow and its total storage capacity (Heinemann, 1981).

Contribution to catchment export Because of losses to flood plain and reservoir deposition, not all suspended sediment delivered to a river network is exported from the catchment outlet. To provide greatest reduction in suspended sediment export, erosion control activities should target the areas of the catchment with the highest rates of contribution to export. Therefore, we calculate the delivery ratio between each link and the catchment outlet. This represents a sediment delivery ratio for the river network (*RSDR*) (equation 4), where n is the number of links between link x and the catchment outlet, and $Sup_x = H_x + G_x + B_x + T_x$, the total suspended sediment supplied to a link from local erosion and upstream tributaries. The contribution of a sub-catchment to suspended sediment export ($Cont_x$; t ha^{-1} year^{-1}), is then given by equation (5), where A_x is the area of the sub-catchment.

$$RSDR_x = \frac{Y_x}{Sup_x} \times \frac{Y_{x+1}}{Sup_{x+1}} \times \ldots \times \frac{Y_n}{Sup_n} \tag{4}$$

$$Cont_x = RSDR_x \times \frac{H_x + G_x + B_x}{A_x} \tag{5}$$

We also want to target the dominant erosion process, since a reduction of given proportion in that erosion process will provide the largest impact on sub-basin sediment supply. Here, we lump gully and riverbank erosion as channel erosion, since both processes are controlled by riparian revegetation. Gullies essentially represent small, incised streams with catchment area smaller than the threshold used to construct the river network. We thus identify the contribution to export, from channel erosion only, of each sub-catchment by removing H_x from equation (5).

Management scenarios

Scenarios were run to assess the effect of various extents of bank and gully stabilization. Sub-catchments were selected in descending order of channel erosion contribution (t ha^{-1} year^{-1}). In each selected sub-catchment, revegetation of 50% of gullies and presently un-vegetated banks in the sub-catchment was simulated by assuming that revegetated gullies and banks contributed no sediment. Scenarios of random revegetation were also run. To ensure the predicted benefit of targeting was conservative, the random scenarios were constrained to areas of the catchment downstream of reservoirs.

RESULTS AND DISCUSSION

The suspended sediment budget for the Upper-Middle Murrumbidgee basin is given in Table 1. The model predicts a suspended sediment load at Wagga Wagga of ~580 kt year^{-1}. This compares favourably with the observed load of 599 kt year^{-1}, calculated from daily turbidity and discharge measurements at Wagga Wagga (Olive *et al.*, 1996). Using fallout radionuclide tracers, Wallbrink *et al.*, (1998) demonstrated that approximately 15 ± 5% of the suspended sediment in the Middle Murrumbidgee basin is sourced from surface erosion, on hillslopes. Over the Middle Murrumbidgee basin only, the predicted proportion of total suspended sediment derived from hillslope erosion is 25%. This discrepancy indicates that either hillslope erosion or the *HSDR* is

Table 1 Upper-Middle Murrumbidgee basin suspended sediment budget.

Inputs (kt year^{-1})		Outputs (kt year^{-1})	
Pre-European			
Hillslope erosion $\sum H_x$	13	Flood plain deposition $\sum F_x$	14
Gully erosion $\sum G_x$	0	Reservoir deposition $\sum R_x$	0
Riverbank erosion $\sum B_x$	56	Export	55
Total	69	Total	69
Present day			
Hillslope erosion $\sum H_x$	453	Flood plain deposition $\sum F_x$	220
Gully erosion $\sum G_x$	286	Reservoir deposition $\sum R_x$	475
Riverbank erosion $\sum B_x$	538	Export	582
Total	1277	Total	1277

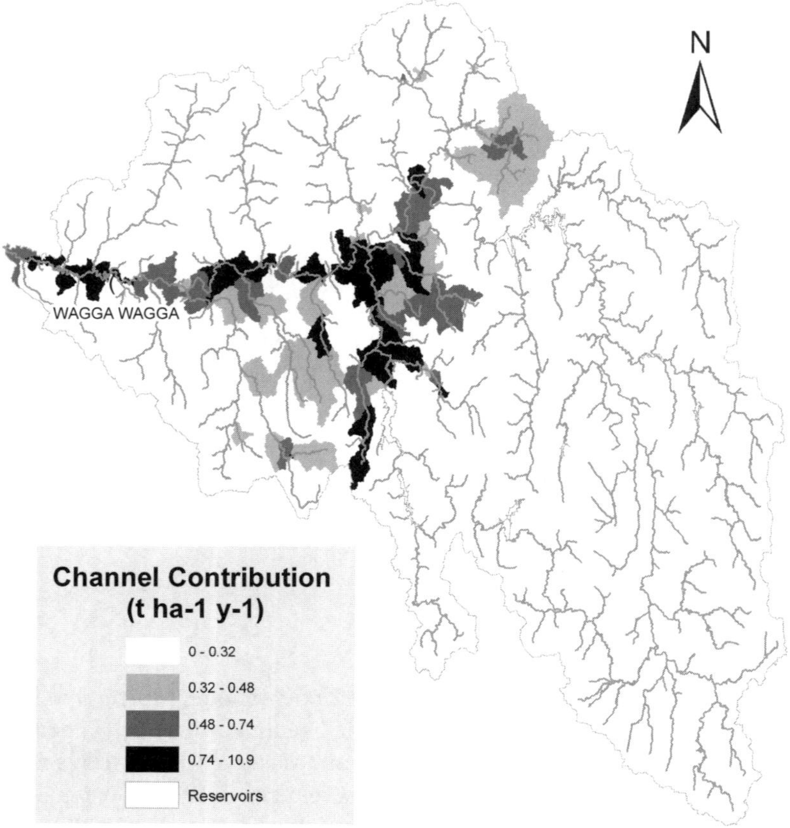

Fig. 4 Rates of specific suspended sediment contribution from channel erosion to export from the Middle and Upper Murrumbidgee catchment (t ha^{-1} $year^{-1}$).

overestimated in SedNet, or that gully and bank erosion rates are under-estimated. Wallbrink *et al.* (1998) also estimated the mean residence time for fine grain sediments within the Middle-Murrumbidgee River network as being a few years, and this justifies neglecting storage of fine sediment in the mean-annual sediment budget.

Bank and gully erosion are the largest sources of suspended sediment (Table 1), supporting a focus on riparian revegetation. The observed suspended sediment load at Wagga Wagga is ~10 times greater than the predicted pre-European load, indicating the need for reducing sediment export. Spatial variation in the contribution of bank and gully erosion to sediment export (Fig. 4) shows that, due to sediment deposition in the reservoirs, the Upper Basin contributes very little to the sediment load at Wagga Wagga. Mapping $Cont_x$ indicates that 70% of suspended sediment exported comes from 20% of the catchment area. If riparian revegetation and erosion control is targeted to sub-catchments in decreasing order of their contribution to export, then export reduces non-linearly as the length of riparian revegetation increases (Fig. 5). In contrast, if riparian revegetation is done at random in the middle basin, the response is approximately linear. For 600 km of revegetation, the reduction in export from random revegetation is less than one third that achieved by targeted revegetation.

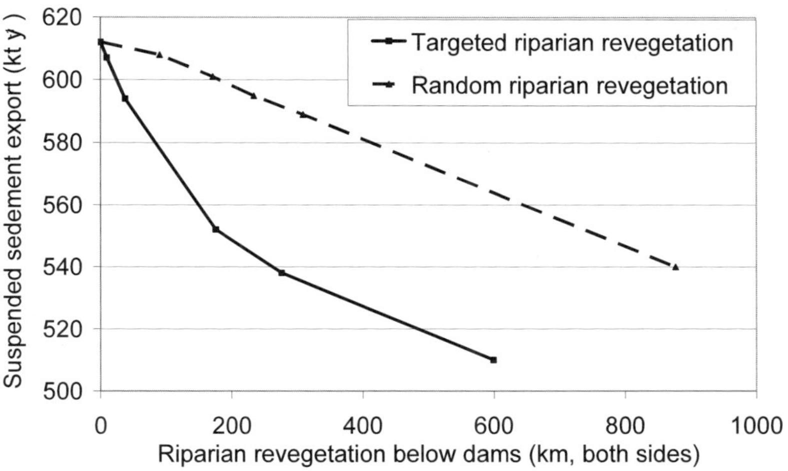

Fig. 5 Response of suspended sediment export to targeted and random channel erosion control.

CONCLUSIONS

The study has used a set of GIS-based models (SedNet) to spatially predict the major sediment sources and sinks within a large temperate river network. The results identify channel erosion as the dominant source of sediment; this is supported by sediment tracing data. The modelled sediment loads compare favourably with those from water quality and flow data. Targeted erosion control is shown to provide more than three times the reduction in suspended sediment export than would occur if the actions were implemented randomly. The results demonstrate how a spatially distributed sediment budget can provide a rational basis for planning erosion control or other rectification measures.

Acknowledgements This research was supported by grant CLW44 from Land & Water Australia. J. Carles Balasch is thanked for a thorough and insightful review of the original manuscript. We also acknowledge the support of Heinz Buettikofer and Tristram Miller (CSIRO Land and Water) for preparing Figs 1 and 4, respectively.

REFERENCES

Barson, M. M., Randell, L. A. & Bordas, V. (2000) *Land Cover Change in Australia*. Bureau of Rural Sciences, Canberra, Australia. ftp://ftp.brs.gov.au/outgoing/longterm/lw/lccreport.pdf

Caitcheon, G. (2004) Data on sediment yield from gully networks in the Murrumbidgee catchment. CSIRO Land and Water, Canberra, Australia (unpublished).

Heinemann, H. G. (1981) A new sediment trap efficiency curve for small reservoirs. *Water Resour. Res.* **17**, 825–830.

Hughes, A. O. & Prosser, I. P. (2003) Gully and riverbank erosion mapping for the Murray-Darling Basin. *Tech. Report 3/03, CSIRO Land and Water, Canberra, Australia.*

Lu, H., Prosser, I. P., Moran, C. J., Gallant, J., Priestley, G. & Stevenson, J. (2003) Predicting sheetwash and rill erosion over the Australian continent. *Australian J. Soil Res.* **41**, 1037–1062.

Olive, L. J., Olley, J. M., Wallbrink, P. J. & Murray, A. S. (1996) Downstream patterns of sediment transport during floods in the Murrumbidgee River, NSW, Australia. *Z. Geomorphol.* **105**, 129–140.

Olley, J. M., Murray, A. S., McKenzie, D. M. & Edwards, K. (1993) Identification of sediment sources in a gullied catchment using natural and artificial radioactivity. *Water Resour. Res.* **29**, 1037–1043.

Olley, J. M. & Wasson R. J. (2003) Changes in the flux of sediment in the Upper Murrumbidgee catchment, SE Australia, since European settlement. *Hydrol. Processes* **17**, 3307–3320.

Page, K. J. (1988) Bankfull discharge frequency for the Murrumbidgee River, New South Wales. In: *Fluvial Geomorphology of Australia*, 267–281. Academic Press, Australia

Pickup, G. & Marks, A. (2001) Identification of floodplains and estimation of floodplain flow velocites for sediment transport modelling. *Tech. Report 14/01, CSIRO Land and Water, Canberra, Australia.*

Prosser, I. P. & Winchester, S. J. (1996) History and processes of gully initiation and development in eastern Australia. *Z. Geomorphol.* **105**, 91–109.

Prosser, I. P., Rutherfurd, I. D., Olley, J. M., Young, W. J., Wallbrink, P. J. & Moran, C. J. (2001a) Large-scale patterns of erosion and sediment transport in rivers networks, with examples from Australia. *Marine Freshwater Res.* **52**, 81–99.

Prosser, I., Young, B., Rustomji, P., Moran, C. & Hughes, A. (2001b) Constructing River Basin Sediment Budgets for the National Land and Water Resources Audit. *Tech. Report 15/01, CSIRO Land and Water, Canberra, Australia.*

Renard, K. G., Foster, G. A., Weesies, D. K., McCool, D. K. & Yoder, D. C. (1997) Predicting soil erosion by water: A guide to conservation planning with the Revised Universal Soil Loss Equation. In: *Agriculture Handbook 703.* United States Department of Agriculture, Washington, DC, USA.

Rustomji, P. (2001) Data on river channel widths in the Murrumbidgee River basin. CSIRO Land and Water, Canberra, Australia (unpublished).

Rutherfurd, I. D. (2000) Some human impacts on Australian stream channel morphology. In: *River Management: The Australasian experience.* (ed. by S. Brizga & B. Finlayson), 11–49. John Wiley and Sons, Chichester, UK.

Shreve, R. L. (1966) Statistical law of stream numbers. *J. Geol.* **74**, 17–37.

Wallbrink, P. J., Murray, A. S., Olley, J. M. & Olive, L. (1998) Determining sediment sources and transit times of suspended sediments in the Murrumbidgee River, NSW, Australia using fallout Cs-137 and Pb-210. *Water Resour. Res.*, **34**(4), 879–887.

Wasson, R. J., Clark, R. L. & Nanninga, P. M. (1987) 210Pb as a chronometer and tracer, Burrinjuck Reservoir, Australia. *Earth Surf. Processes and Landf.* **12**, 399–414.

Wasson, R. J., Mazari, R. K., Starr, B. & Clifton, G. (1998) The recent history of erosion and sedimentation on the Southern Tablelands of southeastern Australia. *Geomorphology* **24**, 291–308.

Wilkinson, S. N., Young, W. J. & DeRose, R. C. (2005) Regionalising mean annual flow and daily flow variability for sediment transport modelling. *Hydrol. Processes* (In press).

Young, W. J., Prosser, I. P. & Hughes, A. O. (2001) Modelling nutrient loads in large-scale river networks for the National Land and Water Resources Audit. *Tech. Report 12/01, CSIRO Land and Water, Canberra, Australia.*

2 Catchment Hydrology and Water Quality

The effects over time of an arterial drainage scheme on the flood magnitude–return period relations

KESHAV P. BHATTARAI[1,2] & KIERAN M. O'CONNOR[1]

1 *Department of Engineering Hydrology, National University of Ireland, Galway, Ireland*
keshav.bhattarai@nuigalway.ie

2 *Environmental Change Institute, National University of Ireland, Galway, Ireland*

Abstract Arterial Drainage Schemes (ADSs) were implemented on Irish rivers for the purposes of improving the land drainage and reducing the frequency and extent of overland flooding. Such ADSs mainly involved the deepening and widening of the river channels in order to increase their discharge-carrying capacity, thereby, affecting not only the basins' response to rainfall but also their peak discharge (Q_T)–return period (T) relation. Earlier Irish studies involving the effects of ADS on a catchment's Q_T–T relation found that the post-drainage Q_T of an ADS-catchment was always larger, for the same T, in comparison to its pre-drainage counterpart. The current study focuses on investigating whether such effects have persisted in the ADS-catchments, or died out with the passage of time. Results of analysis on 16 ADS and 6-non-ADS catchments show that the ADS (post-1945)-catchments have experienced major changes in the Q_T–T relation over time, whereas such changes are less pronounced in other catchments.

Key words Arterial Drainage Scheme (ADS); flood frequency analysis; GEV distribution; method of L-moments; Q_T–T relationship

INTRODUCTION

Arterial Drainage Schemes (ADSs) were implemented in Ireland, in an unplanned *ad hoc* basis, in the years immediately following the famine of 1845–1846, but a more systematic programme of ADS projects was undertaken following the promulgation of the Arterial Drainage Act (ADA) of 1945. The main purpose of such ADSs was to improve the land drainage, thereby reducing both the frequency and the extent of overland flooding. Various activities, such as embankment construction, river straightening and diversion, lake storage development, etc. are involved in ADSs but, for the greatest part, the concentration is on the deepening and widening of the river channels. Clearly, such activities affect not only the hydrological response of these catchments to rainfall, but also their flood characteristics such as peak flood–return period (abbreviated as Q_T–T) relation. Note that in the Q_T–T relation, Q_T is the flood peak discharge (m^3 s^{-1}), which is deemed to have a mean return period of T years if that peak value (Q_T) is exceeded, on average, once in T years.

The problem of evaluating the changes in the hydrological behaviour following the drainage of a natural channel has been the subject of attention of numerous Irish hydrologists and engineers. O'Kelly (1955) sought to establish relations between parameters of the post-drainage unit hydrographs (UHs) and physical characteristics of the catchments. Nash (1955, 1958) analysed the pre- and post-drainage UHs of the Brosna catchment and showed that the peak of the post-drainage UH was virtually twice as large as that of the pre-drainage UH. Bree *et al.* (1979) analysed the unit hydrographs for the pre- and post-drainage conditions of 10 Irish catchments, using the constrained least-squares method. Just recently, taking advantage of the availability of a much longer post-drainage database of River Brosna catchment, Bhattarai & O'Connor (2004) carried out a preliminary study on the pre- and post-drainage response to rainfall, and found that the response has not been consistent over the entire post-drainage period, but has changed over time.

Bree & Cunnane (1979), and Bailey & Bree (1981), investigated the effect of an ADS on the catchment's Q_T–T relation, and found that for each return period of T years, the peak discharge (Q_T) has been increased significantly in the post-drainage condition, but the degree of such change varied from one catchment to another. Lynn, (1981) carried out an investigation of 12 drained catchments in order to identify improved methods of estimating flood magnitudes, especially for the purpose of catchment drainage design. These studies involved Peaks Over Threshold (POT) series for the flood frequency analyses, in preference to the Annual Maximum Series (AMS), the POT being perceived to provide a more accurate estimate of the Q_T–T relationship when only short records are available (NERC, 1975).

The study reported here was initiated with the objective of investigating whether the effects of an ADS on the Q_T–T relations, as reported in the earlier Irish studies, have persisted in the catchments under study, or if these effects died out with the passage of time. The investigation is primarily focused on the catchments drained after 1945.

STRUCTURE OF THE STUDY

Data used in the study

Four categories of catchments (totalling 22), each having an AMS of at least 34 years, up to and including the year 2000, have been considered. These include: (a) seven ADS (post-1945) catchments, which were drained after the promulgation of ADA of 1945, and for which the AMSs are available from immediately following the ADS; (b) five ADS (post-1925) catchments, which were drained after the ADA of 1925, but for which the AMSs are available only from the 1950s onward; (c) four ADS (post-1842) catchments, which were drained after the ADA of 1842, but for which the AMSs are available only from the 1950s onward; and (d) six non-ADS catchments, which were never drained, but for which the AMSs are also available only from the 1950s. The details of the 22 catchments and year of ADSs (where applicable) are shown in Table 1.

Selection of different analysis scenarios

The ADS-catchments considered in this study were drained over different periods, and the lengths of their available post-drainage AMS data also differ. In order to

Table 1 Details of catchments and annual maximum series (AMS) used in the study.

SN	Station no.	River	Gauging station	CA (km^2)	ADS (year)	AMS (from – to)	AMS (years)
A	*ADS (1945) catchments with Arterial Drainage Scheme after 1945*						
1	25006	Brosna	Ferbane	1207	1948–1955	1955–2000	46
2	30004	Clare	Corofin	695	1954–1964	1964–2000	35
3	6021	Glyde	Mansfieldstown	321	1950–1957	1955–2000	46
4	23001	Galey	Inch Bridge	196	1951–1959	1960–2000	41
5	24012	Deel	Grange Bridge	359	1962–1968	1967–2000	34
6	22003	Maine	Riverville	272	1959–1963	1962–2000	39
7	25029	Nenagh	Clarianna	301	1955–1960	1957–2000	42
B	*ADS (1925) catchments with ADS during 1925–1945*						
1	14005	Barrow	Portarlington	398	1925	1955–2000	44
2	36019	Erne	Belturbet	1501	1925	1958–2000	43
3	36010	Annalee	Butler's Bridge	774	1925	1955–2000	46
4	18004	Awbeg	Ballynamona	324	1925	1955–2000	42
5	15005	Erkina	Durrow Foot Bridge	387	1925	1954–2000	46
C	*ADS (1842) catchments with ADS between 1842 and 1925*						
1	26002	Suck	Rookwood	626	1863	1952–2000	49
2	26019	Camling	Mullagh	260	1842	1953–2000	47
3	6012	Fane	Clarebane	167	1842	1957–2000	42
4	25001	Mulkear	Annacotty	646	1863	1953–2000	45
D	*Non–ADS catchments (never underwent an ADS)*						
1	20001	Bandon	Bandon	406	–	1960–2000	41
2	15006	Nore	Brownsbarn	2388	–	1954–2000	45
3	16002	Suir	Beakstown	512	–	1954–2000	47
4	15003	Dinin	Dinin Bridge	298	–	1954–2000	47
5	28001	Inagh	Ennistimon	168	–	1957–2000	42
6	12001	Slaney	Scarawalsh	1036	–	1955–2000	46

investigate the effects over time of an ADS on the Q_T–T relations over the same time frame, the following seven analysis scenarios were considered: (1) *the by-2000-scenario* (consisting of the entire AMS up to and including the year 2000); (2) *the by-1995-scenario* (consisting of the AMS up to and including the year 1995, i.e. excluding the last 5 years of the AMS record); (3) *the by-1990-scenario* (consisting of the AMS up to and including the year 1990, i.e. excluding the last 10 years); (4) *the by-1985-scenario* (consisting of the AMS up to and including the year 1985, i.e. excluding the last 15 years); (5) *the last-30-years scenario* (consisting of the AMS for the period of 1971–2000); (6) *the last-25-years scenario* (consisting of the AMS for the period of 1976–2000); and (7) *the last-20-years scenario* (consisting of the AMS for the period of 1981–2000).

The by-2000-scenario for a catchment gives Q_T–T relation for the un-categorized case for which, in the present study, the AMS length varies from n = 34 to 49 (see Table 1). The by-1995-, by-1990- and by-1985-scenarios give the Q_T–T relation as if the flood frequency analyses had been carried out in the years 1995, 1990 and 1985, respectively, by considering the AMS available by these years. The last-30-year, last-25-years, and last-20-years scenarios are considered to investigate the form the Q_T–T relation would have if, instead of taking the entire available AMS, only those of the last 30-years, the last 25-years, and the last 20-years AMS, respectively, had been available for analyses.

Flood frequency analyses and comparison of Q_T–T relations

The Q_T–T relations for the seven different AMS analysis scenarios, for all 22 catchments, were computed by fitting the Generalized Extreme Value (GEV) distribution to the corresponding 7 sets of AMS. The three parameters of the GEV distribution (i.e. the location parameter α, the scale parameter ξ, and the shape parameter k), for each scenario, were estimated using the relations suggested by Hosking (1990), in which, the sample L-moments were calculated using Wang's (1996) method.

In this study, the last-20-year scenario consists of a set of AMS of length $n = 20$ years, whereas all the other scenarios have longer data sets. According to the Flood Studies Report (NERC, 1975), satisfactory "at-site" estimates can be achieved for up to 40 years return period flood (i.e. Q_{40}), from an AMS set of length of $n = 20$ years. But according to the UK Flood Estimation Handbook (IOH, 1999), an AMS of length $n = 20$ years can provide satisfactory estimates for up to a 10-year return period, i.e. Q_{10}, only. In this study, peak floods of up to 50 years return period (i.e. Q_{50}) have been estimated for each scenario. Thus the peak flood estimates for the last-20-year scenario may not as acceptable. However, as the resulting Q_T values are used here only for comparison purposes, the generality of the overall results is deemed not to be affected by the length of the AMS record used to obtain them.

The Q_T–T relations computed using flood frequency analyses, for each of the seven scenarios, are first compared individually and then collectively for each of the four categories of catchments. For example, the results of the ADS (post-1945)-catchments were compared with those of the ADS (post-1925), ADS (post-1842), and also with those of the non-ADS catchments, in terms of their averages of the normalized Q_T values, written as $(Q_T/\overline{Q})_{(average)}$. These average values for each scenario, and for each category of catchments, were calculated as:

$$(\frac{Q_T}{\overline{Q}})_{(average)x,j} = \frac{1}{N_x}\sum_{i=1}^{N_x}(\frac{Q_{T,i}}{\overline{Q}_i})_j \qquad (1)$$

where the j are the 7 scenarios, x is the category of catchments with N_x number of catchments in that particular category. For example, for the ADS (post-1945) catchments (x), $N_x = 7$, for non-ADS catchments (x), $N_x = 6$, and so on (see Table 1).

RESULTS AND DISCUSSION

As noted above, flood frequency analyses were carried out to fit the GEV distribution to all the sets of the AMS data for all 22 catchments (i.e. 7 ADS (post-1945), 5 ADS

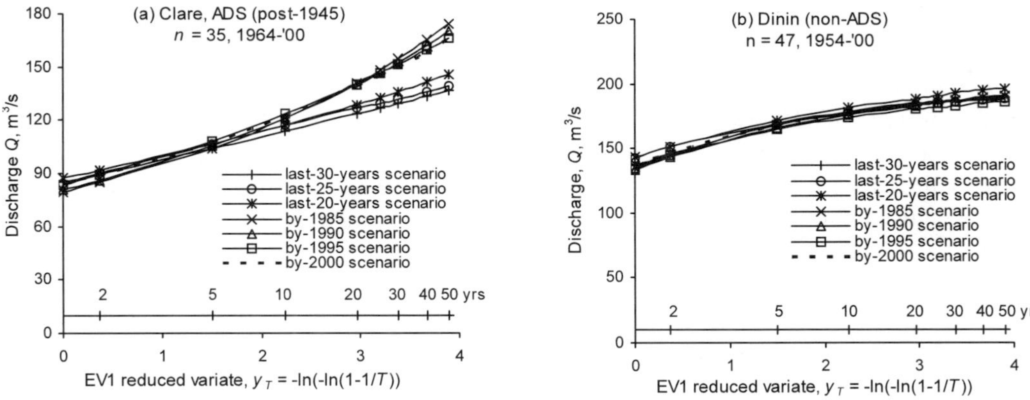

Fig. 1 Comparison of Q_T–T relations for 7 analysis scenarios (subset of annual maximum series based) for (a) ADS (post-1945) and (b) non-ADS catchments.

(post-1925), 4 ADS (post-1842), and 6 non-ADS)), and the resulting Q_T–T relations are plotted in Fig. 1 as flood frequency curves, i.e. the plots of Q_T against the EV1 reduced variate y_T (and also against T), where $y_T = -\ln-\ln(1-1/T)$. From these plots, it is observed that, for the ADS (post-1945) catchments category, the Q_T–T relations are not the same in all scenarios, but vary from one scenario to another, the degree of such variation being different for each basin. For the other three categories of basins, such changes in the Q_T–T relations are less marked. Figure 1(a) shows that, for the case of an ADS (post-1945) catchment, the by-1985-, by-1990- and the by-1995-scenarios (which contain annual floods of early post-drainage periods), produced higher Q_T values than those of the by-2000-scenario, whereas, in contrast, the Q_T values for the last-30-years, last-25-years and last-20-years scenarios (which do not contain early post-drainage annual floods) are smaller than those of the by-2000-scenario. This shows that for the ADS (post-1945) catchments, the effect of the ADSs might have died out after one/two decades of ADS implementation. Furthermore, for the scenario containing the very recent data (i.e. the last-20-years scenario), the value of Q_T has increased above those of the last-30-years and the last-25-years scenarios, but not quite to the level of those scenarios containing the early post-drainage data.

In the case of the non-ADS catchments, a somewhat opposite pattern of changes in the Q_T–T relations was noticed, as the scenarios containing only the very recent annual floods (the last-20-years scenario) produced values of Q_T somewhat higher than those of other scenarios (see Fig. 1(b)). However, no such specific trends of increase/decrease were observed in the case of ADS (post-1925) and ADS (post-1842) catchments. The reason for this may be that the AMS data for such catchments are available only after 25/30 years (for post-1925) and 100 years (post-1842) after the completion of the ADSs, and by that time the effects of the ADSs might have died out. This higher value of Q_T produced by last-20-years scenarios, in comparison to the last-25-years and last-30-years scenarios, in the case of ADS (post-1945) catchments, and the highest values of Q_T produced by last-20-years scenario in comparison to all seven scenarios, in the case of non-ADS catchments, might be attributable to the recent warming trend in Ireland, one that is particularly marked since the beginning of the 1980s (Sweeney *et al.*, 2002).

Comparisons of the averages of the normalized Q_T for the different categories of catchments also follow the trends shown by the comparison of individual catchments. Figure 2(a) shows that for the last-25-year scenario (i.e. 1976–2000), the $(Q_T/\overline{Q})_{average}$ values for all three individual categories of ADS catchments (post-1945, post-1925 and post-1842) are lower than that for the average of the entire 22 catchments, whereas such values for the non-ADS catchments category are higher than that for the entire 22 catchments. Figure 2(b) shows that, for the by-1985-scenario (which contains early post-drainage floods), such average values for the ADS (post-1945) catchments are higher than those for the entire 22 catchments, whereas these values of the ADS (post-1925) and ADS (post-1842) categories of catchments are still lower. The plot of $(Q_T/\overline{Q})_{(average)}$ for the ADS (post-1945) *vs* non-ADS catchments (Fig. 3) shows that,

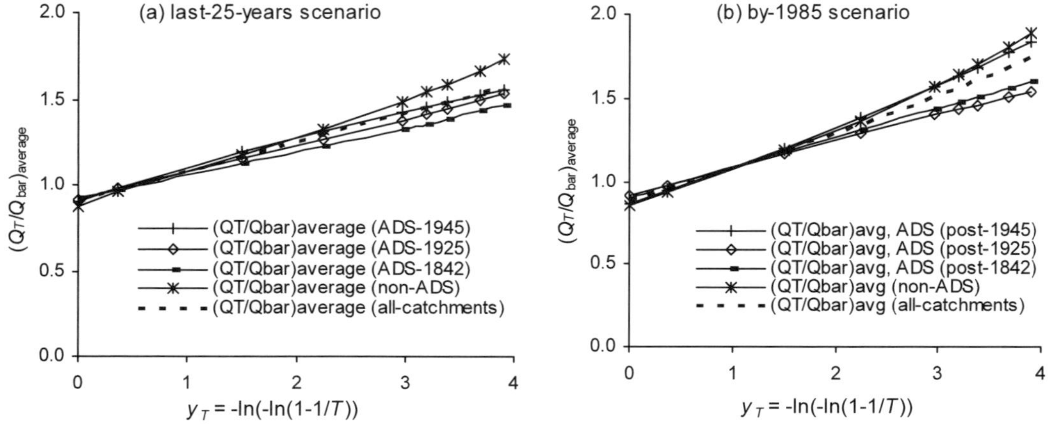

Fig. 2 Comparisons of the averages of the normalized Q_T for the different categories of catchments, for (a) last-25-years scenario and (b) by-1985-scenario.

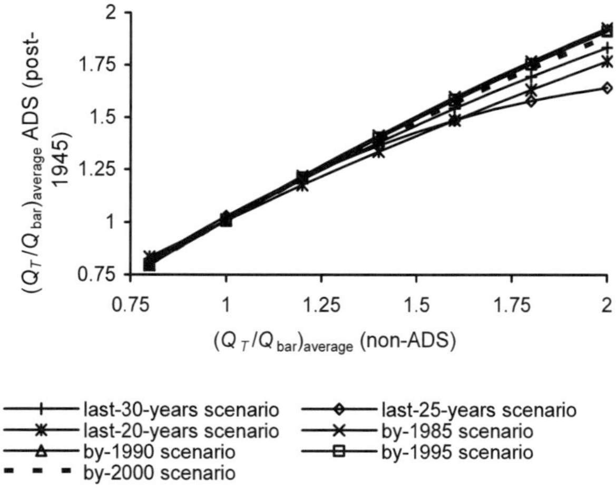

Fig. 3 Comparisons of the averages of the normalized Q_T of different scenarios: ADS (post-1945) *vs* non-ADS catchments.

such $(Q_T / \overline{Q})_{(average)}$ values for the ADS (post-1945) catchments are higher for the by-1985-, by-1990-, and by-1995 scenarios than that for the by-2000-scenario, whereas such $(Q_T / \overline{Q})_{(average)}$ values for the last-30-years, last-25-years and last-20-years scenarios are lower than that of the by-2000-scenario, thus following the above trends for individual catchments.

SUMMARY AND CONCLUSIONS

Comparison of the flood frequency curves for seven AMS subset-based analysis scenarios indicates that there have been considerable changes over time in the Q_T–T relations, for the ADS (post-1945) catchments. However, for the available AMSs, such changes are less pronounced in the ADS (post-1925), ADS (post-1842), and non-ADS catchments, indicating as a general conclusion that the ADSs effects are relatively short term. Moreover, for the ADS (post-1945) catchments, the three scenarios consisting of annual floods of the early post-drainage periods produce somewhat higher Q_T values than the by-2000-scenario (consisting of the entire AMS), whereas the other three scenarios consisting of the most recent subsets of AMS produced lower Q_T values than that for the by-2000-scenario. In contrast, the non-ADS catchments produce slightly higher Q_T values with the most recent subsets of AMS than that for the by-2000-scenarios. Comparison of average Q_T values, for each category of catchments, also showed similar trends. Within the last-30-, last-25-, and last-20-years scenarios, the last-20-years scenario produced higher values, both for ADS (post-1945) and non-ADS catchments, which might be attributable to the effects of the recent warming trend in Ireland, a suggestion that needs further investigation. These results suggests that, in a flood frequency analysis exercise, it would be advantageous to check the Q_T values estimated for entire AMS set against those based on the most recent 20/30 years subset of the entire AMS, as these latter sub-sets might best reflect the current form of the Q_T–T relationship of such catchments.

Acknowledgements The AMSs were downloaded from the official web site of the Irish Office of Public Works (http://www.opw.ie/hydro). Special thanks are due to Mr Peter Newport (Headford) and Mr Vincent Hussey (Trim) of the OPW, for providing additional data and information. The first author is pleased to acknowledge the research support provided by the HEA (of Ireland) through the ECI of the NUI, Galway. Comments made by an anonymous reviewer greatly improved this paper.

REFERENCES

Bailey, A. D. & Bree, T. (1981) Effect of improved land drainage on river flow. In: *Flood Studies Report—5 Years On.*, 131–142. Thomas Telford, London, UK.

Bhattarai, K. P. & O'Connor, K. M. (2004) The effects over time of an arterial drainage scheme on the rainfall–runoff transformation in the Brosna catchment. *Phys. Chem. Earth* **29**, 787–794.

Bree, T., Cunnane, C. & Lynn M. A. (1979) Effect of arterial drainage works on the unit hydrograph. In: *International Symposium on Scientific Aspects of Hydrological Computation for Water Projects* (UNESCO, WMO, IAHS and USSR National IHP Committee, 3–7 September 1979, Leningrad, USSR, 3–11).

Bree, T. & Cunnane, C. (1979) The effect of arterial drainage on flood magnitude. Casebook of scientific project 5.1 of the International Hydrological Programme: Assessment of changes in the hydrological regime of river basins due to human activities. UNESCO, 116–126.

Hosking, J. R. M. (1990) L-moments: analysis and estimation of distributions using linear combinations of order statistics. *J. Royal Statis. Soc.* **52**(2), 105–124.

Institute of Hydrology (1999) *Flood Estimation Handbook*. Wallingford, UK.

Lynn, M. A. (1981) *Estimating Flood Magnitude/Return Period Relations and the Effect of Catchment Drainage*. The Office of Public Works, Dublin, Ireland.

Nash, J. E. (1955) The relation of stream flow to rainfall. ME Degree Thesis, National University of Ireland, University College, Galway, Ireland.

Nash, J. E. (1958) Determining runoff from rainfall. *Proc. ICE* **10**, June 1958.

Natural Environment Research Council (1975) *Flood Studies Report*. NERC, Swindon, UK.

O'Kelly, J. J. (1955) The employment of Unit Hydrographs to determine the flows of Irish arterial drainage channels. *Proc. Inst. Civ. Eng.* (London), August 1955.

Sweeney, J., Donnelly A., McElwain, L. & Jones, M. (2002) *Climate Change Indicators for Ireland*. Environmental Protection Agency, Ireland.

Wang, Q. J. (1996) Direct sample estimators of L-moments. *Water Resour. Res.* **32**(12), 3617–3619.

Modelling the hydraulic regime and the water quality of Sebou River (Morocco): first results

MOHAMMED IGOUZAL[1], JEAN MARIE MOUCHEL[2], KARIM TAMOH[1] & ABDELLATIF MASLOUHI[1]

1 *Ibn Tofail University, BP 2047 Al Massira Kénitra, Morocco*
igouzal@cereve.enpc.fr

2 *Cereve, ENPC, 6–8 Avenue Blaise Pascal, MarneLaVallée, Paris, France*

Abstract In this study, a first attempt is performed to study and to model the hydraulic regime and water quality of the Sebou River in the Gharb agricultural zone. A 90 km long reach situated upstream a reservoir dam is studied during the base flow period. During this period, water resources are scarce and much solicited by agriculture while water quality is very problematic. The dynamic water quality model WASP5 is used. Results give more information about the hydraulic regime which is influenced by a variable morphology. Water depth, flow and velocity evolution along the river and the water reserve available in the dam reservoir are simulated. In the water quality modelling, scenarios of excessive salinity propagation were first simulated. The time of arriving and the amplitude of the salinity were estimated at agricultural pumping stations. This model can play the role of an alert system against this natural phenomenon. Dissolved oxygen (DO) and biological oxygen demand (BOD) were then simulated. The impact on the oxygen depletion of sugar production plant wastewater was estimated. This information gives more suggestions for the envisaged waste load treatment programmes.

Key words hydraulic regime; salinity; Sebou River; WASP5 water quality model

INTRODUCTION

The studied area is situated between Mechraa Belksiri town and Lalla Aicha dam (Fig. 1). River waters are used intensively for agricultural activity and many land-use changes are planned. Flow regime is influenced by considerable natural variations and by the presence of many dams (constructed upstream of the study area) for electricity production. During base flow periods, the Lalla Aicha dam is partially closed in order to facilitate pumping water for agriculture inside the river reach, and to avoid salt intrusion from the estuary. At the same time, two sugar production plants situated near Belksiri town and the third near Sidi Allal Tazi village start dumping their non treated biodegradable wastewater into the river. DO concentrations below the minimum necessary for fish (5 mg l^{-1}) are sometimes measured, which upsets the natural equilibrium of the river and kills a lot of fish. A few modelling studies have already been developed in the Sebou River (Igouzal *et al.*, 1997). The WASP5 model (Ambrose & Martin, 1993) is applied here to help to understand the hydraulic regime and water quality for 1997. Hydraulic simulations provide flows, depths, water level and the water reserve available in the whole area. Water quality data are very scarce, so only DO and the BOD are simulated in order to study the system tendencies and to have a whole view of it. The influence of the dam downstream and the impact of each

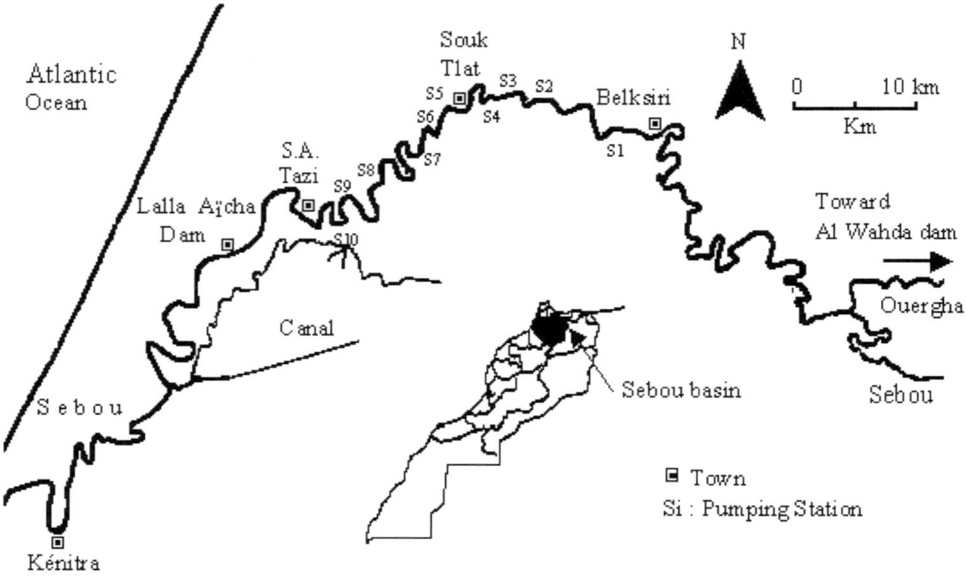

Fig. 1 Position of the studied river reach in the Gharb agricultural zone.

sugar production plant were analysed. In many periods, high salinity concentrations were measured. This affects agriculture and forces some important pumping stations to stop their activities for many weeks each year (Zeraouli & Chakor, 1998). The mean salinity of Sebou waters is 1.15 g l^{-1}. Salinity comes from some specific high-salt soils areas at the upstream zones of Sebou basin. In this context, model simulations can help for the prevention against this natural phenomenon.

MODEL DESCRIPTION

The one-dimensional (1-D) hydraulic submodel called DYNHYD5 was modified in order to get a better morphologic representation of the river (Igouzal & Maslouhi, 2003). DYNHYD5 uses the two Saint-Venant continuity and movement equations. Flow resistance is quantified using the Manning friction coefficient, which depends on the nature of the river bed and on the river morphology. Sebou morphology varies a lot along the studied area, so a variable Manning friction coefficient is used along the river. It ranges from 0.02 to 0.04, with a mean value of 0.037. The discretization of the studied reach was performed using aerial photographs taken during a dry period when the river was nearly dry. A total of 528 grid meshes were adopted. River morphology was evaluated using data available from limited cross sectional areas. Near Souk Tlat village, morphology was quantified using "Numeral photogrammetry" (Igouzal *et al.*, 2003). River photos of a 1/20 000 scale taken in 1983 during a very base flow period were used. Then, the RESTIT.2 Photogrammetry's software (Ejels, 1999) enabled the morphology's stereoscopic measurements on aerial photos to be scanned and calibrated. A correlation coefficient of 0.96 was obtained between measured and calculated cross section areas. The intensive pumping of waters along the river via 10 principal stations (Fig. 1) and many smaller stations is considered in the model. The total pumped flow ranges from 10 to 25 $m^3 s^{-1}$. Also, boundaries and initial time step conditions were needed for the simulations. The upstream boundary is given as values of water level as function of time. The downstream boundary is a discharge as a function of time. The water quality model used is called EUTRO5, based on the advection–dispersion equation. It represents the biochemical water quality submodel of WASP5. The longitudinal dispersion coefficient used in the model (13 $m^2 s^{-1}$) was calculated using salinity propagation measurements (Zeraouli & Chakor, 1998).

RESULTS

Calibration of the hydraulic model is achieved by varying the Manning coefficient (with the same magnitude along the river) in order to minimize the difference between measured and simulated water level. The only available measured water level was at S2 station. The months of July and June 1997 were used for the calibration and validation, respectively. The results show a good agreement between measurements and predictions (Fig. 2). The hydraulic model also calculates the spatial and temporal evolution of velocity and water depth. The water reserve available in the whole reach is also simulated (Fig. 3). This information can allow rapid interventions when this variable begins to decrease dramatically. The fate of salinity intrusion with an amplitude similar to that measured during high salinity periods (Zeraouli & Chakor, 1998) was simulated. Figure 4 gives its theoretical evolution at five locations: Belksiri, S2, S7 and S10 pumping stations and at the dam, for a water flow of 10 $m^3 s^{-1}$. Salinity arrives more dispersed to the stations downstream and with lower peak values. When the flow is increased to 25 $m^3 s^{-1}$, salinity arrives quickly downstream, but with large amplitude (Fig. 5). The S2 station (the most important regarding the quantity of water pumped) is more affected by the salinity because it is situated at the upstream part. Salinity arrives quickly at this station and with high amplitude. Thus, to increase the time of alert against high salinity intrusion for this station, the upstream condition for the model must be moved more upstream of Belksiri hydrological station. Such a coupled model (hydraulics plus salt transport) could prove extremely useful for improved management of the pumping stations.

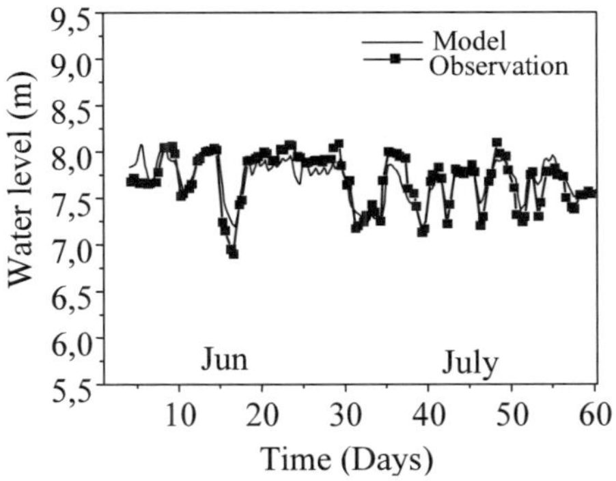

Fig. 2 Calibration of the hydraulic model, June-July 1997 (Igouzal & Maslouhi, 2003).

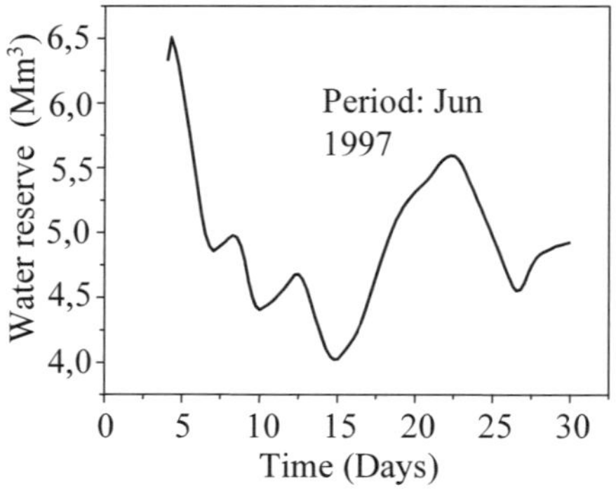

Fig. 3 Evolution of the water reserve, June 1997.

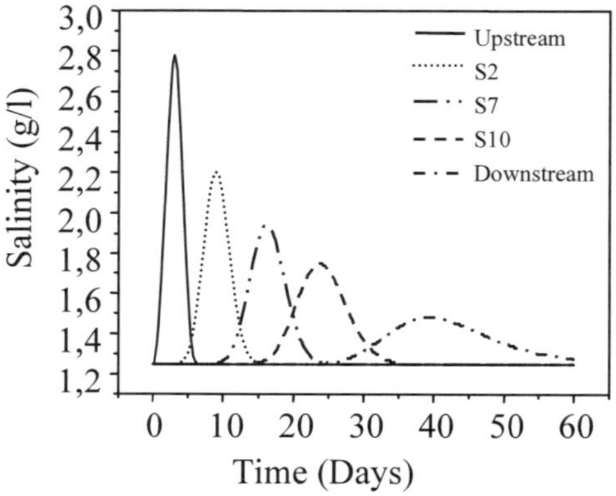

Fig. 4 Evolution of a salinity intrusion at different stations for a 10 $m^3 s^{-1}$ flow.

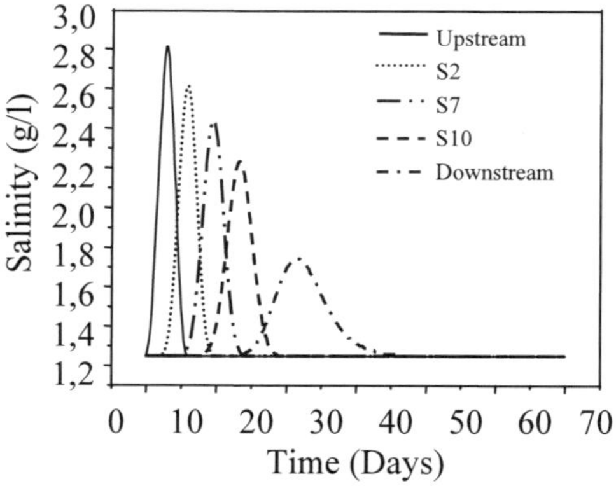

Fig. 5 Evolution of a salinity intrusion at different stations for a 25 $m^3 s^{-1}$ flow.

In biochemical modelling, the reacting terms in EUTRO5 are re-aeration and oxidation for dissolved oxygen, and oxidation for the BOD. Information about the oxidation coefficient for the sugar wastewater was not available. A mean oxidation coefficient of 0.2 day^{-1} was used. Also, continuous measurements of DO and BOD were not found to calibrate the water quality model. The existing data have been collected on an occasional basis. So, different scenarios were simulated to estimate the influence of the dam downstream and the impact of each sugar plant. First, simulations without the presence of wastewater were elaborated to quantify the effect of the dam closure. It showed that even when no industrial wastewater is dumped into the river, the dam closure can contribute to the water quality degradation by reducing the natural re-aeration. With the presence of the dam downstream, velocity decreases downstream and the depth increases, so re-aeration must decrease downstream which reduces the DO concentration. Other simulations considered only the Belksiri wastewater (upstream), only the Tazi wastewater (downstream) and both the Belksiri and Tazi wastewater. These simulations show first that the Belksiri wastewater has the greatest impact on water quality because it increases the BOD and decreases the DO concentrations so much in waters. However, simulations show that the conjunction of the two wastewaters increase the negative influence on the water quality of the river (Figs 6 and 7). Thus, if a treatment must be done, we think that Belksiri wastewater must be treated first. This wastewater is the most important and its travel time along the reach is the longest.

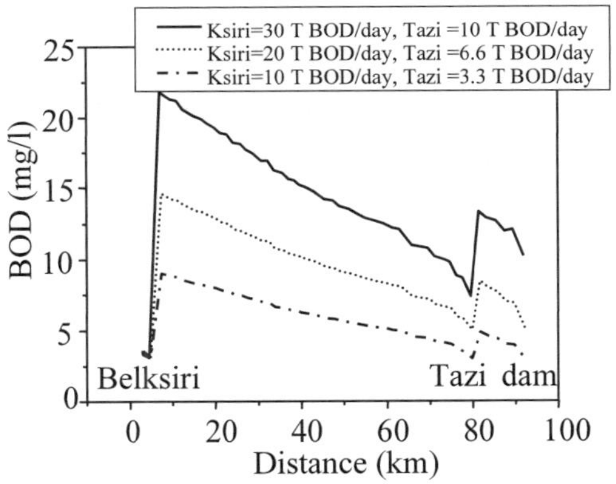

Fig. 6 BOD evolution for three outflows in t day^{-1}.

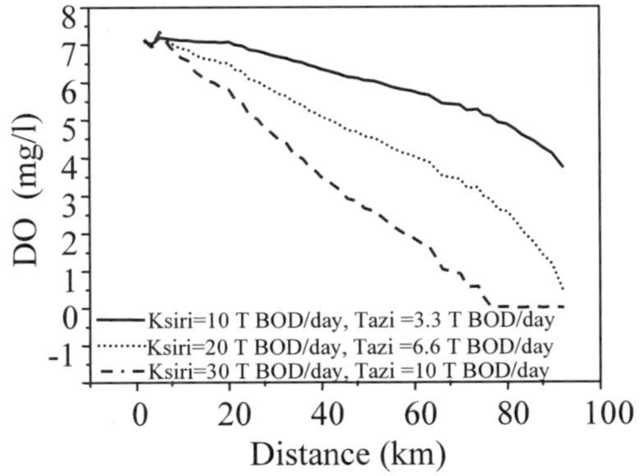

Fig. 7 DO evolution for three outflows in t day^{-1}.

CONCLUSIONS

In this study, a first attempt was performed to study and model the Sebou River waters in a 90 km long reach characterized by human quantitative and qualitative impacts. A base flow period is studied, during which water is very scarce and many water standards are violated. Results permit a good understanding of the hydraulic regime. Despite the lack of adequate field data, salinity and water quality simulations give a lot about the element of response. The time alert against high salinity concentration will be increased more using this mathematical model. Also, simulation of DO and BOD variables provides a lot of suggestions if a wastewater treatment is envisaged. These results are not recommended as a substitute for field studies, but are believed to provide reasonable estimates of water quality variables in a short time to help with many environmental and health safety considerations.

REFERENCES

Ambrose, R. B. & Martin, J. L. (1993) WASP5, a hydrodynamic and water quality model. Model theory, user's manual guide and programmer's guide. *US EPA, Georgia 30613.*

Egels, Y. (1999) *Le logiciel de photogrammétrie RESTIT, Version 5*. Manuel d'utilisation. l'IGN, L'ENSG, France.

Igouzal, M. & Maslouhi, A. (2003) Contribution à la gestion de la retenue d'un barrage réservoir sur la rivière Sebou (Maroc) à l'aide d'un modèle hydraulique. *Rev. Sci. Eau* **16**(4), 443–458.

Igouzal, M., Maslouhi, A., Schelstraete, D. & Chan, J. (2003) Reconstitution de la morphologie d'une rivière à méandres (le Sebou, Maroc) par l'utilisation de la photogrammétrie numérique. *Bull. Soc. Fr. de Photogramm. et de Télédétec.* **171**(3), 16–21.

Igouzal, M., Maslouhi, A., Lahlou, F. & Daniane, M. (1997) Modélisation de la qualité des eaux de l'oued Sebou à l'aval d'un rejet industriel, In: *Fresh Water Contamination.* (ed. by B. Webb) (Proc. Rabat Symp., April–May 1997), 349–356. IAHS Publ. 243. IAHS Press, Wallingford, UK.

Zeraouli, M. & Chakor, B. (1998) Les sources salées: causes de la salinisation de l'Oued Sebou même en périodes pluvieuses. *ORMVAG. Ministère of Agriculture, Morocco.*

Phosphorus concentrations of brook runoff in a Swiss agricultural catchment: longitudinal variability and subsurface drainage impact

RAINER WEISSHAIDINGER[1], BERND HEBEL[2], PETRA OGERMANN[1], PHILIPP SCHNEIDER[1], CHRISTIAN KATTERFELD[1] & RANDY KOCH[1]

1 *University of Basel, Chair of Physical Geography and Landscape Ecology, Klingelbergstrasse 27, CH-4056 Basel, Switzerland*
rainer.weisshaidinger@unibas.ch

2 *Swiss Federal Institute of Technology Zurich, Chair of Forest Engineering, ETH Zentrum HG G21.1, Rämistrasse 101, CH-8092 Zurich, Switzerland*

Abstract Hydrological characteristics and water quality of anthropogenic subsurface drainage and brook runoff were observed in a small agricultural catchment near Basel (NW Switzerland). A two year monitoring programme provided concentration data for soluble Molybdate Reactive Phosphorus (MRP = Dissolved Reactive Phosphorus) for both base flow and storm events. Recurrence frequency and correlation analyses were conducted to evaluate the samples with regard to time–space variability as well as observance of eutrophication limits. Results suggest substantial MRP input into the brook system via subsurface drains, partly originating from private wastewater treatment facilities. Runoff, particularly in upstream brook sections, clearly exceeds eutrophication limits already at MRP concentrations (C_{MRP}) of low recurrence frequency. During non-event periods a decrease in C_{MRP} is obvious from upstream brook sections towards the catchment outlet as a result of dilution as well as hydrobiological and chemical processes.

Key words anthropogenic subsurface drainage; molybdate reactive phosphorus; point and nonpoint sources; recurrence probability; water quality

INTRODUCTION

Phosphorus (P) input to surface water, originating from nonpoint agricultural sources, causes serious eutrophication problems, even within remote off-site areas (Sharpley *et al.*, 1995; Whiters *et al.*, 1999). Surface runoff processes, in particular soil erosion, are usually identified as the most eminent nonpoint sources of P as a result of the strong sorption of P by soil particles (Prasuhn & Grünig, 2001; Quinton *et al.*, 2001). However, several studies attested that anthropogenic subsurface drainage runoff may contribute significantly to eutrophication (Brookes *et al.*, 1997; Haygarth *et al.*, 1998; Stamm *et al.*, 1998; McDowell & Wilcock, 2004).

According to Foy & Whiters (1995) even small concentrations of P may cause eutrophic conditions in certain locations, and consequently standards should be closer to 0.03 ppm. On the basis of a water protection decree (Swiss Federal Assembly, 1998a) the Swiss Agency for the Environment, Forests and Landscape (SAEFL [s.a., Online]) is discussing an eutrophication limit of $C_{MRP,lim} = 0.04$ ppm for receiving waters' runoff. Additionally, Swiss agricultural policy has focused on advancing the ecological orientation of farming since 1993. Ecological measures (Ge.: *Ökomassnahmen*) are promoted in order to reduce sediment and nutrient input originating from nonpoint agricultural sources. Subvention payments for farmers are dependent on ecological efficiency statements (Ge.: *Ökologischer Leistungsnachweis*) requiring a well-balanced fertilizer budget, an appropriate share of ecological compensation areas, well-adapted crop rotations, adequate soil protection and riparian buffer strips (3 m minimum width) along receiving waters (Swiss Federal Assembly, 1998b).

However, the impact of agricultural management and land drainage on phosphorus input to surface waters and their temporal and spatial dynamics are largely unconsidered by agropolitical regulations. This is particularly true for small catchment systems.

Objectives

(a) Statement on the sufficiency of ecological measures with respect to the observance of the discussed eutrophication limits $C_{MRP,lim}$ under the specific conditions of the Swiss Jura Plateau.
(b) Elaboration of methods suited to reflect prevailing MRP dynamics along receiving water courses within a small catchment in the region.
(c) Identification of source areas of P and the quantification of heterogeneous impacts provoked by varying land management and human impacts within sub-catchments.

MATERIAL AND METHODS

Study area

The Länenbach Valley was chosen as representative investigation area (2.61 km^2, first order) for the specific conditions of the Swiss Jura Plateau. These conditions are a pronounced relief, clay soils and an intensive tile draining. The agricultural soils of the region are characterized by high portions of clay fractions (Opalinus clay, partly overlaid by periglacial sediments) particularly concerning the valley's bottom area (Leser *et al.*, 2002).

The majority of agricultural soils possess phosphorus levels below 45 ppm and generally show no P surplus. These soil P conditions are similar to those of the eastern Jura Mountains. However, the conditions are contrary to those of other Swiss landscapes (e.g. grassland dominated regions of the Swiss Plateau), where P has accumulated within surface soil layers during past decades as a result of intensive manure spreading (Stamm *et al.*, 1998). Augmented soil P levels implies increased P transport potential by suspended sediment (Wall *et al.*, 1996) and may result in an increased MRP concentration in subsurface drainage flow (Stamm *et al.*, 1998).

Soil matrix flow and preferential flow paths may generate direct nutrient and sediment input into a widespread tile drain system (Stamm *et al.*, 1998; Lichner *et al.*, 1999; Whiters *et al.*, 1999). This drainage system represents a prevailing transport pathway for both subsurface and surface runoff, carrying suspended sediment (SS) as well as adsorbed and dissolved nutrients towards the brook (Seiberth, 2001). Inlet funnels conduct local surface runoff (originating from thalweg situations, gravel roads

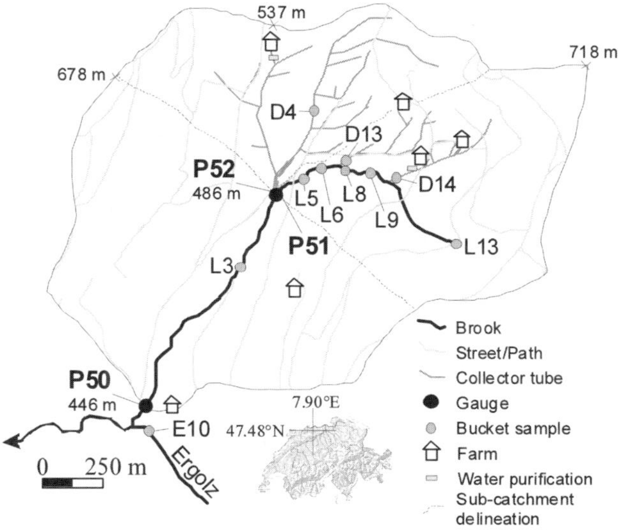

Fig. 1 Map showing the Länenbach catchment (2.61 km^2): artificial drainage system and monitoring sites.

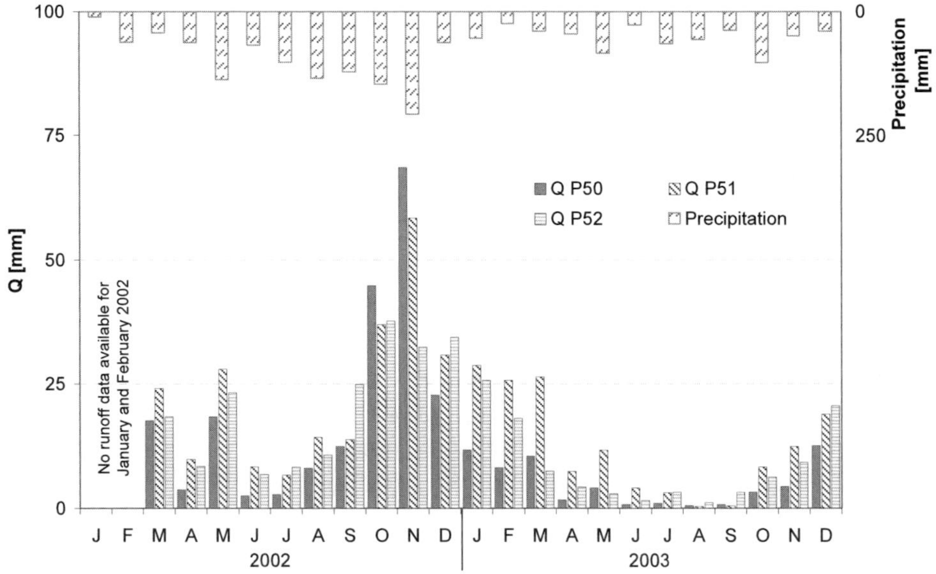

Fig. 2 Monthly precipitation and runoff totals Q [both in mm] at the gauge sites P50 (entire Länenbach catchment) as well as P51 and P52 (sub-catchments) for the period January 2002 to December 2003. Precipitation data source: Swiss Meteorological Institute SMI-MeteoSwiss (Station Rünenberg, TP039; 610 m a.m.s.l.; Long: 7.88°E, Lat: 47.43°N).

and ditches) into the anthropogenic subsurface drainage system. Consequently, remote arable areas are directly connected to the receiving water course (hydraulic connectivity, Fig. 1).

The drainage regime of sub-catchment P51, and similarly the western drainage branch of sub-catchment P52, are additionally influenced by inlets originating from private wastewater treatment facilities (constructed wetlands), whereas the drainage regime of the eastern branch, D4 within P52, is exclusively influenced by agricultural structures.

The investigation period from January 2002 to December 2003 was defined by the effects of two extreme meteorological events (Fig. 2):
(a) Persistent rainfall during the autumn of 2002 (total precipitation 2002: 1 162 mm)
(b) Uncharacteristically warm and dry summer of 2003 (total precipitation 2003: 624 mm).

Monitoring, sampling and evaluation methodology

Characterization of primary investigation methods:
(a) Monitoring of storm event runoff by stage gauges and automated revolving water samplers at high temporal resolution.
(b) Continuous fortnightly bucket sampling of runoff at selected brook profiles as well as drainage pipe inlets into the brook (including collection pipes fed by several drainage pipes before being discharged into the brook), representing non-event periods.
(c) Discharge calculation by means of stage–discharge correlations and Hydras-3 software (Source: OTT Hydrometry).
(d) Colorimetric spectral photometer analysis (at 824 nm) of runoff samples (0.45 μm membrane filter) for C_{MRP} (Murphy & Riley, 1962).
(e) Calculation of recurrence frequencies T_C for C_{MRP} of continuous monitoring as well as storm event runoff samples (available only for sites P50, L3, P51, P52, L8, D13, L9), which was effected according to Linsley *et al.* (1992) using Gumbel distribution and partial series with $n = 13$ considering the highest C_{MRP} as data basis.
(f) Correlation analysis between C_{MRP} of continuous bucket sample series at gauge and pipe inlet sites as well as selected brook profiles taken at identical dates ($n = 31$).

Recurrence frequency and correlation analyses could not be conducted for sites D4, D14 and L13 as the sites were repeatedly dry during the summer months and therefore can only be characterized by long-term average C_{MRP}.

RESULTS AND DISCUSSION

Recurrence frequency of MRP concentrations

As a result of the input from wastewater treatment facilities into the draining system, the MRP concentrations of the inlets P52 and D13 exceed the recommended limits $C_{MRP,lim}$ already at low recurrence frequencies (Fig. 3). An average background load of $C_{MRP} = 0.038$ ppm (continuous bucket samples, $n = 37$, no recurrence analysis available) and a maximum of $C_{MRP} = 0.151$ ppm were recorded originating from the branch D4, which is exclusively dominated by agricultural structures does not

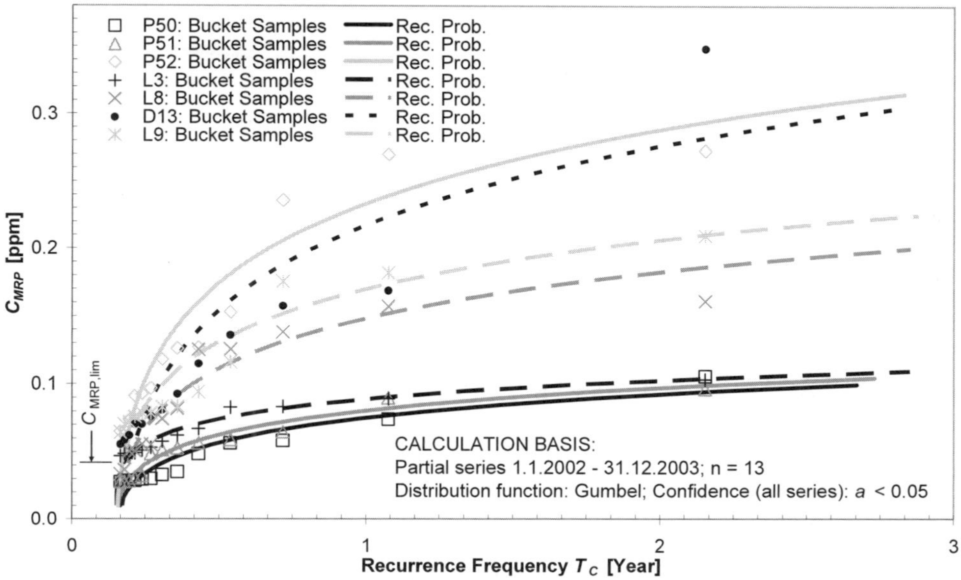

Fig. 3 Recurrence frequency of C_{MRP} in continuous bucket samples during non-event periods of all sampling sites (calculation according to Linsley *et al.*, 1992).

contribute to such high concentrations during non-event periods. Haygarth *et al.* (1998) also identified a small C_{MRP} in artificial drainage flow from grassland on clayey soils (C_{MRP} mean: 0.04 ppm, min: 0.01 ppm, max: 0.30 ppm) in southwest England. Concentrations in subsurface drainage flow investigated by Stamm *et al.* (1998) illustrate much higher levels under intensive grassland use of the Swiss Plateau (C_{MRP} min: 0.05 ppm, max: 4.8 ppm).

C_{MRP} at brook profiles L8 and L9 in upstream brook sections exceed eutrophication limits $C_{MRP,lim}$ already at low recurrence intervals due to low discharge rates. These sites are situated downstream of a drainage pipe inlet D14, contributing high nutrient loads originating from a private constructed wetland (wastewater treatment) into a section of the brook with low discharge rates. Long-term average background load at brook profile L13 originating from the non-agricultural areas is $C_{MRP} = 0.009$ ppm (continuous bucket samples, $n = 28$) with a maximum of $C_{MRP} = 0.042$ ppm. Even at downstream brook profiles, not directly influenced by pipe inlets (P50, P51, L3), C_{MRP} moderately exceeds eutrophication limit $C_{MRP,lim}$ already at low recurrence intervals.

C_{MRP} regime at L3 renders integration processes of P51 and P52. Downstream, differences in C_{MRP} regime of subsequent brook profiles (e.g. L3 and P50) become obvious. Both effects are caused by:

(a) Diluting of the concentrated input from pipe P52 as a result of higher non-event average discharge rates of brook runoff ($Q_{P51,mean} = 4.9$ l s^{-1}) in relation to pipe inlet rates ($Q_{P52,mean} = 3.0$ l s^{-1}).

(b) MRP adsorption and uptake due to hydrobiological and -chemical activity in the brooks cascading sequence (step-pool systems, sediment retaining backwaters above small steps or dead wood locks, sediment remobilizing pools and scours).

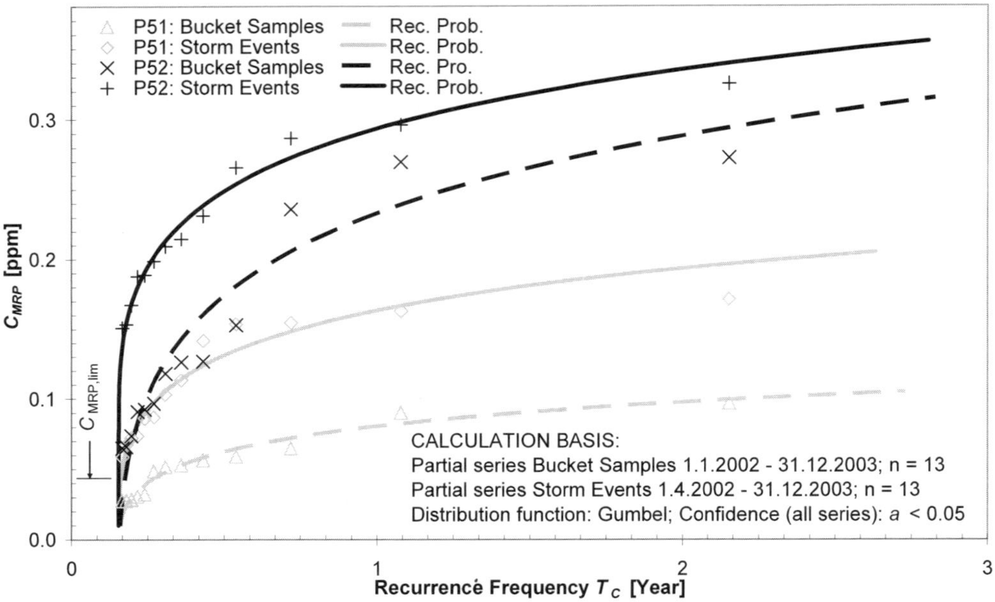

Fig. 4 Recurrence frequency of C_{MRP} in bucket samples during non-event periods opposed to max. C_{MRP} in storm event samples of two gauge stations (calculation according to Linsley *et al.*, 1992).

During storm periods the C_{MRP} of both brook runoff and drain input exceed the eutrophication limit $C_{MRP,lim}$ even for low recurrence intervals (Fig. 4). During storm events C_{MRP} is about 100% (P51) and 20% (P52) higher than during non-event periods.

C_{MRP} in drainage input is approximately two to three times higher than in brook runoff. This is valid during both storm events ($Q_{P51,mean} = 57.0$ l s^{-1}, $Q_{P51,max} = 68.2$ l s^{-1}; $Q_{P52,mean} = 61.1$ l s^{-1}, $Q_{P52,max} = 68.0$ l s^{-1}), as well as non-event periods (see above).

Correlation coefficient between MRP concentrations

Distinct correlations between the C_{MRP} regimes of profiles within the central brook section (L8 down to L3, characterized by low average discharge rates) prove that downstream sections are influenced by upstream impacts, such as concentrated pipe inflow (Table 1). This is especially obvious for pipe inlet D13 with corresponding downstream profiles L5, L6.

The downstream brook section (L3 down to P50 and the confluence E10 with the Ergolz River, characterized by higher average discharge rates compared to Länenbach brook) is however dominated by diluting effects as well as hydrobiological and -chemical processes creating low correlation rates.

C_{MRP} regimes of drainage pipes (P52, D13) generally show little correlation between one another, implying different land management situations have been adopted and a diverting influence of the wastewater treatment facilities within their corresponding sub-basins.

Table 1 Correlation coefficients r^2 between C_{MRP} ($n = 31$, period January 2002 to December 2003) of continuous bucket samples at gauge (P) as well as selected brook (L, E) and drainage pipe inlet (D) sites (entire series, samples taken at identical dates). $0.3 < r^2 < 0.5$ is highlighted with grey backgrounds, $r^2 \geq 0.5$ with black backgrounds.

	E10	P50	L3	P51	P52	L5	L6	L8	D13
P50	0.39								
L3	0.02	0.35							
P51	0.08	0.42	0.28						
P52	0.08	0.39	0.33	0.19					
L5	0.02	0.30	0.45	**0.52**	0.21				
L6	0.01	0.21	0.40	**0.59**	0.20	**0.87**			
L8	0.00	0.14	0.15	**0.52**	0.24	0.33	0.48		
D13	0.02	0.05	0.23	0.06	0.11	0.44	0.45	0.01	
L9	0.00	0.25	0.25	**0.67**	0.28	0.39	**0.52**	**0.81**	0.04

CONCLUSIONS

Methodology The analysis of variations in terms of MRP regime by means of recurrence stochastics appears to be more revealing in contrast to a common indication of arithmetic means. Thereby, an identification of source areas and a quantification of human impacts were feasible as a result of a systematic monitoring of representative spots. The results suggest that conventional point observations of nutrient yield exclusively at catchment outlets are not suited to constitute a reliable quantification of the complex nutrient dynamics within upstream river sections.

MRP limits The artificial drainage system was identified to be the dominating transport pathway for MRP within the investigated catchment. Due to regular and prominent MRP limit excess, the present disregard of subsurface drainage and concentrated point inlet impacts within ecological measures (Ge.: *Oekomassnahmen*) appears to be problematic as far as the specific regional regime of the Swiss Jura is concerned. As long as surface runoff from undrained arable areas remains the prevailing transport process, such ecological measures designated by Swiss legislation may be sufficiently effective. However, in the case of anthropogenic subsurface drainage systems—especially in combination with any kind of hydraulic connectivity—the measures taken will have to be supplemented and extended. This is particularly true for storm event periods. Improvements may be created by grass covered thalweg flowpaths, grass buffer strips surrounding drainage inlets, and extended grass strips next to arable land, especially at the end of inclinated furrows.

Longitudinal variability Only during non-event periods may the impact of concentrated point sources on the downstream brook system be considered low. Within short distances along the brook reduction of C_{MRP} was marked as a result of diluting effects and hydrobiological and chemical processes. However, during major storm events the impact of concentrated MRP input from drainage pipe inlets was measurable along the entire downstream brook sections. Considering large artificially drained areas, the re-establishment of natural wetlands may be an appropriate way to decrease MRP inputs into river systems (Petry *et al.*, 2002; Jordan *et al.* 2003). Additionally, drainage pipe inlets influenced by private wastewater treatment facilities should enter a brook section only where brook discharge can guarantee an adequate dilution.

Acknowledgements The authors appreciate the financing of the covering research programme by Swiss National Science Foundation (SNF). Compliments to MeteoSwiss as well as the technical and laboratory personnel of the Chair of Physical Geography and Landscape Ecology (University of Basel) for their vital support. Comments of Dr Volker Prasuhn and language support of Anna Squires are gratefully acknowledged. Comments by Xavier Úbeda were helpful in improving the manuscript.

REFERENCES

Brookes, P. C., Heckrath, G., De Smet, J., Hofman, G. & Vanderdeelen, J. (1997) Losses of phosphorus in drainage water. In: *Phosphorus Loss from Soil to Water* (ed. by H. Tunney, O. T. Carton, P. C. Brookes & A. E. Johnston), 253–271. CAB International, Wallingford, UK.

Foy, R. H. & Withers, P. J. A. (1995) The contribution of agricultural phosphorus to eutrophication. *Fertilizer Soc. Proc.* **365**, 1–32.

Haygarth, P. M., Hepworth, L. & Jarvis, S. C. (1998): Forms of phosphorus transfer in hydrological pathways from soil under grassland. *Europ. J. Soil Sci.* **49**, 65–72.

Jordan, T. E., Whigham, D. F., Hofmockel, K. H. & Pittek, M. A. (2003) Nutrient removal by a restored wetland receiving agricultural runoff. *J. Environ. Qual.* **32**, 1534–1547.

Leser, H., Meier-Zielinski, S., Prasuhn, V. & Seiberth, C. (2002) Soil erosion in catchment areas of northwest-Switzerland: Methodological conclusions from a 25-year research program. *Z. Geomorphol.* **46**, 35–60.

Lichner, L., Mészároš, I., Germann, P. F., Mdaghi Alaoui, A., Šír, M. & Faško, P. (1999) Impact of land-use changes on nutrient fluxes in a structured clay-loam soil. In: *Impact of Land-Use Change on Nutrient Loads from Diffuse Sources* (ed. by L. Heathwaite) (Birmingham, UK, July 1999), 171–177. IAHS Publ. 257, IAHS Press, Wallingford, UK.

Linsley, P. K., Franzini, J. B., Freyberg, D. L. & Tchobanoglous, G. (1992) *Water-Resources Engineering*. McGraw-Hill, New York, USA.

McDowell, R. W. & Wilcock, R. J. (2004) Particulate phosphorus transport within stream flow of an agricultural catchment. *J. Environ. Qual.* **33**, 2111–2121.

Murphy, J. & Riley, J. P. (1962) A modified single solution method fort the determination of phosphate in natural water. *Anal. Chim. Acta* **27**, 31–36.

Petry, J., Soulsby, C., Malcolm, I. A. & Youngson, A. F. (2002) Hydrological controls on nutrient concentrations and fluxes in agricultural catchments. *Sci. Total Environ.* **294**, 95–110.

Prasuhn, V. & Grünig, K. (2001) Evaluation der Ökomassnahmen: Phosphorbelastung der Oberflächengewässer durch Bodenerosion. (Evaluation of ecological measures: Phosphorus contamination of surface water systems by soil erosion. Schriftenreihe FAL, vol. 37. Swiss Federal Research Station for Agroecology and Agriculture, Zurich-Reckenholz, Switzerland.

Quinton, J. N., Catt, J. A. & Hess, T. M. (2001) The selective removal of phosphorus from soil: Is event size important? *J. Environ. Qual.* **30**, 538–545.

Seiberth, C. (2001) Relation between soil erosion and sediment yield in catchment scale. In: *Sustaining the Global Farm* (ed. by D. E. Scott, R. H. Mohtar & G. C. Steinhardt) (soil cons. org. meeting, West Lafayette, USA, May 1999), 725–731.

Sharpley, A. N., Hedley, M. J., Sibbesen, E., Hillbricht-Ilkowska, A., House, W. A. & Ryszkowski, L. (1995) Phosphorus transfers from terrestrial to aquatic ecosystems. In: *Phosphorus in the Global Environment: Transfers, Cycles and Mangement* (ed. by H. Tiessen), 171–199. Wiley, Chichester, UK.

Stamm, C., Flühler, H., Gächter, R., Leuenberger, J. & Wunderli, H. (1998) Preferential transport of phosphorus in drained grassland soils. *J. Environ. Qual.* **27**, 515-522.

Swiss Federal Assembly (1998a) Gewässerschutzverordung (GSchV) from the 28 Oct. 1998. (Decree on water protection). SR: 814.201, Bern, Switzerland.

Swiss Federal Assembly (1998b) Verordnung über Direktzahlungen an die Landwirtschaft (DZV) from the 7 Dec. 1998. (Decree on subvention payments to agriculture). SR: 910.13, Bern, Switzerland.

Wall, G. J., Bos, A. W. & Marshall, A. H. (1996) The relationship between phosphorus and suspended sediment loads in Ontario watersheds. *J. Soil Water Conserv.* **51**(6), 504–507.

Whiters, P. J. A., Dils, R. M. & Hodgkinson, R. A. (1999) Transfer of phosphorus from small agricultural basins with variable soil types and land use. In: *Impact of Land-Use Change on Nutrient Loads from Diffuse Sources* (ed. by L. Heathwaite) (Birmingham, UK, July 1999), 41–50. IAHS Publ. 257, IAHS Press, Wallingford, UK.

3 Suspended Sediment Transport and Yield

Comparison of a neural network and a regression model to estimate suspended sediment in a semiarid basin

SUSANNE SCHNABEL & MARCO MANETA

Departmento de Geografía y Ordenación del Territorio, Universidad de Extremadura, Avda. de la Universidad, 10071 Cáceres, Spain

maneta@unex.es

Abstract In small semiarid basins with ephemeral flows most of the sediment conveyance takes place during runoff peaks exceeding a certain discharge. Sediment load is commonly modelled using rating curves fitting the water–sediment discharge relationship. The performance of a feed-forward back-propagation artificial neural network (ANN) and a multiple quadratic regression (MQR) model are tested using data from the Parapuños Catchment, a wooded rangeland located in SW Spain. Both models were calibrated using rainfall and discharge time series and derived variables such as rainfall intensity, runoff coefficient and rate of change of discharge. The final set of variables used in the analysis was done based on sensitivity analysis for the ANN model and based on an analysis of statistical significance of parameters in the MQR model. The performance of ANN and MQR were similar but better than rating curves of a single variable. In addition, ANN and MQR can reproduce the hysteretic loop of the sediment–discharge relationship.

Key words modelling; neural network; semiarid environments; suspended sediment

INTRODUCTION

Assessing the sediment yield at a basin outlet is of major concern for land degradation studies in small catchments. It is also important when evaluating sedimentation in estuaries, rivers and reservoirs. Traditional methods to estimate suspended sediment transport relate limited sediment concentration data to flow discharge. The nature of sediment dynamics and the inter-related processes that act in the soil detachment and movement towards the outlet of a basin are highly nonlinear and make sediment concentrations difficult to forecast. So far, the most commonly used theoretical functions trying to absorb part of this nonlinearity in the flow–sediment concentration relationship are quadratic and cubic polynomials and potential or exponential functions (Cordova & Gonzalez, 1997). In spite of the high order polynomials used, the scatter plots commonly show a large dispersion indicating that flow alone is not enough to explain suspended sediment discharge. When dealing with small semiarid basins with ephemeral streams the problem of nonlinearity between runoff and soil loss on an event basis is more acute. The relationship between soil loss and runoff at hillslopes in semiarid environments is generally poor and soil loss is highly related to rainfall intensity and the state of the vegetation cover. Channel erosion is a further sediment source and it has been shown that high losses are related to high discharge peaks produced by low frequency rainfall events (e.g. Schnabel *et al.*, 1999). We may expect that most of the available (loose or detached) sediment is conveyed in the first runoff peak and sediment concentration will eventually become less strong at later stages, even if the subsequent runoff peaks are higher.

Within this context, Artificial Neural Networks (ANNs) have proven to be a very effective technique in nonlinear time series forecasting and could be regarded as an alternative approach to other regression techniques as they provide a robust and reliable solution with high levels of generalization, even if trained with noisy or small data sets. There are numerous examples of successful use of ANN to solve nonlinear problems in hydrology and related fields (e.g. Hsu *et al.*, 1995; Sivakumar *et al.*, 2002). Despite the intensive and successful use in time-series prediction and forecasting, there are only a few examples of ANNs applied to suspended sediment transfer, with variable results (e.g. Abrahart & White, 2001; Cigizoglu, 2002). Those works have a continuous time series of data, which allow the application of autoregressive neural networks, forecasting the suspended sediment concentration based on previous concentrations at earlier time steps or based on earlier discharges.

The aim of this work is testing whether an ANN is able to mine enough information from discharge and time series to forecast the suspended sediment concentration without a pre-classification of the data set based on typology of event, season or flow regime, and to compare the performance of ANN and multiple quadratic regressions (MQR). Both techniques will be used for the prediction of suspended sediment concentration time series in a small semiarid basin with an ephemeral stream where an autoregressive approach is not possible due to discontinuities in the data.

STUDY AREA AND DATASET

This work has been carried out in the Parapuños experimental watershed located in Extremadura (SW Spain). The basin has an area of 99.5 ha and belongs to the wooded rangelands called *dehesas*. It forms part of an undulating erosion surface formed in schist and graywacke, with some residual pediments found in the highest parts of the basin. Most of the soils are shallow, have low organic matter content and a silty texture. The main channel is a second order stream, which in the lower part of the catchment is incised into alluvial sediments of approximately 1 m thickness, reaching the underlying schist. The channel and its tributary can be classified as gullies due to the active erosion taking place. Continuous water flow is most frequent from November to January under wet soil conditions. During the dry season, Hortonian overland flow is only generated during high rainfall intensities (Ceballos & Schnabel, 1998). The climate is Mediterranean with continental and Atlantic influences. Mean annual precipitation is about 500 mm. Sheep and pig grazing is the main land-use. Holm oaks (*Quercus rotundifolia*) grow in the area, together with an herbaceous layer. Data used in this study was obtained between September 2001 and March 2004. Rainfall is measured at three points in the catchment with a tipping bucket device, and discharge is determined at the outlet of the catchment using a V-notch gauging station with a maximum capacity of 4 m^3 s^{-1}. All data is collected in 5-min intervals. Suspended sediment concentrations are obtained with an ISCO water sampler, which takes 0.5-l samples every 5 min when discharges exceed 3.7 l s^{-1}. This sampling system may leave part of the event unsampled. In that case calculating the total sediment yield of the event is difficult (e.g. Alexandrov *et al.*, 2003).

CONCEPTS, VARIABLES AND NEURAL NETWORK SET UP

An artificial neural network (ANN) is a structure of parallel computation where many small operating units called neurons are massively interconnected, storing, transferring and receiving information from other neurons. The information distributed through all the neurons of the network makes it able to reproduce complex nonlinear patterns learnt in the training process (Fausset, 1994). The type of ANN used in this study is of the type feed forward with a back-propagation learning algorithm (Rumelhart *et al.*, 1986). This type of network architecture has been found to be the most common and most efficient type of network to deal with time-series and input–output function approximations such as rainfall–runoff modelling (Hsu *et al.*, 1995; Furundzic, 1998). This type of ANN (Fig. 1) is composed by a first layer (input layer) with as many neurons as variables have the information vector to be introduced. This input layer is usually connected to one or more hidden layers composed of one or more neurons which get the information from the previous layer. The incoming information is multiplied by a set of correlation weights and filtered through a transform function. The transformed information will finally get to a last layer (output layer) which will bring out the results. The learning of the network is distributed throughout all the neurons by the re-adjustment of the correlation weights. In feed-forward networks, the information is always transferred forward from the input to the output layer and the error is back-propagated from the output layer back to the hidden layers to adjust iteratively those numerical weights. In the training process, the output of the network for every input is compared iteratively with a training set of experimental values. The residual between the observed and calculated is then used to re-adjust the internal network weights so a global objective error function can be diminished.

Together with the ANN model, the forecasting performance of a multiple quadratic regression (MQR) analysis has been tested. The general model for MQR is a polynomial of the form:

$$Y_i = c_0 + a_1 x_{1,i} + a_2 x_{2,i} + \ldots + a_{n-1,i} x_{n-1,i} + b_1 x_{1,i}^2 + b_2 x_{2,i}^2 + \ldots + b_{n-1} x_{n-1,i}^2 + \varepsilon_i \quad (1)$$

in which the subscripts n and i are the number of variables and observation points respectively, Y_i is the suspended sediment concentration at time i, x_n are the input

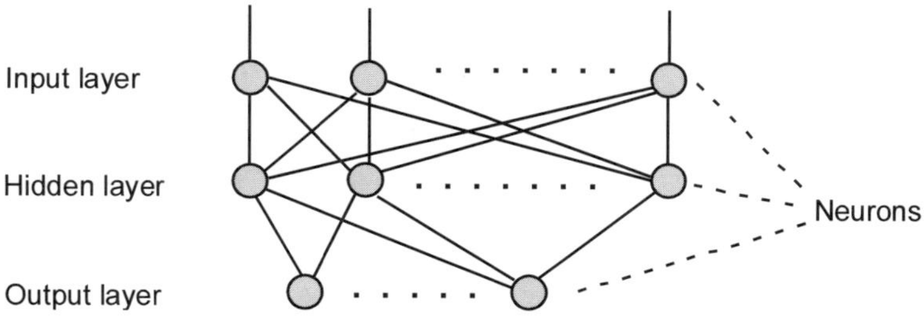

Fig. 1 Generic structure of a feed-forward back-propagated ANN.

variables, c_0 is the independent term, a_n and b_n are constant coefficients adjusted by ordinary least squares and ε_i is a residual term holding the prediction error.

The capabilities of both methods to determine the suspended sediment time-series are assessed using the same variables that are only derived from rainfall and discharge data and that try to give information about the most important factors affecting sediment movement. We let the models decide what the most salient factors explaining the process are, and drop from the analysis those whose partial slope is non-significantly different from zero in the case of MQR, or relatively insensitive in the case of ANN. The sensitivity analysis of each variable is done by assessing the deterioration of the ANN performance by ignoring in turn one variable. The criterion to prune a variable is when the ratio between the network error when all variables are available and the error when the specific variable is ignored falls below 1.05. Once the best set of variables is found, a sensitivity ranking is done, repeating the analysis using the errors of the training set and the verification set separately to test for robustness in the saliency of the variables.

The initial factors taken into account as the most important in the sediment transport process are depletion of sediment sources, hysteretic relationships between discharge and sediment concentration, rainfall intensity, antecedent soil moisture conditions and seasonality.

Sediment yield at a point is highly dependent on sediment availability. Part of this available sediment is conveyed in the beginning of the runoff event, depleting the sediment source and leaving less sediment to be conveyed by the subsequent runoff. The cumulated discharge from the last four hours (*CumQ*) has been introduced to account for this, assuming that the higher the accumulated discharge the less the available sediment.

The discharge–sediment relationship is highly hysteretic. The rate of change of discharge (d*Q*) has been introduced to give information to the neural network about this effect. Positive or negative values will indicate, respectively, whether the current discharge is in the rising or recession limb of the hydrograph. It also gives information on the rate of change of the discharge.

To account for rainfall intensity the cumulated precipitation from the last 30 min (I30) has been introduced. It is also assumed that the antecedent soil moisture conditions of the watershed play a major role in sediment detachability (e.g. Ciampalini & Torri, 1998). To account for this, the discharge/precipitation ratio (*Q*/*R* and *Q*/*CumR*) has been used as a surrogate to introduce information on antecedent moisture conditions (the higher the ratio the wetter the basin). When precipitation is zero during any of the time steps of the event, an interpolated value of the ratio has been used.

In addition to those derived variables, the number of the month (*M*) (1 January; 12, December) is used as a surrogate for seasonality and hence of the soil cover. Those five variables together with the current discharge (discharge at lag 0) (*Q*), precipitation lagged one time interval (5 min) (R_{t-1}) and the cumulated precipitation (*CumR*), which were found to be the most correlated lags of those variables with sediment concentration after a cross-correlation analysis, form the nine input variables to the neural network and MQR models (Table 1).

Table 1 Variables and identification symbols used in the analysis.

Input / dependent variables	Number of month (ordinal 1–12)	M
	$Rainfall_{t-1}$ (mm)	R_{t-1}
	Cumulated rainfall (mm)	*CumR*
	Discharge (l s^{-1})	*Q*
	Discharge/rainfall ratio (-)	*Q/R*
	Discharge/cumulated rainfall ratio (-)	*Q/CumR*
	Rate of change of discharge (-)	d*Q*
	4 h cumulated discharge (l s^{-1})	*CumQ*
	30 min, cumulated rainfall (mm)	I30
Output / independent variables	Suspended sediment concentration (g l^{-1})	Sed

As the variables implicitly maintain information of their positions within the rainfall–runoff time series, every set of seven input elements can be treated independently of the chronological order. A total of 636 records of seven input elements are available after treatment of the raw data. From this complete set 467 records are used as a training set, 40 are randomly segregated for its use as a validating set in order to avoid over-fitting of the network in the training process, and 129 records from the 2003–2004 hydrological year serve as a testing set to assess the error and performance of the models.

RESULTS AND DISCUSSION

Figure 2 and Table 2 present the calculated values for the parameters of the MQR model, together with their estimated standard errors and their significance. Most of them present low sensitivities (low partial slopes) and high standard errors, the latter indicating interaction (high correlation) between the parameters. The uncertainty in their calculated values makes many of them fail in the significance test against the null hypothesis. All this strongly suggests that the MQR model is over-parameterized. To

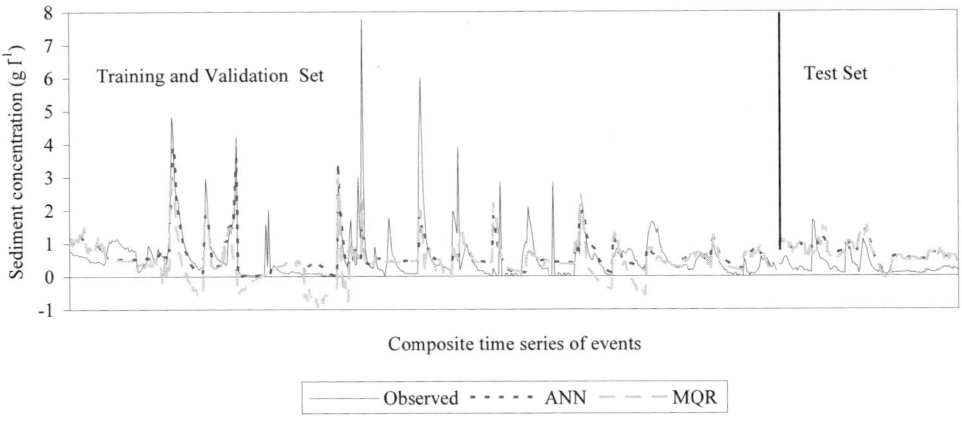

Fig. 2 Composite suspended sediment time series for all events included in the analysis and predictions calculated with MQR and ANN models.

Table 2 Parameter values, standard errors (SE) and *t*-test for the MQR model. See Table 1 for identification of symbols.

Parameter	Value	SE	*t*-test H0: param=0	*p*-value 2 tails
M	–0.00629	0.051365	–0.12	0.90
CumR	–0.003138	0.047059	–0.07	0.95
R_{t-1}	–1.010164	0.22995	–4.39	0.00
Q	0.001074	0.000784	1.37	0.17
Q/*R*	0.000266	0.000209	1.27	0.20
Q/*CumR*	0.000315	0.000268	1.17	0.24
d*Q*	–0.005897	0.004805	–1.23	0.22
CumQ	–0.00027	4.80E–05	–5.61	0.00
I30	0.259708	0.035635	7.29	0.00
M^2	0.000288	0.003825	0.08	0.94
$CumR^2$	0.000302	0.004451	0.07	0.95
R_{t-1}^2	0.22156	0.10761	2.06	0.04
Q^2	1.00E–06	1.00E–06	1.48	0.14
$(Q/R)^2$	0	0	–1.99	0.05
$(Q/CumR)^2$	0	0	–0.65	0.51
dQ^2	2.70E–05	9.00E–05	0.31	0.76
$CumQ^2$	0	0	4.17	0.00
$I30^2$	–0.006362	0.002215	–2.87	0.00
Constant	0.552429	0.111316	4.96	0.00

Table 3 Ranking of sensitivity of the chosen variables in explaining test and validation sets. See Table 1 for identification of symbols.

	Variables					
	Q	*Q*/*R*	*CumR*	d*Q*	*CumQ*	I30
Ranking test set	1	6	4	5	3	2
Ranking validation set	1	5	6	4	3	2

solve this and following the methodology, the terms that have passed a 2-tail significance test at 0.05 have been chosen and the model parameters recalculated, yielding the following more parsimonious final model:

$$Sed = -1.058 \cdot R_{t-1} + -0.000133 \cdot \mathrm{CumQ} + 0.325 \cdot \mathrm{I30} + 0.29 \cdot R_{t-1}^{\ 2} + 0.0073 \cdot I30^2 + 0.495$$

in which all the parameters are statistically significant at 0.05 (2-tails test) and include the most sensitive parameters in the initial model (R_{t-1} and I30).

For the ANN model, the best architecture found has a hidden layer with 10 neurons, and after performing the sensitivity analysis described earlier, six variables were chosen to be the most explicative ones, ranked in importance as shown in Table 3 for the training and for the test set. Note that the three variables that seem to carry more information in explaining the data are quite robust, being the same for both training and validating sets (Fig. 2). It is interesting noting the fact that the most explicative variable for the ANN model is discharge, which has a low partial sensitivity and was non-significant in the initial MQR model. It is also worthwhile noting that mainly 30 min intensity but also the 4 h cumulated discharge are important to both models, suggesting the saliency of those variables in the process of explaining suspended sediment concentration in small basins. It agrees with the idea that rainfall intensity and sediment availability is more important than runoff in explaining the

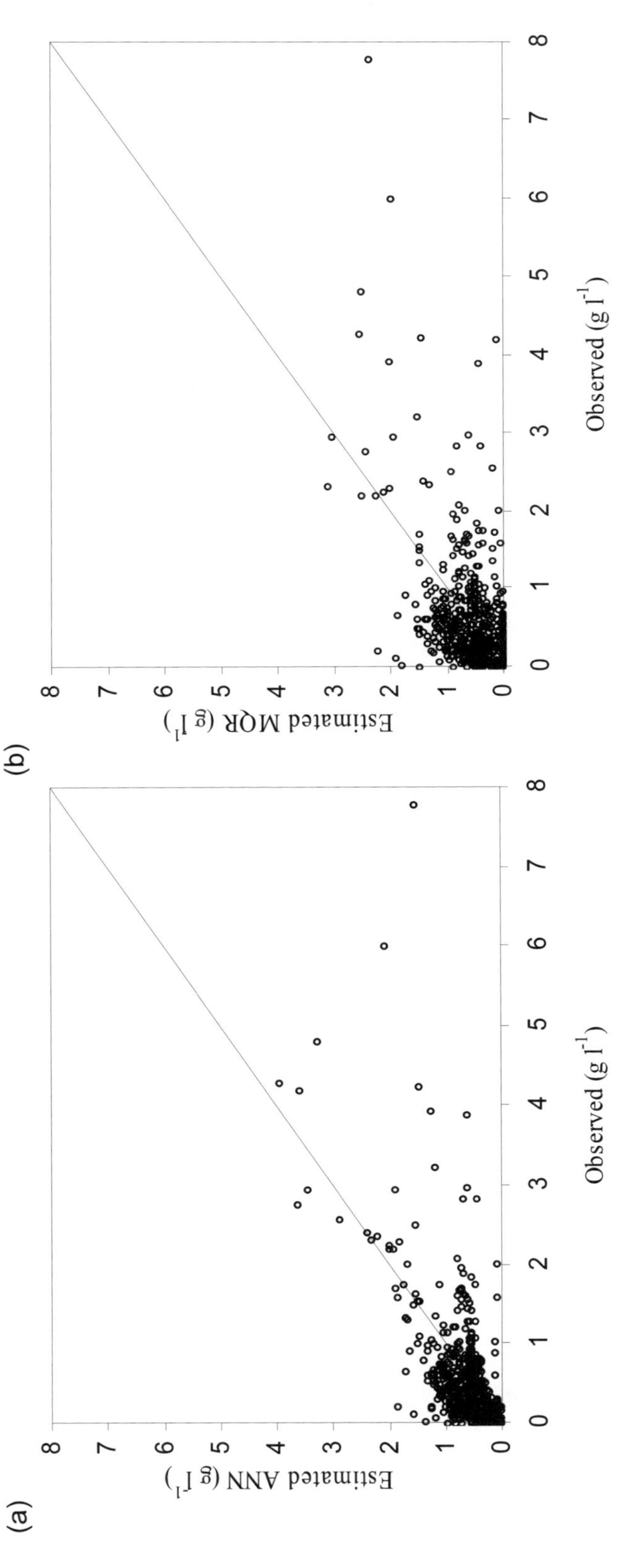

Fig. 3 Predicted *vs* observed sediment concentration (whole dataset) with line of perfect agreement for (a) ANN (artificial neural network) and (b) MQR (multiple quadratic regression) with negative predictions truncated to zero.

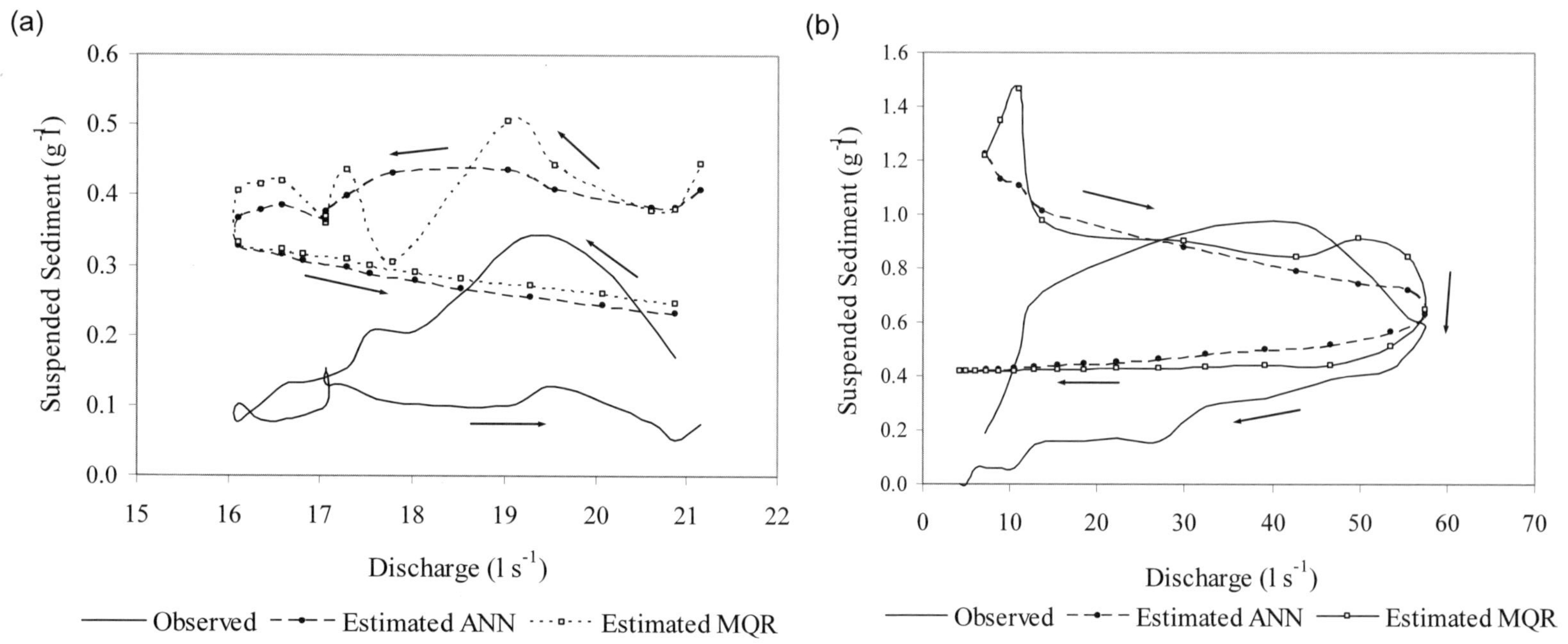

Fig. 4 Observed and simulated hysteretic behaviour (a) event 18/01/2001, counter-clockwise and (b) event 14/11/2002, clockwise.

erosion process for small basins, which will be the controlling factor for suspended sediment in large basins (e.g. Klein, 1984).

Figures 3 and 4 show the model predictions and scatterplots observed *vs* estimated for the whole data set including training, validation and test subsets. Following the behaviour of both models, they seem to react and be insensitive to nearly the same peak events, suggesting that the information supplied is insufficient for both models. With exceptions, the general trend is to overestimate low sediment concentrations and underestimate the peaks. The test set, which is composed by events of relatively lower sediment concentrations, is mostly overestimated by both models.

As the MQR model tends to estimate sediment concentrations as negative for low flows after high peaks, the negative values have been truncated to zero when evaluating its performance giving a more fair comparison with ANN performance. Table 4 shows the Root Mean Square Error (RMSE) for both models. While the ANN model performs slightly better in the training and validation set, when it comes to estimate data not included in the training process, both models have similar generalization performance.

The statistical relations between suspended sediment concentrations and *Q* and I30 yield a maximum explained variance of 22% and 28%, respectively (note how I30 is again more explicative than *Q*) while MQR with three variables accounts for up to 33% of the total variance and ANN up to 39%. Although the estimations given by MQR and ANN are better than those simple rating curves against a single variable, there is still a high amount of unexplained variance left, seen in the wide scatter of points around the line of perfect agreement. The reason remains in the complex nature of the erosion processes and the varying flow conditions. In the studied catchment there is sheetwash at the hillslopes and gully erosion in the valley bottom, meaning that the factors driving the sediment erosion and their relative importance is different, depending on the process involved and the type of event. This different events and processes of runoff and sediment production have been classified and studied through a classification of hysteretic loops. Seeger *et al.* (2004) associate sediment hysteresis with the antecedent wetness conditions of the basin and the intensity of the event, which indicate different processes of runoff generation and sediment transport. Furthermore, the hysteretic behaviour of the discharge–suspended sediment relationship has been shown to carry certain information about the sediment sources. Klein (1984) comments that the phenomenon of hysteresis as a function of the sediment travel time and its source with respect to the water contributing area. Those different processes are known to occur in the study area at different stages of the year. Most of the events in the study area show a clockwise hysteretic loop indicating that most of the sediment conveyed is due to gully erosion, which is a known feature of the watershed (Fig. 4(b)). When the basin is at its wettest stage (usually in January), there is saturation excess overland flow reaching the outlet and sediment incorporated in the

Table 4 Performance of the models as expressed by the root mean square error (RMSE).

	ANN	MQR*
Overall RMSE	0.59	0.68
Test set RMSE	0.52	0.53

* RMSE calculated with negative estimates truncated to zero.
ANN, artificial neural network; MQR, multiple quadratic regression.

flow from the slopes. In those cases there are events presenting eight-shaped and counter-clockwise hysteresis (Fig. 4a). While hysteresis carries some important structural information about the runoff and sediment transport process and has helped the research on suspended sediment transport to separate and classify the typology of events accordingly (Seeger *et al.*, 2004), it also means that, even using sophisticated regression techniques such as MQR and ANN, the scatter of data in the relation between suspended sediment and the predictors is large, let alone simple rating curves (Alexandrov *et al.*, 2003). Given the inability of ANN to mine out enough information about the events from the raw data supplied, further studies to improve the predictive capacity of suspended sediments with ANN and MQR may consider explicitly different classified sets, maybe even treating the data of each sort of event as samples of different populations. A classification based on hysteretic loops would probably substantially improve the performance of ANN and MQR by implicitly introducing new information about sediment sources and antecedent conditions, information which could not be properly introduced by the variables derived from discharge and rainfall time-series, at least in the form used in this study.

Acknowledgements Funding for research was provided by the Spanish Ministry of Science under contract REN2001-2268-C02-02. Authors wish to thank an anonymous referee for the helpful comments during the review process.

REFERENCES

Abrahart, R. J. & White, S. M. (2001) Modelling sediment transfer in Malawi: comparing backpropagation neural network solutions against a multiple linear regression benchmark using small data sets. *Physics and Chemistry of the Earth, Part B: Hydrology, Oceans and Atmosphere* **26**(1), 19–24.

Alexandrov, Y., Laronne, J. B. & Reid, I. (2003) Suspended sediment concentration and its variation with water discharge in a dryland ephemeral channel, northern Negev, Israel. *J. Arid Environments* **53**(1), 73–84.

Ceballos, A. & Schnabel, S. (1998) Hydrological behaviour of a small catchment in the dehesa landuse system (Extremadura, SW Spain). *J. Hydrol.* **210**, 146–160.

Ciampalini, R. & Torri, D. (1998) Detachment of soil particles by shallow flow: Sampling methodology and observations. *Catena* **32**(1), 37–53.

Cigizoglu, H. K. (2002) Suspended sediment estimation for rivers using artificial neural networks and sediment rating curves. *Turkish J. Eng. Env. Sci.* **26**, 27–36.

Cordova, J. R. & Gonzalez, M. (1997) Sediment yield estimation in small watersheds based on streamflow and suspended sediment discharge measurements. *Soil Tech.* **11**(1), 57–65.

Fausset, L. (1994) *Fundamentals of Neural Networks: Architectures, Algorithms and Applications.* Prentice Hall, New York, USA.

Furundzic, D. (1998) Application example of neural networks for time series analysis: Rainfall–runoff modeling. *J. Signal Processing* **64**, 383–396.

Hsu, K. l., Gupta, H. V. & Sorooshian, S. (1995) Artificial neural network modeling of the rainfall–runoff process. *Water Resour. Res.* **31**(10), 2517–2530.

Klein, M. (1984) Anti clockwise hysteresis in suspended sediment concentration during individual storms: Holbeck catchment; Yorkshire, England. *Catena* **11**(2–3), 251–257.

Rumelhart, D. E., Hinton, G. E. & Williams, R. J. (1986) Learning internal representations by error propagation. In: *Parallel Distributed Processing* (ed. by D. E. Rumelhart & J. L. McClelland), vol. 1. MIT Press, Cambridge, Massachusetts, USA.

Schnabel, S., Gómez Amelia, D. & Ceballos, A. (1999) Extreme events and gully erosion. In: *Proceedings of the International Seminar on Land Degradation and Desertification* (ed. by C. Coelho) (International Geographical Union, Lisbon), 17–26.

Seeger, M. Errea, M. -P., Begueria, S., Arnaez, J., Marti, C. & Garcia-Ruiz, J. M. (2004) Catchment soil moisture and rainfall characteristics as determinant factors for discharge/suspended sediment hysteretic loops in a small headwater catchment in the Spanish Pyrenees. *J. Hydrol.* **288**(3–4), 299–311.

Sivakumar, B., Jayawardena, A. W. & Fernando, T. M. K. G. (2002) River flow forecasting: use of phase-space reconstruction and artificial neural networks approaches. *J. Hydrol.* **265**(1–4), 225–245.

Suspended sediment transport and trapping processes in the regulated lower Durance River, southeastern France

EMMANUEL RIBES & MIREILLE PROVANSAL

Department of Geography, University of Provence, CEREGE, Europôle de l'Arbois, BP 80, 13545 Aix-en-Provence, Cedex 04, France

eribes@cerege.fr

Abstract Since 1994 the lower Durance River has been experiencing a notable increase of hydro-sedimentary inputs, which have diverse morphological impacts on its bankfull channel, depending on water discharges and suspended sediment concentrations. At discharges of less than 300 $m^3 s^{-1}$, the suspended sediments are effectively transported within the active bed, and deposited in downstream gravel pits of the system. At discharges greater than 300 $m^3 s^{-1}$, the suspended sediments are deposited predominantly on vegetated and elevated bars, while the less cohesive sediments trapped in the gravel pits are removed. Results highlight the difficulties of restoring sediment transit in a highly disturbed and vegetated river channel.

Key words clearing operation; Durance River; gravel pit; regulated river; suspended sediments; water restoration programmes

INTRODUCTION

Several studies have already investigated the drastic decrease of solid fluxes in the Durance River and the consequent narrowing and plant colonization of the channel during the 20th century (Juramy & Montfort, 1985; Warner, 2001). The decrease of suspended sediment (SS) fluxes entering the lower Durance River after the 1950s is partly explained by the construction of dams since 1959 by Electricité de France (EDF), the diversion of 92% of the river discharge into an hydroelectric canal toward the Etang de Berre (Fig. 1(a)), and the decrease of the flood frequency from the 1960s to the 1980s (Fig. 2). However, since 1994, eutrophication problems in the Etang de Berre have led EDF to restore water and SS fluxes into the lower Durance channel downstream of Mallemort dam (Fig. 1(b)). The same period (1994–2003) is characterized by an important increase in the number of floods (Fig. 2). Consequently, the lower Durance channel has experienced a notable increase of SS inputs after the 1950s: 720 000 t $year^{-1}$ during the 1960s, rising to 1 250 000 t $year^{-1}$ during the 1990s and 2 300 000 t $year^{-1}$ since 2000 (Fig. 2). This paper summarizes field observations carried out to quantify sedimentation in the bankfull channel produced by this renewed increase of hydro-sedimentary activity.

STUDY AREA AND HYDRO-SEDIMENTARY REGIME

The study area is made up of two morphologically distinct reaches that extend from the Mallemort dam (Kilometric Point 0 or KP 0) to the confluence with the Rhône (KP 45)

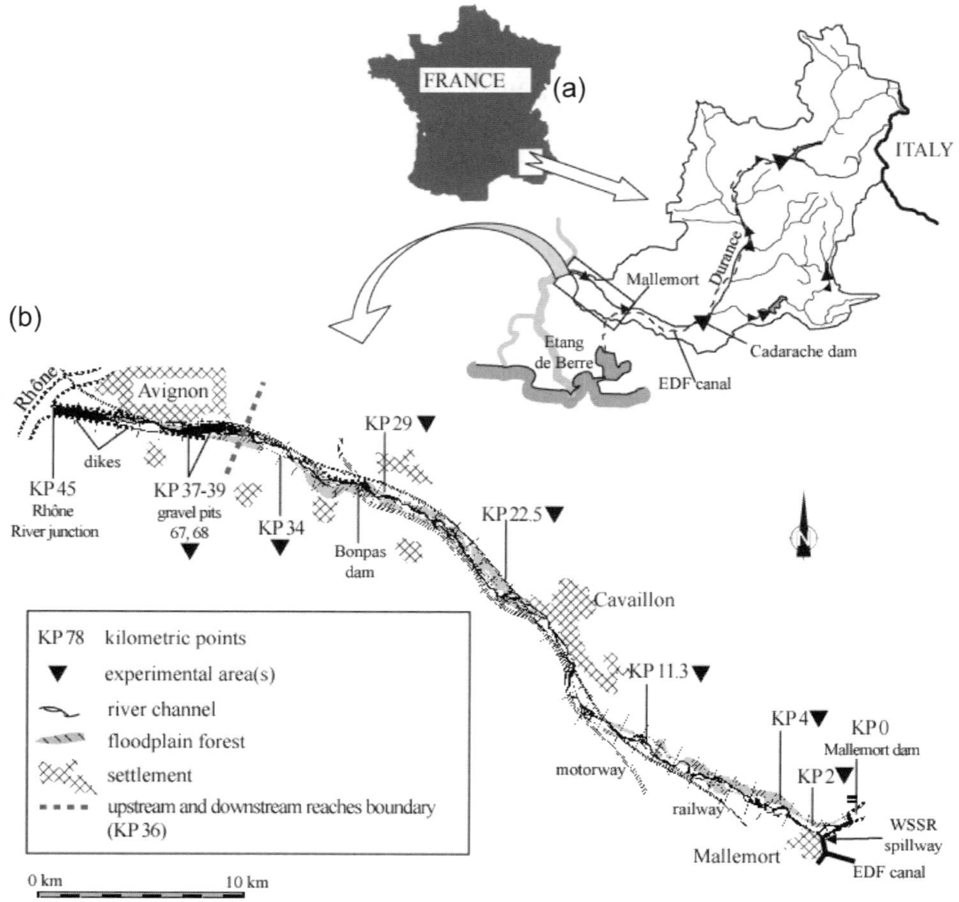

Fig. 1 (a) Durance basin map showing EDF hydroelectric canal, major dams and study area; (b) Location of study reaches and measurement area at KPs, downstream of Mallemort dam.

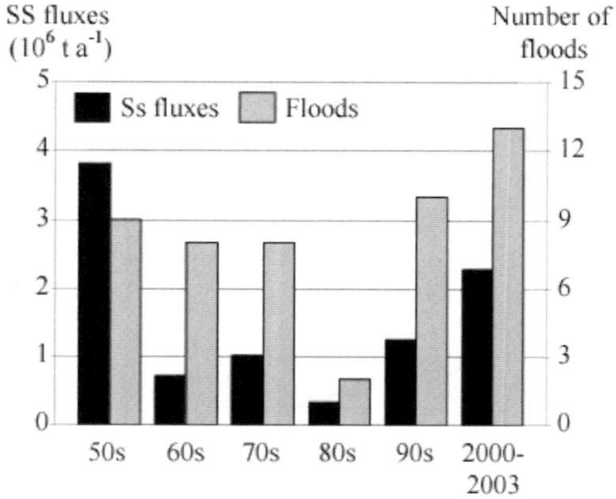

Fig. 2 Number of floods by decade from the 1950s to 2003, and SS inputs into the lower Durance River (based on EDF data).

(Fig. 1b). The channel reach upstream of KP 36, characterized by a mean bed slope of 0.24% and a mean bankfull width of 380 m, is weakly meandering (sinuosity index: 1.09). The bankfull channel here is composed of an entrenched active gravel bed (AB) with elevated and vegetated bars covering 1/3 of the bankfull width. The channel reach downstream of KP 36, characterized by a mean bed slope of 0.11% and a mean bankfull width of 345 m, is straighter (sinuosity index 1.01) because of the almost complete embankment of the bankfull channel. From KP 37 to KP 39, deep and extensive gravel pits (numbered 67 and 68) are bound by rock-fill transverse weirs crossing the entire width of the low-flow channel. In operation from 1973 to 1998, these pits are each about 500 000 m^2 in area, with a maximum depth of 10 m.

Floods transport 89% of SS fluxes during 5% of the time, whereas water and SS restoration (WSSR) by EDF, characterized by maximal discharges of 275 $m^3 s^{-1}$, transport only 9% of SS fluxes during 16% of the time (Fig. 3). The remaining 2% of SS fluxes are transported during minimum required discharge, released upstream of the dams for maintaining healthy conditions in the river (mean discharge = 10 $m^3 s^{-1}$). These different contributions to SS fluxes may be partly due to higher mean SS concentrations during floods (2.45 g l^{-1}) than during WSSR (0.35 g l^{-1}) and minimum required discharge (0.05 g l^{-1}).

METHODOLOGY AND LOCATION OF MEASUREMENT AREAS

The fine-grained sedimentation was measured along the river, at several KPs (Fig. 1(b)), on experimental areas considered as representative of various types of alluvial environments.

Upstream reach (KP 0 to KP 36)

Upstream of KP 36, measurements were carried out after different events occurred between 2001 and 2003 (Fig. 4 and Table 1): (a) after floods in November 2002, November 2003 and December 2003 (surveys no. 3, 9 and 10 in Fig. 4), with mean SS

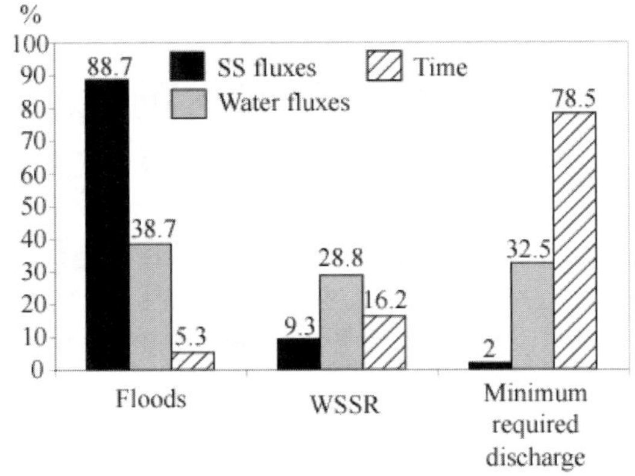

Fig. 3 Hydro-sedimentary regime of the lower Durance River since 2001 (based on EDF data).

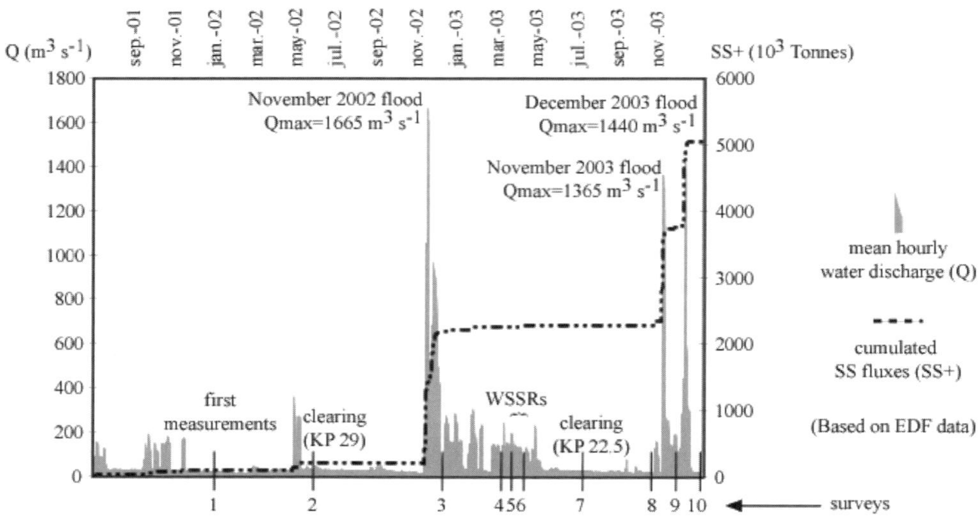

Fig. 4 Hydrograph, cumulated SS fluxes and surveys of the study period (2001–2003).

Table 1 Variations of fine-grained sedimentation and grain-size distribution in different experimental areas after 10 events.

Areas	KP	$D_{50}(3)$	$D_{50}(10)$	Thickness of fine-grained deposits (cm)										cm
AB	2	10	25	no data	no data	0.05	0.05	0	0	0	0.05	no data	0.1	0.3 0.2 0.1 0.0
AB	4	15	18	0.1	0.1	0	0	0	0	0	0.1	0.15	0	0.3 0.2 0.1 0.0
AB	11.3	06	09	no data	no data	0.1	0.05	0.05	0	0	0.1	0.3	0.2	0.3 0.2 0.1 0.0
AB	22.5	12	23	0.2	0.2	0.3	0.3	0.2	0.2	0.2	0.3	0	0.1	0.3 0.2 0.1 0.0
AB	29	16	19	0.05	0.05	0	0	0	0	0	0.1	0.25	0.1	0.3 0.2 0.1 0.0
AB	34	08	23	0.1	0.1	0.05	0.05	0.05	0.05	0.05	0.1	0.3	0.2	0.3 0.2 0.1 0.0
T1	22.5	97	94	no data	no data	20	20	20	20	18	18	no data	11	20 10 0
T1	29	88	78	22	22	18	18	18	18	18	18	no data	24	20 10 0
T2	4	66	61	42	42	55	55	55	55	55	55	no data	57	60 40 20 0
OX	2	92	86	no data	no data	41	41	41	41	41	41	no data	50	60 40 20 0
surveys				**1**	**2**	**3**	**4**	**5**	**6**	**7**	**8**	**9**	**10**	
previous event measured during survey				none; first survey	clearing KP 22.5	Flood	clear WSSR	clear WSSR	clear WSSR	clearing KP 29	turbid WSSR	Flood	Flood	
Q_m/Q_{max} (*)				X	X	590/1665	135/145	135/145	135/245	X	120/160	840/1365	625/1440	
$d_{>300}/d_{<300}$ (*)				X	X	237/227	0/59	0/60	0/67	X	0/72	29/5	71/53	
SSC_m/SSC_{max} (*)				X	X	1.5/8.5	0.04/0.1	0.04/0.25	0.01/0.07	X	1.8/5.2	10.5/18.5	3.5/11	

concentrations of 1.5 g l^{-1}, 10.5 g l^{-1} and 3.5 g l^{-1}, respectively; (b) after clear WSSRs with SS concentrations always less than 0.3 g l^{-1} (surveys 4, 5 and 6); (c) after a turbid WSSR with a mean SS concentration of 1.8 g l^{-1} (survey n°8); (d) the surveys 2 and 7 occurred after clearing operations at KP 29 and 22.5. The clearing consists of pulling up the vegetation with bulldozers, so that floods can then remove the deposits. After each event, topographic and fine-grained deposit thickness profiles were surveyed in each experimental area (Figs 5 and 6), using a tachometer and a graduated metal rod (Fig. 6(a) and 6(b)) (or millimetre ruler for weak deposits on active gravel bed). Surfaces of experimental areas vary from 8000 m^2 to 16 000 m^2 and their measurement point densities vary from 0.1 m^{-1} to 0.2 m^{-1} (Fig. 7). The data are then spatially

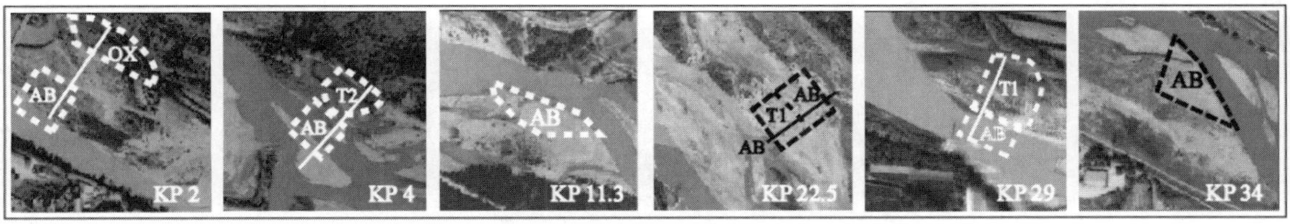

Fig. 5 Photographs of experimental areas (contoured by line with dashes).

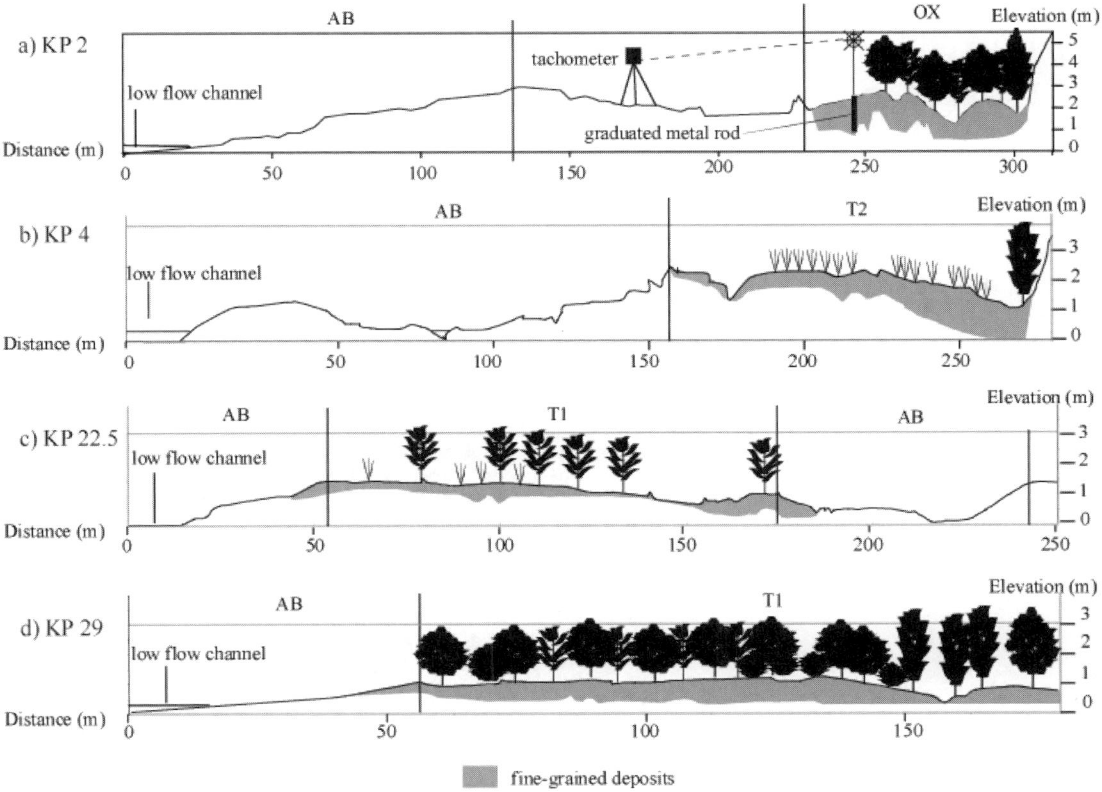

Fig. 6 Examples of topographic profiles (symbolized by a continuous line on Fig. 5): (a) at KP2, and measurement methods; (b) at KP 4, (c) at KP 22.5, (d) at KP29.

interpolated using Surfer 7.0 software to obtain a digital elevation model (DEM) of the surface and the deposit thickness (Fig. 7). Mean thickness deposit of each area, after each event, is obtained dividing the volumes calculated from the DEM of each experimental surface by their respective surface. The experimental areas are located on (Figs 5 and 6): (a) non-vegetated bars of the active gravel bed (AB); (b) vegetated and regularly cleared bars (T1); (c) stable elevated lateral bars (T2) situated above the active bed; (d) vegetated oxbows (OX). The growth of vegetation and types of management vary according to the different locations: at KP 2, the OX area is colonized by willows and black poplar shrubs, which have never been cleared; at KP 4, the T2 area is covered by herbs and willows; the T1 areas, currently including some willow groves, were cleared in 1996 and 2002 at KP 22.5, and in 1997 and 2003 at KP 29.

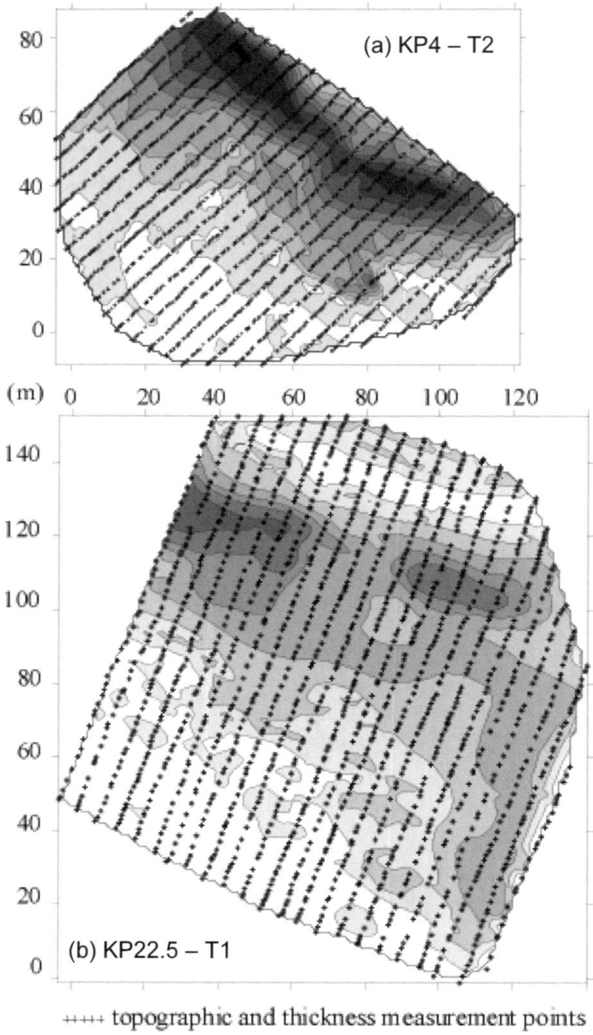

Fig. 7 DEM planimetric maps of the deposit thickness of experimental areas with topographic and thickness measurement points; (a) T2 area at KP 2; (b) T1 area at KP 29.

Downstream reach: gravel pits (KP 37 to KP 39)

Bathymetric surveys in 1999, 2002 and 2003 were carried out using a differential GPS with an echo-sounder (XYZ error = ±5 cm) (Fig. 8(a)). Comparisons between profiles using BMAP software allow us to calculate sedimentation rates in the gravel pits (Fig. 8(b)). Granulometric analyses of six sedimentary cores (4 m depth), using a Mastersizer S Malver Coulter counter were used to characterize the textural composition of the deposits. The hydraulic dynamics were studied in collaboration with the Laboratoire National d'Hydraulique et Environnement (LNHE-EDF), using a two-dimensional flow modelling software, TELEMAC-2D (Fig. 9). Flow models were run for different water discharges, varying from 15 to 2000 $m^3 s^{-1}$.

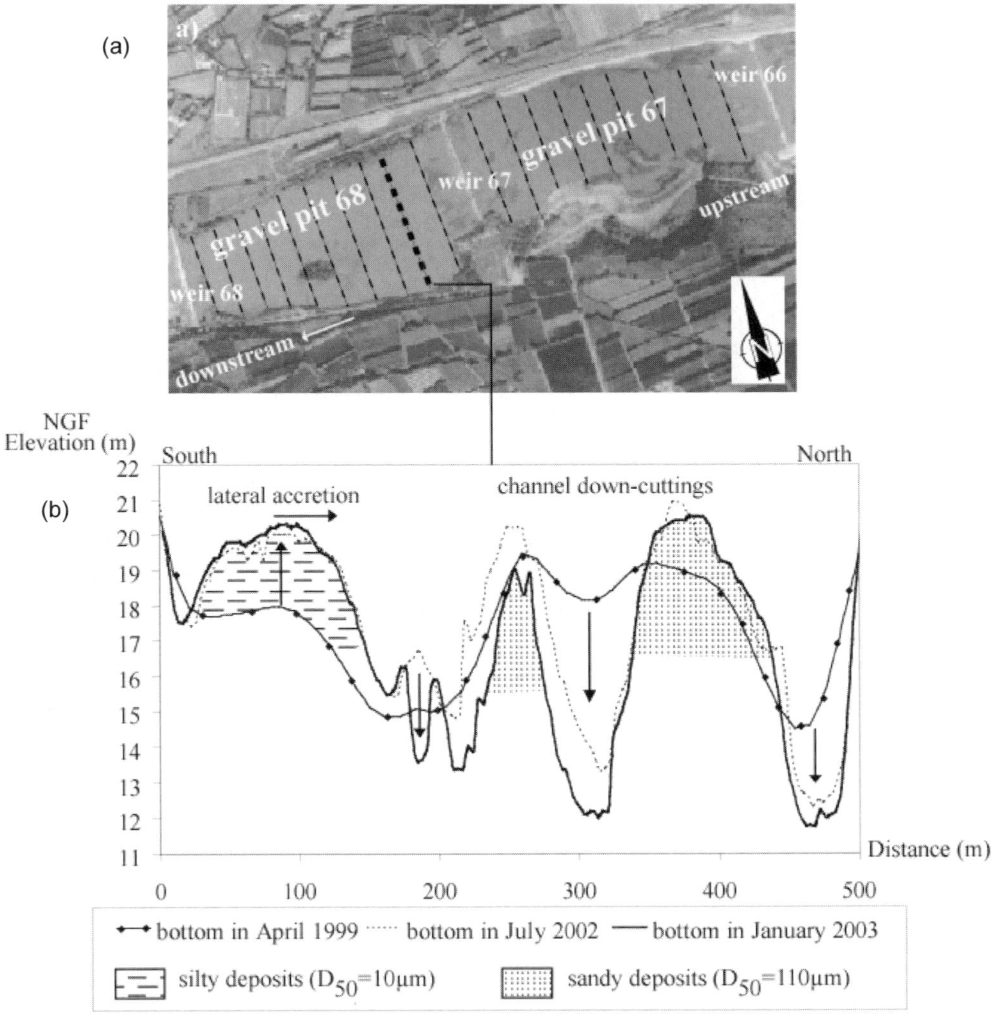

Fig. 8 Photograph of the gravel pits and location of cross-section profiles (black transverse lines); (b) example of cross-section profile in the gravel pit 68 (symbolized by a bold line on (a), showing typical evolution of the bottom between 1999 and 2003.

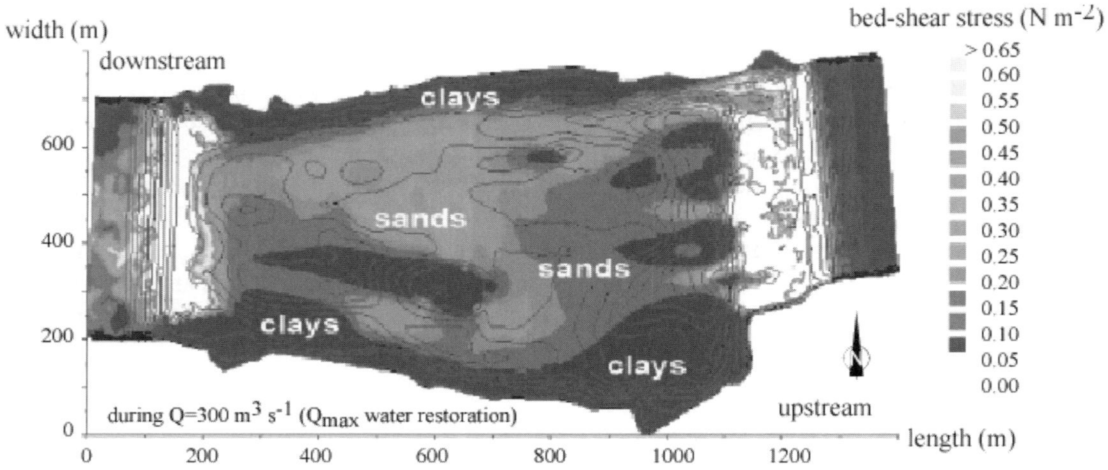

Fig. 9 Mapping of bed-shear stresses in gravel pit 68 during water discharge at 300 $m^3 s^{-1}$ (TELEMAC-2D simulation).

RESULTS

Upstream reach (KP 0 to KP 36)

On the active gravel bed (AB), a few mm of silts were deposited after the turbid WSSR (survey 8) and after the November 2003 flood (survey 9) (Table 1). On the contrary, clear WSSRs (surveys 4, 5 and 6) as well as floods of November 2002 (survey 3) and December 2003 (survey 10) partially removed fine sediment layers deposited during previous events. These differences may be partly due to higher SS concentrations during the turbid WSSR and the November 2003 flood (Table 1). In addition, the duration of the falling limb of the hydrograph during the November 2003 flash-flood was not sufficiently long to allow thorough washing of the active bar (too few days, after the peak flow, with discharge less than 300 $m^3 s^{-1}$, Table 1). On regularly cleared bars (T1), the deposits are made up of fine loose sands (D_{50} = [78–97] μm) with a thickness of about 20 cm. The floods immediately following the clearing of vegetation (KP 79: 2002, survey 3; KP 72.5: 2003, survey 10) removed these sands (KP 29: –4 cm; KP 22.5: –7 cm). However, the 2003 floods (surveys 9–10) had a positive effect on sedimentation at KP 29 (+6 cm), linked to the rapid and spontaneous growing up of the vegetation after clearing operations in 2002. On the elevated bars and vegetated oxbows (T2 and OX), the fine-grained deposits (D_{50} = [61–92] μm) show a greater thickness (>40 cm) resulting from high sedimentation rates after floods: the increase is +13 cm at KP 4 after the 2002 flood (survey 3) and +9 cm at KP 2 after the 2003 floods (survey no 10) in the OX area. However, only +2 cm was recorded in the T2 area after the 2003 floods, which reached a lower level than in November 2002. Finally, the stability of sediment thickness measured after WSSRs (surveys 4, 5, 6, 8) shows that T1, T2 and OX areas are not submerged during WSSRs (events with maximal discharge always less than 300 $m^3 s^{-1}$).

Downstream reach: gravel pits (KP 37 to 39)

Diachronic analysis of the bathymetric profiles shows a global accretion of the bottom (15 cm year^{-1}). It is the result of the important aggradation and progradation of the outer parts of the pit toward the centre of the channel (100 cm year^{-1}) (Fig. 8(b)). The decrease of the channel capacity caused by the pit infilling occurs as deep incisions in the central part of the channel during floods such as the November 2002 flood (Fig. 8(b)). Sedimentological analysis shows that the marginal deposits are composed of fine cohesive silts (D_{50} = 10 μm), whereas the eroded central deposits are made up of loose fine sands (D_{50} = 110 μm). These results suggest a granulometric sorting of SS in relation with variations of the sediment-carrying capacity of the flow. TELEMAC-2D models, indicating a reasonable decrease of bed-shear stresses from the channel centre toward the banks, are in agreement with these sedimentological observations. Bed-shear stress calculations were performed considering both types of sediment. However, the existence of bedforms, steep gradients and helicoidal currents lead to very complex calculations that cannot be presented in more detail here. For the sands, we propose a more simplistic approach based on Shields (1936) function of critical bed-shear stress for the initiation of motion and deposition over a plane bed, as described by Van Rijn (1984). According to the parametric equations used by this author, we obtain a value of 0.14 N m^{-2} for sands characterized by D_{50} = 110 μm. For the silts, we used predetermined values of 1 N m^{-2} for critical bed-shear stress for the initiation of deposition (Mehta & Partheniades, 1975) and 13 N m^{-2} for the critical bed-shear stress for initiation of motion, according to Bouchard (2003) on silty bank erosion in middle Durance reservoirs (2003). TELEMAC-2D flow models show that the initiation of motion and deposition of sands is around 300 m^3 s$^-$1 (Fig. 9). On the other hand, while silts begin to deposit at discharges less than 500 m^3 s^{-1}, silty outer parts of the pit are never eroded whatever the modelled water discharge (maximum discharge = 2000 m^3 s^{-1}) because of insufficient bed-shear stress, i.e. always less than 13 N m^{-2}. These results are in accordance with bathymetric data (Fig. 8(b)). The calculated thresholds (300–500 m^3 s^{-1}) suggest that WSSR, for which water discharges are always less than 300 m^3 s^{-1}, make little or no contribution to the flushing out of deposits.

CONCLUSIONS

In the lower Durance River, the transit and trapping of fine sediments in the bankfull channel depends on hydrological, topographic and vegetation factors. Upstream, the sedimentation rates are conditioned by the height and distance from the active gravel bed, as well as the development of vegetation cover. Restorations of water flow do not reach the gravel bars (T1, T2) and channel margins (OX). They do not create permanent deposits, and their suspended load is mainly transported directly into the downstream gravel pits. Deposits in T1 areas (vegetated bars) are reworked immediately after vegetation clearing, but it is important to stress the limited temporal effect of this management option (<1 year). Despite the partial upstream retention of sediment, the downstream gravel pits are infilled rapidly during floods and water restoration (WSSR). Only sands are partially flushed out at discharges greater than 300–500 $m^3 s^{-1}$ and even the maximum simulated discharge (2000 $m^3 s^{-1}$) does not lead to the reworking of silty banks. To conclude, the sedimentary evolution of the lower Durance River is characterized by the predominance of depositional processes and the infilling of the fluvial environment. This highlights the difficulties of restoring sediment transit in a disturbed river.

Acknowledgements EDF, SMAVD, Région PACA are acknowledged for their collaboration and financial support as well as Dr Carpenter for assisting in post-editing the English text. Authors are grateful to the anonymous reviewer for his revision of the manuscript that has greatly improved the final version of the paper.

REFERENCES

Bouchard, J. P. (2003) Modélisation des processus sédimentaires dans les retenues de l'Escale et St Lazare. LNHE EDF R&D (unpublished report).

Mehta, A. J. & Partheniades, E. (1975) An investigation of the depositional properties of flocculated fine sediment. *J. Hydraul. Res.* **13**(4), 361–381.

Juramy, S. & Montfort, I. (1985) L'évolution des lits fluviaux : l'exemple d'une rivière aménagée: la Durance. PhD Thesis, University of Aix-Marseille II, France.

Shields, A. (1936) Anwendung der Aehnlichkeitsmechanik und der Turbulenzforschung auf die Geschiebebewegung. Mitteilung der Preussischen Versuchsanstalt fur Wasserbau und Schiffbau, 26, Berlin, Germany.

Van Rijn, L. C. (1984) Sediment transport—part I: bed load transport. *J. Hydraul. Eng. ASCE* **110**(10), 1431–1456.

Warner, R. (2001) Relevance of geomorphology in exploitive and sustainable management of water resources in the Durance River, France. In: *Integrated Water Resources Management* (Proceedings of Davis symposium, California, April 2000), 272–284. IAHS Publ. 272. IAHS Press, Wallingford, UK.

Seasonal suspended sediment concentrations and their relationship with discharge in the Anoia River basin, NE Spain

JOAQUIM FARGUELL & MARIA SALA
Department of Physical Geography, University of Barcelona, Baldiri Reixac s/n, 08028 Barcelona, Spain
jfarguell@hotmail.com

Abstract The relations between suspended sediment concentration and discharge in the Anoia River basin are reported. The scatter is mainly a consequence of seasonality of events and the hysteretic loops that take place within them. The seasonal concentration–discharge relationship improves the complete data set relationships reinforcing the seasonality of events. Concentrations are greater during autumn than during spring. For instance, at the upper site mean suspended sediment concentration in autumn is 480 mg l^{-1} while in spring it is 200 mg l^{-1}. At the downstream site, maximum concentration during autumn was 15 400 mg l^{-1} and during spring was 10 700 mg l^{-1}. The rating curve has a regression coefficient of $r^2 = 0.2$ whereas with seasonal subdivision the coefficients are $r^2 = 0.6$ for the autumn period and 0.5 for the spring period. Events exhibit hysteretic clockwise and anticlockwise loops increasing the degree of scatter in the concentration–discharge relationships.

Key words Anoia catchment; hysteresis; seasonal variation; sediment rating curves; suspended sediment concentration

INTRODUCTION

Suspended sediment concentrations show considerable scatter when plotted against discharge. The scatter occurs because increasing discharge is not always accompanied by an increase in suspended sediment concentration, due to seasonal contrasts, hysteresis and antecedent soil moisture conditions (Walling & Webb, 1987). These facts make it difficult to establish reliable rating curves, which are used in many studies to estimate suspended loads, and should be taken into account to interpret the rating relationships (Walling, 1977; Reid & Frostick, 1994; Asselman, 2000). Specific studies undertaken in Mediterranean environments show a clear seasonal pattern of hydrological processes (Gallart *et al.*, 2002) and suspended sediment concentration (Batalla & Sala, 1994). In addition, suspended sediment is mainly transported during flood events, which involve differences in concentration during the rising and falling stage. Furthermore, there are seasonal differences between rising and falling stages. This paper highlights the seasonal differences between concentrations measured in the Anoia River basin and the role of the events in the scatter of the concentration–discharge relationships.

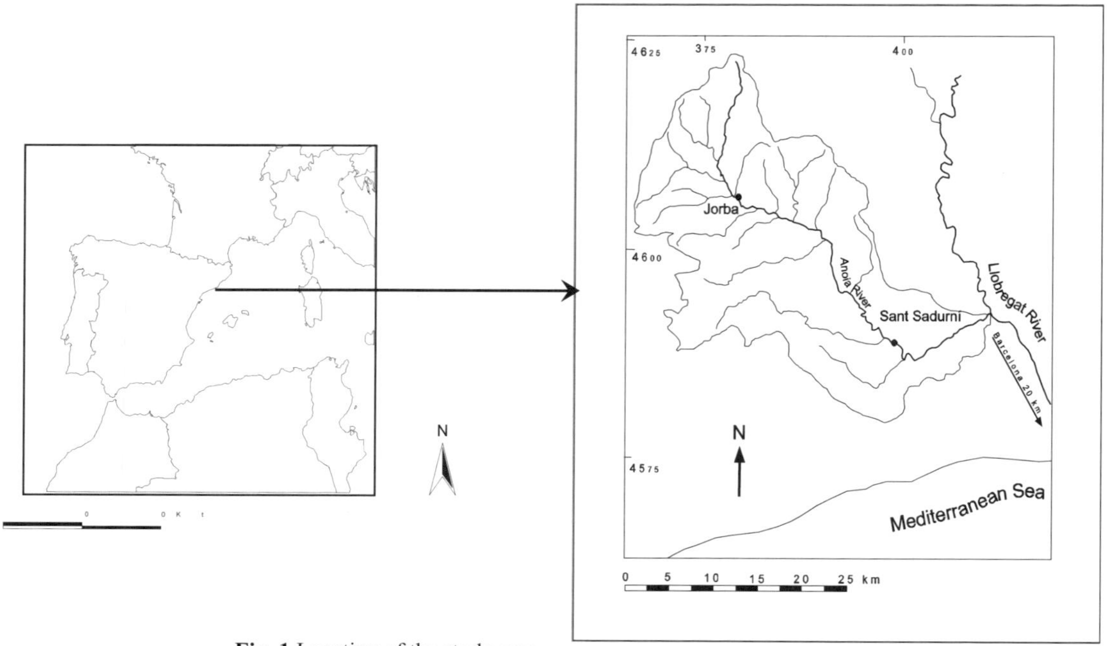

Fig. 1 Location of the study area.

STUDY AREA

The Anoia River basin is located in the northeast of the Iberian Peninsula and drains an area of 926 km^2 (Fig. 1). The upper part of the basin is Jorba, which drains an area of 220 km^2. Elevation ranges from 900 m a.m.s.l. down to 60, at the junction with the Llobregat River. This junction is 20 km west of its outlet to the Mediterranean Sea at Barcelona. The underlying materials in the upper part of the basin are mainly sedimentary rocks, sandstone and marls, whereas, in the lower part, silt and clay are widespread. The Catalan Coastal Ranges, formed by limestone and small areas of granites, divide the upper and lower part of the basin. Agriculture covers 45% of the basin's surface with mainly winter cereal crops in the upper part of the basin and vineyards in the lower part; forests and scrubland cover 51%.

MATERIALS AND METHODS

Samples were taken using a US DH48 integrated sampler during low flows at a weekly interval. Additionally, during flood events, two automatic pump samplers, ISCO 3700 type, were used in order to collect hourly samples. These samplers were sited at Jorba and Sant Sadurní. A comparison between manual and automatic samples shows a good correlation, with $r^2 = 0.98$ and 0.99 for Jorba and Sant Sadurní, respectively. Sediment concentration was measured from a 250 ml sample which was filtered using a Millipore vacuum pump through a 0.45 µm pore size filter, which was previously weighed to within 0.01 mg. Filters were then air-dried for a week and re-weighed.

RESULTS

Suspended sediment concentration and its relationship with discharge during the study period shows considerable scatter for both sampling sites. The rating curves differ significantly (p-value < 0.01) between sites. Discharge explains 17% of the variance of concentration at Jorba and 56% at Sant Sadurní. These coefficients reflect the influence of other variables on the suspended sediment concentration values obtained, especially in the upstream site, where the regression coefficient is low. Seasonal concentrations and the hysteresis loops during events are key elements contributing to the scatter of the concentration–discharge relationships at both sampling sites. The seasonal rating curves greatly improve the regression coefficients, with $r^2 = 0.60$ and 0.46 at Jorba during autumn and spring respectively, and $r^2 = 0.85$ and 0.61 at Sant Sadurní (Fig. 2). Concentrations are higher during autumn than during spring at both sites (Fig. 3). The range of concentrations is also greater during autumn, increasing the mean and median values with respect to concentrations in spring (Table 1). Greater autumn concentrations are attributed to higher rainfall intensities and tillage of

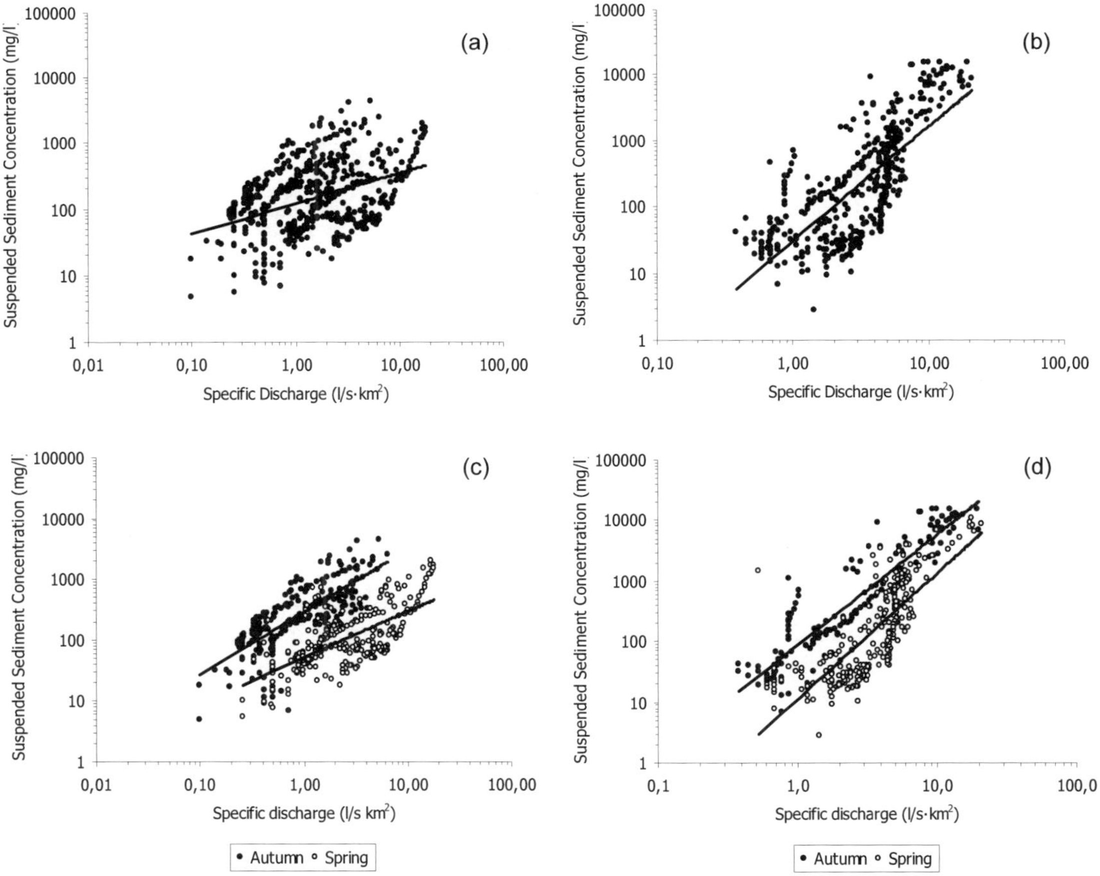

Fig. 2 Complete suspended sediment rating curves for Jorba (a); and Sant Sadurní (b). Seasonal rating curve at Jorba (c) and seasonal rating curve at Sant Sadurní (d).

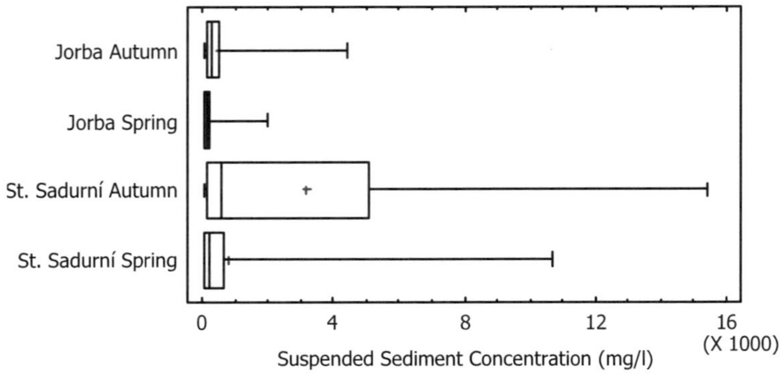

Fig. 3 Seasonal distribution of concentration at Jorba and Sant Sadurní.

Table 1 Distribution of suspended sediment concentrations at the sampling sites.

	Lower Quartile		Median		Upper Quartile		90%		Maximum(mg l^{-1})	
	Autumn	Spring	Autumn	Spring	Autumn	Spring	Autumn	Spring	Autumn	Spring
Jorba	112	50	226	75.5	485	176	1090	470	4400	1960
Sant Sadurní	68	32	274	140	2725	609	10100	1775	15400	10700

Table 2 Suspended sediment concentrations (mg l^{-1}) during rising and falling stages at the sampling sites.

	Rising stage			Falling stage		
	General	Autumn	Spring	General	Autumn	Spring
Jorba	680	900	473	230	359	132
Sant Sadurní	2707	4172	2060	1240	2545	362

agricultural fields during autumn. The concentration during events exhibits hysteresis, being generally greater during the rising stage than during the falling stage. Seasonality also affects this relationship and concentrations during the rising stage in autumn are greater than concentrations during spring (Table 2). The concentration pattern is analogous during the falling stage. The fact that peak sediment concentration is produced either before or after the peak discharge leads to clockwise or anticlockwise loops. The single-peaked events exhibit clockwise loops at both sampling sites (Fig. 4) whereas multiple peak events show both clockwise and anticlockwise loops. The anticlockwise loops are generated by the arrival of different upstream tributary peaks, with a small delay between sediment peak in relation to the peak of discharge. The delay is usually one hour and it is enough to create complex hysteresis loops (Fig. 5). In addition, multiple peak events show that the second discharge peak exhibits little increase in concentration, indicating an exhaustion of the sediment supply. In small basins, however, the clockwise loop is produced when sediment is coming from the river bed and anticlockwise loops are produced when the sediment comes from the slopes (Klein, 1984).

Events that take place during late summer are usually single-peaked storms, with high rainfall intensity, although brief in duration. These events affect only part of the basin, especially the lower part. Maximum concentrations are recorded during autumn

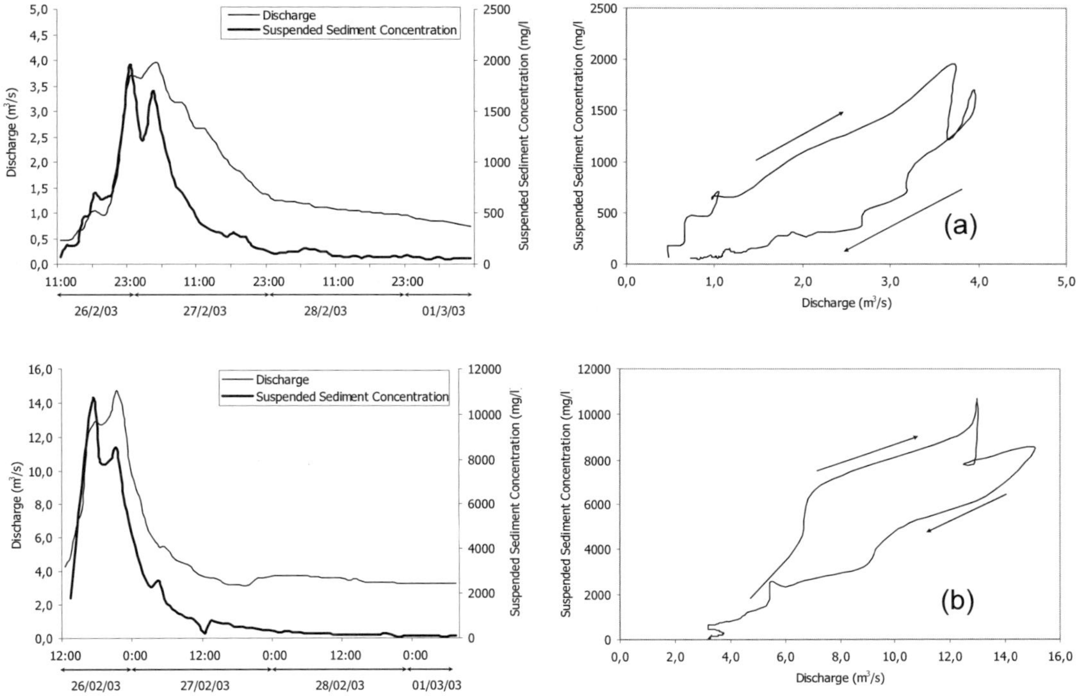

Fig. 4 Single peak event and hysteresis at Jorba (a) and Sant Sadurní (b).

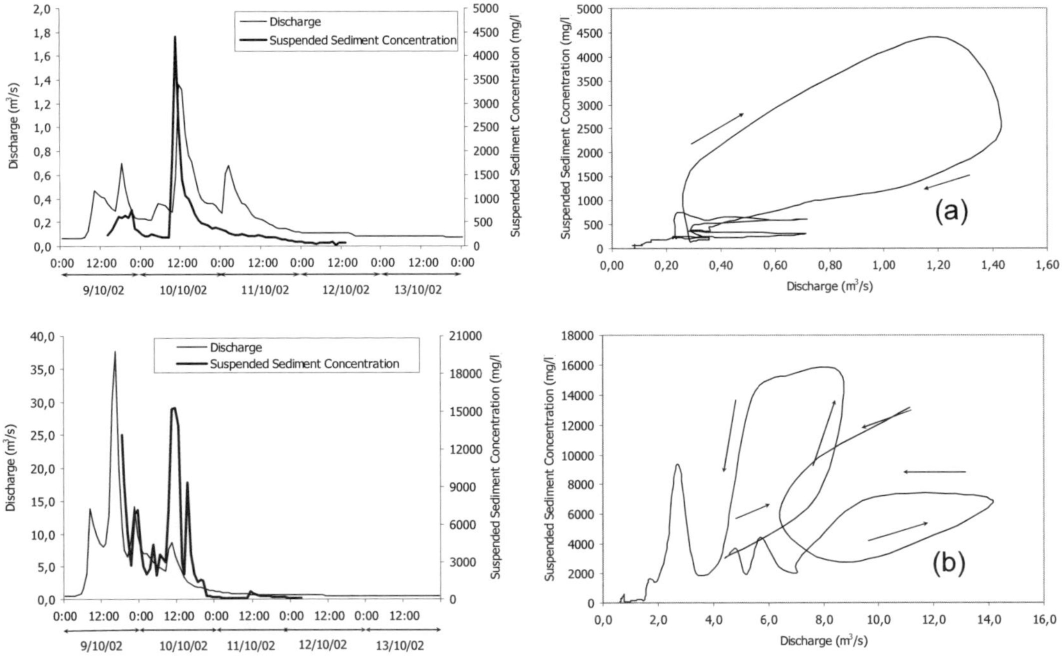

Fig. 5 Multiple peak event and hysteresis at Jorba (a) and Sant Sadurní (b)

events, which are longer in duration and record higher mean rainfall intensities. Rainfall intensity is lower during spring and the land is protected by the growth of cereal crops, which result in lower sediment concentrations. However, the fact that base flow is greater during this season than during autumn suggests that there may also be a dilution effect on concentrations.

CONCLUSIONS

Results from the Anoia River basin for 2001–2003 show that there are marked differences between suspended sediment concentrations according to season, which can be seen in mean, median and maximum values. Several factors help to understand the seasonal contrasts observed, but their relative importance has not been determined. Seasonal separation of the concentration–discharge relationship considerably improves the overall relationships. The events play a key role in the scatter of the rating curves, as they introduce hysteretic clockwise and anticlockwise loops according to single peak events or multiple peak events. This pattern is consistent at both sites. However, the values are one order of magnitude greater in the lower part than in the upper part.

Acknowledgements The authors wish to express their gratitude to the University of Barcelona for funding the PhD fellowship, to Sara Bernia, whose help in the field is gratefully appreciated and to the Water Catalan Authorities for the discharge data supply. Authors want to thank Mike Kirkby for the helpful comments on the paper.

REFERENCES

Asselman, N. (2000) Fitting and interpretation of sediment rating curves. *J. Hydrol.* **234**, 228–248.

Batalla, R. J. & Sala, M. (1994) Temporal variability of suspended sediment transport in a Mediterranean sandy gravel-bed river. In: *Variability in Stream Erosion and Sediment Transport* (ed. by L. J. Olive, R. J. Loughran & J. A. Kesby), 299–305. IAHS Publ. 224. IAHS Press, Wallingford, UK.

Gallart, F., Llorens, P., Latron, J. & Regüés, D. (2002) Hydrological processes and their seasonal controls in a small Mediterranean mountain catchment in the Pyrenees. *Hydrol. Earth System Sci.* **6**(3), 527–537.

Klein, M. (1984) Anti clockwise hysteresis in suspended sediment concentration during individual storms: Holbeck Catchment, Yorkshire, England. *Catena* **11**, 251–257.

Reid, I. & Frostick, L. E. (1994) Fluvial sediment transport and deposition. In: *Sediment Transport and Depositional Processes* (ed. by K. Pye), 89–155. Blackwells, Oxford, UK.

Walling, D. E. (1977) Assessing the accuracy of suspended sediment rating curves for a small basin. *Water Resour. Res.* **13**, 531–538.

Walling, D. E. & Webb, B. W. (1987) Suspended load in gravel-bed rivers: the UK experience. In: *Sediment Transport in Gravel-Bed Rivers* (ed. by C. R. Thorne, J. C. Bathurst & R. D. Hey), 691–732. Chichester, UK.

Monitoring suspended sediment fluxes in the Na Borges basin, Mallorca, Spain

JOAN ESTRANY & CELSO GARCIA

Department of Earth Sciences, Building Beatriu de Pinós, University of the Balearic Islands, 07122 Palma de Mallorca, Spain

joan.estrany@uib.es

Abstract Suspended sediment fluxes have been monitored since October 2002 within the Na Borges basin, a carbonate lowland agricultural basin (327 km^2) in the northeast of Mallorca. Over the study period, between October 2002 and March 2004, four flood events were selected and analysed. Mean suspended sediment concentrations ranged from 46.4 mg l^{-1} to 166.6 mg l^{-1} with a maximum suspended sediment concentration of 4000 mg l^{-1}. The analysis of the relationship of water discharge and suspended sediment concentration for the four events shows different behaviour in the availability and depletion of the sediment according to the magnitude and extension of each event, and according to the mixing and routing of water from different sources. This study reports for the first time suspended sediment data for the Balearic Islands.

Key words hysteresis; Mallorca; Mediterranean; sediment availability; suspended sediment concentrations; urban floods

INTRODUCTION

Mediterranean environments have undergone a profound transformation as the result of human activity. Historically, anthropic pressure on Mediterranean drainage basins has modified the landscape clearing fields for agriculture with the result of accelerated land degradation (Conacher & Sala, 1998). Contemporaneously, there has been a progressive loss of agricultural land due to increased urbanization that has modified the hydrological response of Mediterranean coastal basins (Sala, 2003). Moreover, the increasing use of the Mediterranean coastline by tourism has severely impacted the region, especially the islands (Garcia & Servera, 2003), and are responsible for environmental degradation. These land use changes affect runoff and sediment load of drainage basins, but make it difficult to discern between natural or human sources of sediment load (Woodward, 1995). Another complication is the influence of drainage area. Although there have been numerous studies on suspended sediment dynamics in small basins (e.g. Walling & Webb, 1983), fewer studies exist for medium and large river basins. These systems are more complex because their drainage basins may span multiple geologic or land-use settings, complicating the discharge and suspended sediment relationship, where mixing and routing of water from different sources plays an important role (e.g. Asselman, 1999; Hudson, 2003). In Mediterranean river basins there is a general lack of this type of data and only a few studies have been reported (e.g. Batalla *et al.*, 1995; Pont *et al.*, 2002).

This paper reports on the preliminary results of an investigation of suspended sediment transport within an agricultural basin on the island of Mallorca, where urban expansion of one of its tributaries has caused a change in suspended-sediment dynamics. In particular, it aims to quantify the magnitude and duration characteristics of suspended sediment concentrations within the Na Borges basin.

STUDY AREA

The Na Borges basin (327 km^2) extends over the northeast of Mallorca (Fig. 1) and is the second largest basin on the island. Limestone, marl and dolomite are the main rock type, although the lower areas are covered by Quaternary alluvial sediments. Altitudes range from 300 to 500 m in the headwaters to 0 m where the basin discharges to the sea. The average slope is 2.5% along 39 km (7.2% the first km and 0.4% the remaining 38 km).

The Na Borges basin has a Mediterranean climate and is affected by high-intensity rainstorms. The mean annual precipitation lies in the range 500 to 700 mm. The hydrological regime is essentially ephemeral with flash floods, although in the lower part of the basin the stream flows most of the year fed by groundwater. Historically, floods have affected the town of Manacor. In order to avoid this risk, the tributary of Sa Cabana was diverted to the north of the town in 2004. The mean annual discharge (1970–1990) at B05 station (264 km^2) was 0.097 m^3 s^{-1}, although during 2003 it was 0.265 m^3 s^{-1}. The basin is predominantly rural with a population of 47 458 inhabitants. More than 80% of land is in dry agricultural use. Agriculture has modified the lower

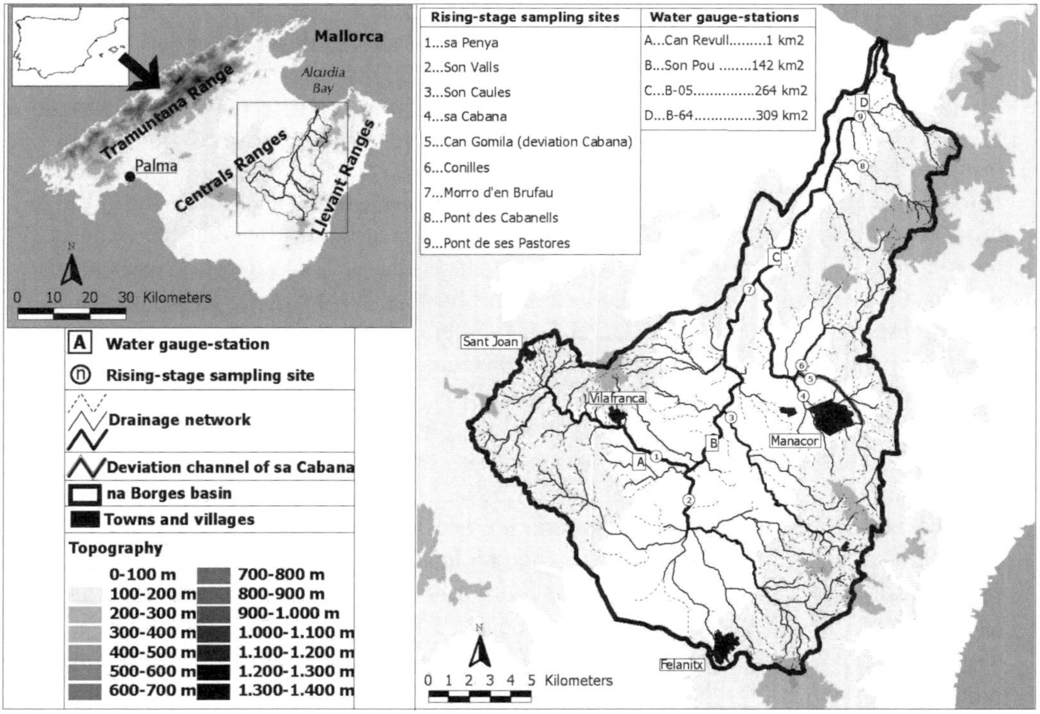

Fig. 1 Location map of the Na Borges basin showing the water gauge stations and the sampling sites.

areas by subsurface tile drainage, and the steepest areas with dry-stone walls. *Garrigue* and forest occupy 8 and 5.3%, respectively, irrigated agriculture represents 4.5%, and urban areas 2%.

METHODS

Suspended sediment concentration was determined by water samples obtained using a programmed pump sampler containing 24 bottles. The pump is programmed to sample at hourly intervals. Depth-integrated suspended sediment samples were also collected over a range of low and medium flow conditions. These samples were filtered to determine the suspended sediment concentration. The initial monitoring programme was established at the discharge gauge station B05 (264 km^2) between October 2002 and February 2003. Later on, and to assemble detailed suspended sediment flux data for the basin, turbidity monitoring equipment was installed at three discharge gauging stations within the Na Borges basin (Fig. 1):

(a) Can Revull (1 km^2) is a small sub-basin at headwaters where herbaceous agriculture and subsurface tile drainage occupy 91% of the land.
(b) Son Pou (142 km^2) closes headwaters, where dry agricultural land uses represent 82.8%.
(c) B05 (264 km^2) provides a location towards the medium end of the basin and receives flows from the Sa Cabana tributary (58 km^2). This tributary has an important urban flow influence caused by spills from Manacor's urban drainage system and its wastewater treatment plant. During rainstorms the capacity of the system is often exceeded and sewers overflowed into watercourse via combined sewer overflows (CSOs) (e.g. Old *et al.*, 2003).

At each gauging station, turbidity is continuously measured using a turbidity sensor (McVan Analyte NEP-9516-G) linked to a data logger (Campbell CR10X and Data-Taker D50) which records the output every 15 mins. Turbidity has been used successfully to investigate the short-term dynamics of suspended sediment transport (i.e. Gippel, 1989; Foster *et al.*, 1992) and to improve the reliability of estimates of suspended sediment load (i.e. Walling, 1977; Lewis, 1996). To convert turbidity measurements (NTUs) a calibration relating suspended sediment concentration to the measured turbidity was required. Samples collected manually and with the automatic pump sampler between December 2003 and March 2004 were used to generate a linear relationship between the turbidity and suspended sediment concentration data (Fig. 2). Additionally, nine rising-stage samplers, of a type modified from Finlayson (1981) with seven samples and 1 m stage, were installed in tributaries and along the main stream in order to provide more information about suspended sediment concentrations and dynamics in the basin.

RESULTS

Since the initial monitoring period, four events recorded in gauge station B05 have been selected. Two events are the first floods of the winter season (February 2003) that

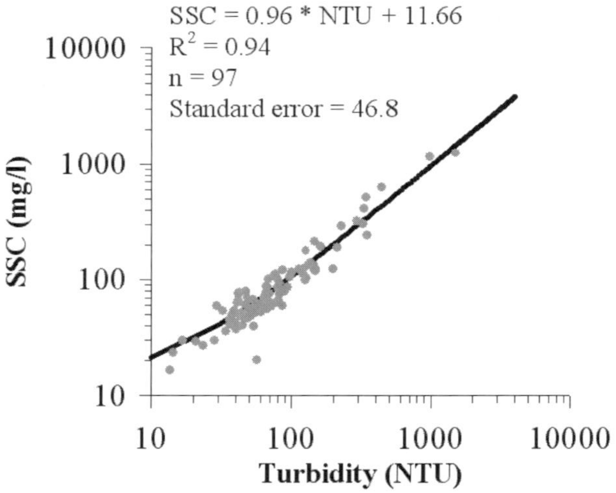

Fig. 2 Calibration relating suspended sediment concentration to the measured turbidity. Manual and automatic samples collected between December 2003 and March 2004.

affected to the whole basin. Both floods were bankfull events, the first being an exceptional flood with an instantaneous peak of 12.3 $m^3\ s^{-1}$ that in some areas overflowed the main channel. The third event occurred at the beginning of March 2004 and was produced by the tributary of Sa Cabana, coming from the town of Manacor (26 698 inhabitants). The fourth event was recorded at the end of March 2004 and came mainly from the urban tributary and a small part from the headwaters. This was proved by the flow height reached at the rising-stage samplers in Sa Cabana tributary (1 m) and Morro den Brufau (0.33 m) in the main stream (points 4 and 7 respectively, Fig. 1). The last two events were of lower magnitude than the two first floods, but had higher suspended sediment concentrations. Suspended sediment concentration (SSC) for the four flood events was recorded using the automatic pump sampler and, additionally, the last two were recorded with the turbidity sensor.

Figure 3 provides information relating to the water discharge, rainfall, turbidity and SSC during the four flood events. Figure 4 shows the relationship between SSC and water discharge and illustrates the patterns of hysteresis for each event:

(a) First event (Fig. 3(a)). Flood event with a peak discharge of 12.3 $m^3\ s^{-1}$ and a mean SSC of 103.5 mg l^{-1}. The exponent of the SSC/discharge relationship is high (1.46). The hysteresis of the event is characterized by a clockwise and tight loop. Patterns, exponent and hysteresis, suggest a hydraulic control that caused a rapid increment of SSC followed by a very fast sediment exhaustion with the peak discharge (Fig. 4(a)).

(b) Second event (Fig. 3b). Flood event with a lower mean SSC (46.4 mg l^{-1}) than the previous event. The low exponent of the SSC/discharge relationship (0.59) indicates a decrease in sediment availability, a fact that can be related to a lower availability of fine sediment in the channel, likely caused by the short time interval between the two floods. The hysteresis shows different initial loops that directly respond to sediment supplied by the flushing of material disturbed that had been left by the first event into temporary storage in the stream channel. The result is a very low coefficient of determination (0.28) on the relationship between SSC and discharge (Fig. 4(b)).

(c) Third event (Fig. 3c). Flood with a mean SSC of 166.6 mg l^{-1} and a peak discharge of 0.4 m^3 s^{-1}. It shows a high mobility of the fine sediment delivered by the combined sewage system entering in the main channel through the urban sub-basin. This fact provides high sediment availability for the event (2.07) with a figure-eight hysteresis relationship consisting of a clockwise loop at high flow and a counterclockwise loop at low flow caused by high values of SSC at low discharge on the falling limb compared with those on the rising limb (Williams, 1989) (Fig. 4(c)).

(d) Fourth event (Fig. 3(d)). Flood with a mean SSC of 185 mg l^{-1} and two peaks discharge. The first, reached 1.2 m^3 s^{-1} and the second 0.36 m^3 s^{-1} 20 h after coming from headwaters. A very low coefficient of determination (0.32) on the relationship between SSC and discharge denotes different sediment sources. The total sediment availability for the event was 0.68. The first peak is responsible for

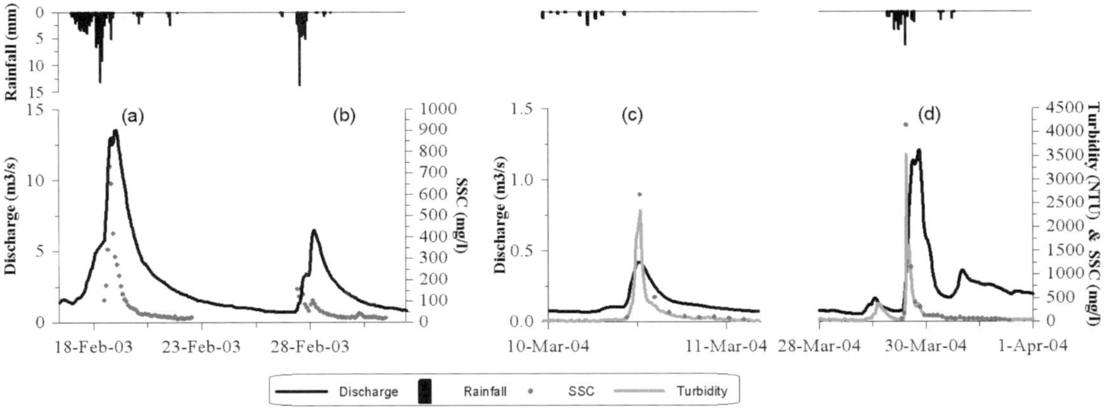

Fig. 3 Water discharge, rainfall, SSC and turbidity in the events: (a) first; (b) second; (c) third; (d) fourth.

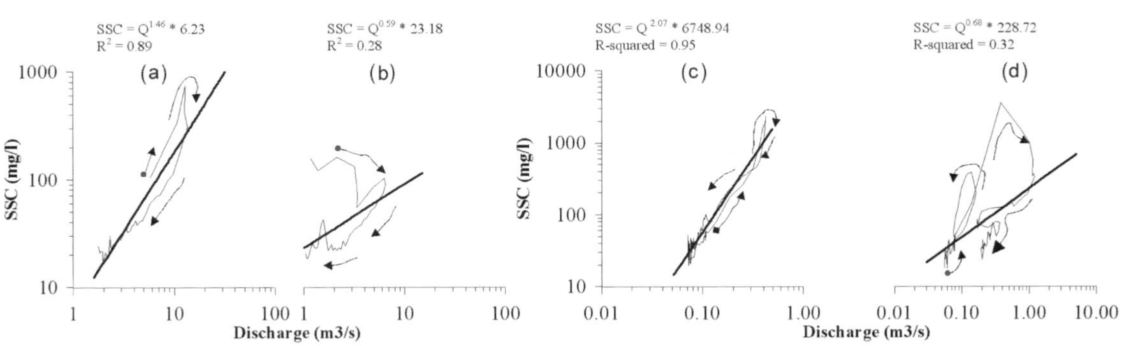

Fig. 4 SSC/water discharge relationships and hysteresis for the events: (a) first; (b) second; (c) third; (d) fourth.

high suspended sediment concentrations coming from the combined sewage system. The second, coming from headwaters, is responsible for a large sediment depletion, as the last hysteresis loop shows. This is due to the low slope between the headwaters and the gauging station (0.4%) that allows large temporary channel storage for low-flow volumes like this event (Fig. 4(d)).

CONCLUSIONS

The findings presented provide a valuable insight into suspended sediment fluxes and dynamics of a Mediterranean basin. The results demonstrate that sediment availability and its relationship with water discharge is a complex process in which many variables take part. Analysis of suspended sediment concentrations measured for four flood events, with different flow magnitude, allow an initial scheme of suspended sediment behaviour through the hysteresis loops to be identified. This has provided a first idea for analysing the significance of temporary storage of sediment according to the magnitude and the extension of the flood. The results show the importance of the Sa Cabana sub-basin (58 km^2), a tributary partially urbanized and affected by the combined sewage system of Manacor that produced urban floods having lower sediment depletion than floods affecting the whole basin due to its better connection with the main channel.

The integrated approach to establishing a basin sediment budget, especially in larger basins as the Na Borges, requires high quality data that can only be obtained through an intensive monitoring programme (Walling & Collins, 2000). Further research is being carried out to provide and improve understanding of suspended sediment dynamics in the study basin, where suspended sediment concentration is measured continuously at three sampling sites and the different sediment source areas are being identified in order to determine the relative contributions derived from those sources.

Acknowledgements This work has been funded by the Spanish Ministry of Science and Technology (Research project REN2001-0281) and by an agreement between the Spanish Ministry of Environment (LUCDEME project) and the University of the Balearic Islands. We are indebted to Joan Miquel Carmona and Andreu Ramon for their technical assistance during the field work. We thank Ian Reid for the constructive criticism that helped focusing the paper.

REFERENCES

Asselmann, N. E. M. (1999) Suspended sediment dynamics in a large drainage basin: the River Rhine. *Hydrol. Processes* **13**, 1437–1450.

Batalla, R. J., Sala, M. & Werrity, A. (1995) Sediment budget focused in solid material transport in a subhumid Mediterranean drainage basin. *Z. Geomorphol.* **39**(2), 249–264.

Conacher, A. J. & Sala, M. (1998) *Land Degradation in Mediterranean Environments of the World. Nature, Extent, Causes and Solutions*. John Wiley & Sons, Chichester, UK.

Finlayson, B. (1981) The analysis of stream suspended loads as a geomorphological teaching exercise. *J. Geogr. Higher Education* **5**, 23–35.

Foster, I. D. L., Millington, R. & Grew, R. G. (1992) The impact of particle size controls on stream turbidity measurements: some implications for suspended sediment yield estimation. In: *Erosion and Sediment Transport Monitoring Programmes in River Basins* (ed. by J. Bogen, D. E. Walling & T. J. Day), 51–62. IAHS Publ. 210. IAHS Press, Wallingford, UK.

Garcia, C. & Servera, J. (2003) Impacts of tourism development on water demand and beach degradation on the island of Mallorca (Spain). *Geogr. Ann.* **85A**(3–4), 287–300.

Gippel, C. J. (1989) The use of turbidimeters in suspended sediment research. *Hydrobiologica* **176/177**, 354–379.

Hudson, P. F. (2003) Event sequence and sediment exhaustion in the lower Panuco Basin, Mexico. *Catena* **52**, 57–76.

Lewis, J. (1996) Turbidity-controlled suspended sediment sampling for runoff-event load estimation. *Water Resour. Res.* **32**(7), 2299–2310.

Old, G. H., Leeks, G. J., Packman, J. C., Smith, B. P., Lewis, S., Hewitt, E. J., Holmes & M. & Young, A. (2003) The impact of a convectional summer rainfall event on river flow and fine sediment transport in a highly urbanised catchment: Bradford, West Yorkshire. *Sci. Total Environ.* **314–316**, 495–512.

Pont, D., Simonnet, J. P. & Walter, A. V. (2002) Medium-term changes in suspended sediment delivery to the ocean: consequences of catchment heterogeneity and river management (Rhône River, France). *Estuarine, Coastal and Shelf Science* **54**, 1–18.

Sala, M. (2003) Floods triggered by natural conditions and by human activities in a Mediterranean coastal environment. *Geogr. Ann.* **85A**(3–4), 301–312.

Walling, D. E. (1977) Suspended sediment and solute response characteristics of the River Exe, Devon, England. In: *Research in Fluvial Systems*. (ed. by R. Davidson-Arnott & W. Nickling), 58–87. Geo Abstracts, Norwich, UK.

Walling, D. E. & Webb, B. W. (1983) Patterns of sediment yield. In: *Background to Paleohydrology*. (ed. by K. J. Gregory), 69–100. Wiley, New York, USA.

Walling, D. E. & Collins, A. L. (2000) *Integrated* assessment of catchment sediment budgets: a technical manual. Framework of UK Department for International Development Research Project R6868, University of Exeter, Exeter, UK.

Williams, G. P. (1989) Sediment concentration versus water discharge during simple hydrologic events in rivers. *J. Hydrol.* **111**, 89–106.

Woodward, J. C. (1995) Patterns of Erosion and Suspended Sediment Yield in Mediterranean River Basins. In: *Sediment and Water Quality in River Catchments*. (ed. by I. D. L. Foster, A. M. Gurnell & B. W. Webb), 365–389. John Wiley & Sons, Chichester, UK.

Geomorphological Processes and Human Impacts in River Basins (Proceedings of the International Conference held at Solsona, Catalonia, Spain, May 2004). IAHS Publ. 299, 2005.

Assessment of sediment yield of watersheds by reservoir survey and simulation modelling in Brazilian semiarid environment

JOSÉ CARLOS DE ARAÚJO[1,2] **& DONALD W. KNIGHT**[2]

1 *Department of Civil Engineering, The University of Birmingham, Edgbaston, Birmingham, B15 2TT, UK*

2 *Department of Hydraulic and Environmental Engineering, Federal University of Ceará, Campus do Pici, bl.713, 60.451-970, Fortaleza, Ceará, Brazil.*
jcaraujo@ufc.br

Abstract Reservoir survey is a reliable method for assessing sediment yield, which is a major environmental issue concerning impacts on water resources. The paper presents field surveys of one urban and six rural watersheds located in the Brazilian semiarid region, with a wide range of characteristics. Basin areas range from 3 to 1220 km^2; reservoir capacity 0.5–126 Mm^3; silting history 46–94 years; and hydrological regimes range from very dry "sertão" to mild mountain and coastal regions. Sediment yield assessment consisted of topo-bathymetric reservoir surveys; volume reduction computation; dry bulk density evaluation based on core samples; and trap efficiency estimation. Results showed *per capita* urban contribution to be 21 kg year^{-1}; average sediment yield to be 426 t km^{-2} year^{-1}; and reservoir-volume reduction at a rate of 2% per decade. The Hidrosed model was applied and field data were used for its validation. Hidrosed showed good agreement with field data, with Nash and Sutcliffe efficiency coefficient 0.81 for Maner SDR formulation.

Key words Brazil; Ceará; Hidrosed; reservoir silting; sediment yield; semiarid

INTRODUCTION

The Brazilian semiarid region (10^6 km^2) in general and the State of Ceará (1.5·× 10^5 km^2) in particular are characterized by recurrent droughts and water-use conflicts. Due to the lack of reliable groundwater resources and to the intermittency of river flows, Brazilian semiarid water resources policy has historically focused on surface reservoirs as the strategic tool to provide water with a level of confidence. Today such reservoirs provide more than 90% of the water supply in the State of Ceará. If excessive and ungauged, sediment yield is a risk to reservoirs in both quantitative and qualitative aspects.

To enhance the knowledge of sediment yield in the semiarid environment, a research project considering measurement and modelling aspects is in progress. The selected measurement methodology was the morphological survey of outlet reservoirs, which is expected to be representative of the region. This technology has been used and recommended by several researchers (e.g. by Morris & Fan, 1997; Verstraeten & Poesen 2001a,b; Zarris *et al.*, 2002). Erosion and sediment yield modelling was accomplished by running Hidrosed (Araujo, 2003; Araujo *et al.*, 2003), a parametric model applied to watersheds.

METHODOLOGY

Field survey

Watersheds representative of the Brazilian semiarid State of Ceará were selected. The selected watersheds represented a wide range of basin areas, reservoir storage capacities, hydrological and climatological regimes, geographic positions in the State, and environmental conservation status (see Table 1). Within each selected basin surface water is controlled by a reservoir. The original reservoir designs were recovered so that actual topographic surface before dam construction (and the respective storage capacity) could be assessed.

Updated topo-bathymetric surveys for the seven selected reservoirs were conducted. Surveys were usually made in the dry seasons due to reduced water volume, which improve accuracy. Volume reduction was estimated for each reservoir by comparing the new surveys with surveys conducted before dam construction. Field campaigns also included core sampling of deposited sediment in the reservoirs to assess dry bulk density and enhance sediment yield estimation accuracy. In fact, Verstraeten & Poesen (2001a,b) performed sensitivity analysis of dry bulk density and trap efficiency on sediment yield assessment by reservoir morphology. The authors recommended intense sampling of sediments to reduce sediment yield estimation error, which was shown to be as high as 72% when based on empirical equations. Core samples of the sedimentation material were usually obtained in the dry season, so that they would be as undisturbed as possible. Sedimentation mass (dM) was simply calculated by considering dM = dV.B, in which dV = volume reduction and B = dry bulk density of deposited sediment.

Trap efficiency was computed using the median curve of Brune (1953, quoted in Morris & Fan, 1997, p.10.27). Monitoring of the sediment budget in S. Mateus reservoir was performed in 1999 by Almeida (2001) so as to coarsely assess the applicability of Brune's method. Daily samples of sediment input and output, as well

Table 1 Characteristics of the seven selected watersheds in Brazilian semiarid environment used in this research.

Watershed	Catchment area (km^2)	Reservoir initial capacity (Mm^3)	Hydro-climatological feature	Geographic position in Ceará	Environment conservation status [(1)]
Acarape	208	34.1	mild, mountain	central–east	relatively high
V. Boi	1221	51.9	dry, sertão	central–west	relatively high
S. Mateus	229	10.3	dry, sertão	central–north	regular
Cedro	220	125.7	dry, sertão	central	low
V. Volta	155	12.5	transition, sertão coast	northwest	relatively high
Canabrava	2.9	1.2	transition, sertão mountain	southwest	regular
S. Anastacio[(2)]	9.3	0.5	mild, coast	northeast	low
Average	292	33.7	(–)	(–)	(–)

[(1)] Environmental conservation status during sedimentation measurement period.
[(2)] Urban watershed.

as daily measurements of evaporation, outflow discharge and reservoir level were taken to develop water and sediment budgets.

Hidrosed model

Hidrosed is a model designed to assess gross erosion, sediment yield and reservoir sedimentation in middle-size watersheds for planning time scales, i.e. decades. The watershed is divided into homogeneous cells with sizes ranging from 1 to 25 km^2. Gross erosion is computed by the Revised Universal Soil Loss Equation (RUSLE, see thorough explanation in Haan *et al.*, 1994) with modification to adjust the erosivity factor from event to monthly time steps. Sediment yield is assessed by use of a sediment delivery ratio (SDR) concept computed either by Maner (1958), Roehl (1962) or Williams & Berndt (1972). The results hereafter presented were based on the SDR formulation by Maner.

Reservoir trap efficiency can be evaluated by the median curve of Brune (1953, quoted in Morris & Fan, 1997) or by a sediment-budget approach. In the latter case, the reservoir is divided into three sub-reservoirs: sub-reservoir (1) is the volume where sediment is suspended in water; sub-reservoir (2) is the volume in which sediment concentration is high but re-suspension is possible; and sub-reservoir (3) is the volume where sediment is deposited and cannot be re-suspended. Monthly time-steps are used and each month is associated with either a wet or a dry season. A more detailed explanation of Hidrosed can be found elsewhere (Araújo, 2003; Araújo *et al.*, 2003). Other models follow an approach analogous to that used in Hidrosed (see e.g. Beasley & Huggins, 1980; Khanbilvardi & Rogowski, 1984; Young *et al.*, 1987; Ferro & Porto, 2000; van Rompaey *et al.*, 2001).

Field parameterization was necessary, since no calibration procedure was used. Monthly rainfall data were acquired and each cell was associated with the closest raingauge station. Soil was sampled to assess erodibility (Wischmeier & Smith, 1978) for predominant soil formations in the watershed, and weighted erodibility values were associated with each cell. Topographic maps were used to estimate slope gradient and length. Aerial-photographic surveys of 1958, 1969 and 1997 (complemented by field inspection excursions) and design reports were the basis for historical land use assessment. A land-use factor was calculated for each cell in each reference year (i.e. whenever data was available) and land-use factors in the remaining years were obtained by linear interpolation of reference-year values.

Validation was assessed by means of *model prediction error* (MPE), given by equation (1) and the Nash and Sutcliffe (1970) *model efficiency coefficient* (NSE), given by equation (2): *Yobs* = observed value; *Ypred* = predicted value; and *Ym,obs* = mean observed value.

$$MPE = (Ypred - Yobs)/Yobs \quad (1)$$

$$NSE = 1 - \frac{\sum (Yobs - Ypred)^2}{\sum (Yobs - Ym,obs)^2} \quad (2)$$

MPE varies from -1 to $+\infty$, once $Ypred \geq 0$, and the closer *MPE* is to null, the higher the model efficiency. *NSE* ranges from $-\infty$ to 1. If $NSE \rightarrow -\infty$, observations tend to constant value (*Yobs* ≈ constant ≈ *Ym,obs*) and modelling is meaningless; if $NSE < 0$, model preview capacity is lower than just admitting mean observed value; if $NSE > 0$, model previews with higher accuracy than mean observed value. The closer *NSE* is to 1, the higher the model efficiency.

RESULTS AND DISCUSSION

Field survey

Data from field survey indicate an arithmetic average sediment yield of 426 t km^{-2} $year^{-1}$ and median 233 t km^{-2} $year^{-1}$ (Table 2), which is of the same order of magnitude of other Brazilian watersheds (see Fig. 1). Although Cedro basin has shown high sediment yield (1303 t km^{-2} $year^{-1}$), it is of the same order of magnitude as other watersheds in Brazil (see Sume and Paraiba do Sul, Fig. 1). Average sediment yield can be as low as 0.005 t km^{-2} $year^{-1}$ as observed in the Galie Lake watershed, Canada (Schiffer *et al.*, 2001), or as high as 6400 t km^{-2} $year^{-1}$ as observed in Yellow River, China (Shensheng, p.377–380 in: Bordas & Walling, 1988). In fact, Gang & Mingquan (1999) estimate average erosion in the Yellow River in the range of 10 000–15 000 t km^{-2} $year^{-1}$.

Table 2 Results of field survey in seven watersheds in Brazilian semiarid environment.

Watershed	Volume reduction (Mm^3)	Silting period (years)	Volume reduction rate (% $year^{-1}$)	Silting rate (m^3 km^{-2} $year^{-1}$)	Dry bulk density (t m^{-3})	Sediment yield (t km^{-2} $year^{-1}$)
Acarape	2.67	74	0.106	173	1.32	233
V. Boi	5.82	47	0.239	101	1.20	124
S. Mateus	1.46	46	0.306	138	1.36	194
Cedro	20.64	94	0.175	998	1.28	1303
V. Volta	1.53	81	0.151	122	1.32	164
Canabrava	0.08	57	0.121	517	1.38	704
S. Anastacio	0.14	75	0.357	194	1.26	257
Average	4.62	68	0.208	320	1.30	426

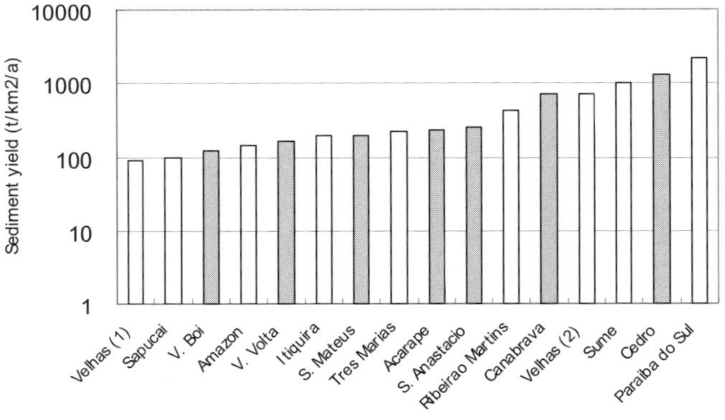

Fig. 1 Sediment yield in several watersheds in Brazil (the seven watersheds investigated in this research are hatched).

From Table 2 a high (2% per decade) volume reduction rate can be seen, This reduction rate demands long-term measures to maintain reservoir capacity. Extrapolation of this rate to the whole State of Ceará leads to an estimation of approximately 30 Mm^3 storage-capacity loss every year, considering present-day conditions.

Urban sediment yield was also assessed. Araujo *et al.* (2003) estimated per capita contribution of the Sao Jorge urbanized watershed in Fortaleza, Brazil, as 0.56 kg $year^{-1}$. Soares (2003) studied the S. Anastacio urban watershed from 1992 to 2002, when another bathymetric survey took place. An average per capita contribution of 21 kg $year^{-1}$ was reported. The higher sediment contribution of the second watershed is due to its lower sanitation level and to the fact that urbanization process in that site was ongoing.

Almeida (2001) monitored sediment budget in S. Mateus watershed and observed lower trap efficiency (86%) than predicted (97%) using the median curve of Brune. The lower trap efficiency of reservoirs in the semiarid environment (compared to those in temperate climate) possibly occurs because of the consistent correlation between high sediment inflow and high sediment outflow in very wet years. In dry years, trap efficiency is very high in semiarid reservoirs (usually over 99%), but inflow is insignificant.

Hidrosed model validation

Model Hidrosed was applied for the investigated watersheds without calibration procedures so as to assess model applicability. The main results of Hidrosed performance (with trap efficiency computed by the method of Brune) can be seen in Table 3. The use of Maner equation for SDR estimation within Hidrosed produced the best results for all watersheds, with average model prediction error of 46%, followed by Roehl's equation (76%) and Williams & Berndt's (303%). Maner's equation also presented the least scatter. Note that all SDR formulations generated low quality results for Varzea da Volta watershed. This is due to its topographic sequence: mountainous–extremely steep–very flat, which cannot be correctly represented by any of the lumped SDR equations.

Table 3 Validation of Hidrosed model (Brune median curve used to assess trap efficiency) for seven watersheds in Brazilian semiarid environment and three SDR formulations.

Watershed	Predicted sediment yield (t km^{-2} $year^{-1}$)			Model prediction error		
	Maner	Roehl	WB	Maner	Roehl	WB (1)
Acarape	345	508	1013	+47.9%	+118.1%	+303.5%
V. Boi	106	71	253	–14.7%	–42.5%	+104.3%
S. Mateus	221	358	1181	+13.7%	+84.4%	+508.6%
Cedro	1001	671	3714	–23.2%	–48.5%	+185.0%
V. Volta	462	478	1710	+181.6%	+191.4%	+942.8%
Canabrava	560	464	1087	–20.4%	–34.1%	+54.4%
S. Anastacio(2)	188	325	189	–23.0%	+23.2%	–23.0%
Average	412	411	1307	46.4% (3)	77.5% (3)	303.1% (3)
NSE (4)				0.81	0.41	–7.84

(1) WB = Williams & Berndt (1972).

(2) Per capita sediment contribution 21 kg $year^{-1}$ according to Soares (2003).

(3) Average considers module of errors.

(4) NSE = Nash & Sutcliffe (1970) model efficiency coefficient.

The Nash and Sutcliffe (1970) efficiency coefficient also showed similar results, leading to the conclusion that the equation by Maner is the best (*NSE* = 0.81), followed by that of Roehl (*NSE* = 0.41). The SDR formulation by Williams & Berndt showed *NSE* = –7.84, which means that using the average observed sediment yield value is better than applying the model. Nevertheless, Williams & Berndt's equation did apply for catchments whose areas were less than 10 km^2 (see Canabrava and S. Anastacio watersheds, Table 3). The performance of the sediment yield model Hidrosed can be enhanced by about 20% if reservoir trap efficiency is computed by the physical modelling proposed by Almeida (2001), instead of using the empirical equation of Brune.

CONCLUSIONS

This paper presents the measured sediment yield for seven watersheds (one of which is in an urban area) in the Brazilian semiarid State of Ceará. The methodology used for sediment yield assessment was the morphologic survey of the reservoirs located in the outlet of the respective basins. Sediment yield in the investigated semiarid watersheds (426 t km^{-2} $year^{-1}$) is comparable with that in other Brazilian environments. Urban sediment production was assessed: the per capita sediment yield for S. Anastacio basin is 21 kg $year^{-1}$, a much higher value than that for S. Jorge basin, also located in Fortaleza, (0.56 kg $year^{-1}$). The different per capita yield is caused by differences in sanitation infrastructure and in urbanization processes.

The model Hidrosed was parameterized and applied to the basins for simulation periods ranging from 46 to 94 years. Good results were achieved when the SDR formulation by Maner (1958) was used. Roehl's (1962) formulation showed lower applicability, whereas Williams & Berndt's (1972) formulation should only be applied for catchment areas below 10 km^2.

Reservoir storage capacity is being reduced at a 2%-per-decade rate, which corresponds to a reduction of 30 Mm^3 every year in the State of Ceará. Considering that sediment deposition not only affects reservoirs quantitatively, but has also impacts on water quality, it is urgent that a long-term policy concerning sediment yield be proposed for the State.

Acknowledgements The authors are grateful to the Brazilian Government (CAPES–Fundação Coordenação de Aperfeiçoamento do Pessoal de Nível Superior, process BEX 1540/03-3) for the post-doctorate scholarship granted to the first author. The authors also acknowledge the University of Kassel, Germany, for the support of bathymetric surveys (WAVES Project). Comments made by an anonymous reviewer greatly improved this paper.

REFERENCES

Almeida, G. B. (2001) *Modelagem da retenção de sedimentos em reservatórios no semi-árido.* MSc Thesis, DEHA, Universidade Federal do Ceará, Fortaleza, Brazil.

Araújo, J. C. (2003) Assoreamento em reservatórios do semi-árido: modelagem e validação. *Rev. Bras. Rec. Hidricos* **8**(2), 39–56.

Araújo, J. C., Fernandes, L., Machado Junior, J. C., Oliveira, M. R. L. & Sousa, T. C. (2003) Sedimentation of reservoirs in semiarid Brazil. In: *Global Change and Regional Impacts* (ed. by T. Gaiser, M. Krol, H. Frischkorn & J. C. Araújo), 205–216. Springer, Berlin, Germany.

Beasley, D. B., Huggins, L. F. & Monke, E. J. (1980) ANSWERS: A model for watershed planning. *Trans. ASAE* **23**(4), 938–944.

Bordas, M. P. & Walling, D. E. (eds) (1988) *Sediment Budgets* (Proc. Porto Alegre Symp.) IAHS Publ. 174. IAHS Press, Wallingford, UK.

Ferro, V. & Porto, P. (2000) Sediment delivery distributed (SEDD) model. *J. Hydraul. Engng.* **5**(4), 411–422.

Gang, W. & Mingquan, X. (1999) Analysis on peculiarities of "92.8" flood event in the lower Yellow River. In: *Sino-US Workshop on Sediment Transport and Disasters* (ed. by Z. Y. Wang, T. W. Soong & B. C. Yen), 205–213. *Special Issue Int. J. Sediment Res.* **14**(2).

Haan, C. T., Barfield, B. J. & Hayes, J. C. (1994) *Design Hydrology and Sedimentology for Small Catchments*, 2nd edn. Academic Press, San Diego, USA.

Khanbilvardi, R. M. & Rogowski, A. S. (1984) Quantitative evaluation of sediment delivery ratios. *Water Resour Bull. AWRA* **20**(6), 865–874.

Maner, S. B. (1958) Factors affecting sediment delivery rates in the Red Hills physiographic area. *Trans. Am. Geophys. Un.* **39**(4), 669–675.

Morris, G. L. & Fan, J. (1997) *Reservoir Sedimentation Handbook: Design and Management of Dams, Reservoirs and Watersheds for Sustainable Use.* McGraw-Hill Professional, New York, USA.

Nash, J. E. & Sutcliffe, J. V. (1970) River flow forecasting through conceptual models–Part I: A discussion of principles. *J. Hydrol.* **10**, 282–290.

Roehl, J. W. (1962) Sediment sources areas, delivery ratios and influencing morphological factors. In: *International Association for Scientific Hydrology Comissions of Land Erosion*, 202–203. IAHS Publ. 59. IAHS Press, Wallingford, UK.

Schiffer, E., Slaymaker, O. & Klinkenberg, B. (2001) Physiographically controlled allometry of specific sediment yield in the Canadian Cordillera: a lake-sediment based approach. *Geogr. Ann.* **A83**(1–2), 55–65.

Soares, F. A. (2003) Assoreamento de bacia urbana: estudo de caso do açude Santo Anastácio, Fortaleza, Ceará. MSc Thesis, DEHA, Federal Univ. Ceará, Brazil.

van Rompaey, A. J. J., Verstraeten, G., van Oost, K., Govers, G. & Poesen, J. (2001) Modelling means annual sediment yield using a distributed approach. *Earth Surf. Processes Landf.* **26**, 1221–1236.

Verstraeten, G. & Poesen, J. (2001a) Variability of dry sediment bulk density between and within retention ponds and its impact on the calculation of sediment yields. *Earth Surf. Processes Landf.* **26**, 375–394.

Verstraeten, G. & Poesen, J. (2001b) The importance of sediment characteristics and trap efficiency in assessing sediment yield using retention ponds. *Phys. Chem. Earth* (B), **26**(1), 83–87

Wischmeier, W. H. & Smith, D. D. (1978) Predicting rainfall erosion losses—a guide to conservation planning. *USDA, Agriculture Handbook 537, Washington DC, USA.*

Williams, J. R. & Berndt, H. D. (1972) Sediment yield computed with universal equation. *J. Hydraul. Div. ASCE* **98**(HY12), 2087–2097.

Young, R. A., Onstad, C. A., Bosch, D. D. & Anderson, W. P. (1987) AGNPS, Agricultural non-point-source pollution model: a large watershed analysis tool. *Conservation Research Report 35, USDA-ARS, Washington, USA.*

Zarris, D., Lykoudi, E. & Koutsoyiannis, D. (2002) Sediment yield estimation of a hydrological basin using measurements of reservoir deposits: a case study for the Kremasta reservoir, Western Greece. In: *Proc. 5th Intern. Conf. European Water Resources Association* (ed. by G. Tsakiris), 338–345. Athens, Greece.

Patterns of suspended sediment transport in two forested Mediterranean mountainous basins (Ribera Salada, Catalan Pre-Pyrenees, NE Spain)

J. CARLES BALASCH[1], RAMON J. BATALLA[1,2], ROSA M. POCH[1,2] & DAMIA VERICAT[1]

1 *Department of Environment and Soil Science, University of Lleida, Rovira Roure 191 E-25198, Lleida, Catalonia, Spain*
cbalasch@macs.udl.es

2 *Forestry Institute of Catalonia, Pujada del Seminari s/n, E-25280, Solsona, Catalonia, Spain*

Abstract Suspended sediment dynamics and trends are analysed in two Mediterranean mountainous basins in the southern Pyrenees. Rainfall, discharge and suspended sediment are monitored in order to establish the long-term water and sediment budgets of the basins and to determine sediment transport dynamics. A nested basins research scheme (2.6 km^2 and 65 km^2) allows determination of the conveyance and routing of sediment through headwaters, tributaries and main channel during flood response. Hysteretic loops show seasonal complex relationships between water response and sediment availability. Clockwise and anticlockwise hysteretic patterns and exhaustion effects are recognized and explained in terms of temporal runoff generation, sediment sources distribution and stream channel sinks, demonstrating the different seasonal patterns of processes involved. Mean sediment yield of the two basins was in the order of 1–3 Mg km^{-2} year^{-1}, well below the regional values.

Key words forested mountainous basin; hysteresis; Mediterranean rivers; sediment transport

INTRODUCTION

In Mediterranean regions, sediment transport and yield are usually measured in very small drainage basins (<1 km^2). As a consequence of this, there is a general lack of this type of data in larger basins, especially in areas with predominant undisturbed forest use.

The objective of this work is to study the patterns of suspended sediment transport in relation to the hydrological response in two forested Mediterranean mountainous basins. Cogulers and Canalda drainage basins (2.6 km^2 and 65 km^2, respectively) are located in the Ribera Salada basin, over molassic gradually varied detritic substrate and with a dominant *Pinus sylvestris* cover. Ribera Salada basin has a mean annual precipitation of 780 mm and drains to the Segre River in the Rialb reservoir (Catalan Pre-Pyrenees) (Table 1, Fig. 1). Preliminary results include the analysis of temporal dynamics of discharge and suspended sediment during individual floods, and an estimation of the annual sediment yield of the two basins.

METHODOLOGY

Water discharge and suspended sediment transport (mainly silt and clay particles) are monitored following a nested basins research scheme. Water level sensors are used to

Table 1 Main characteristics of the two experimental basins.

	Cogulers	Canalda
Area (km^2)	2.65	65
Mean channel slope (%)	6	1.9
Lithology	Conglomerates	Limestone, clay and conglomerates
Time concentration (min)	28	82.6
Altitude interval (m)	442	1700
Woodland (% basin area)	76	60
Bare rock (% basin area)	13.6	11
Crops (% basin area)	0.4	5.8
Pastureland (% basin area)	10	23.5

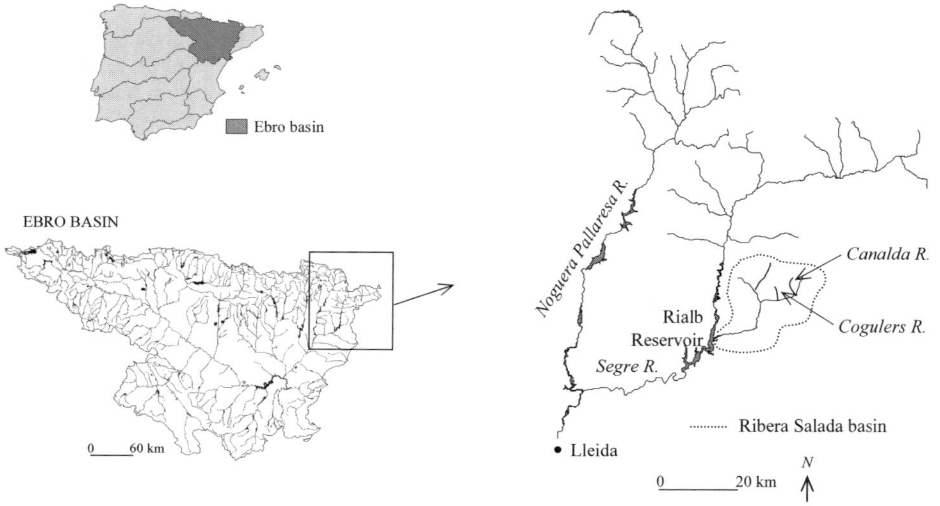

Fig. 1 Location of the Cogulers and the Canalda experimental basins in the Ribera Salada basin and in the whole Ebro basin (NE Iberian Peninsula).

measure flow stage and automatic ISCO-3700 samplers are used to collect samples of water and suspended sediment, at two control river sections on alluvial channels (Table 2, Fig. 2). Rainfall is recorded in an automatic weather station and in a series of rain gauges distributed along the basins (Table 2). A total of eight floods were monitored in Cogulers and six floods in Canalda during the period 1998–2000 (Tables 3 and 4). An estimation of the sediment yield for those two years was done from load-rating relations between discharge and suspended sediment using the flow duration curve method described by Walling (1984).

RESULTS AND REMARKS

Mean discharge of the Cogulers torrent (1998–2000) is 0.0029 m^3 s^{-1} (0.0011 m^3 s^{-1} km^{-2}), while mean discharge of the Canalda river is 0.137 m^3 s^{-1} (0.0021 m^3 s^{-1} km^{-2}), well below the average values obtained in the Catalan Pre-Pyrenees and in the Coastal

Ranges of around 0.01–0.015 $m^3\ s^{-1}\ km^{-2}$ (FEC, 1985; Batalla *et al.,* 1995). Runoff in Canalda is less variable through time in comparison with Cogulers (Fig. 3). Runoff coefficients for the whole period are 5% in Cogulers and 10% in Canalda, also lower than the regional values, whilst runoff coefficients during floods are, in general, smaller than 1% in the two basins.

Table 2 Monitoring sections and sampling instruments.

Instrumentation	Cogulers (2.65 km^2)	Canalda (65.2 km^2)
Meteorological station		1
Raingauges	3	8
Hydrometric station (weir)	1	
Hydrometric station (river bed)		1
Water level sensors	1	1
Water temperature sensors	1	1
Water and sediment samplers	1	1

Fig. 2 Hydrometric and sediment sampling stations at the Cogulers torrent (a) and at the Canalda river (b) monitoring sections.

Table 3 Monitoring of sediment transport in the Cogulers and Canalda basins (1998–2000).

Sampling section and period	Range of discharges ($m^3\ s^{-1}$)*	Nº of samples	Time interval (%)†
Cogulers	0.001–0.240	54	0–65
Canalda	0.300–2.460	158	0–90

*Range of discharges under which suspended load was sampled.

†Time interval covered by samples.

Table 4 Main hydrological and sedimentary characteristics of different runoff events recorded in Cogulers and Canalda basins. Floods have been divided (FT) in relation to both, the magnitude of the hydrological response and the suspended sediment dynamics.

Flood	[1]P (mm)	[2]Q_{peak} ($m^3 s^{-1}$)	Baseflow (m^3)	Direct Runoff (m^3)	Total Runoff (m^3)	[4]SSC_{max} ($g l^{-1}$)	[3]FT
Cogulers Basin							
26 Sep. 1998	51.6	0.028	177 (37)[3]	298 (63)[3]	475	0.27	I
18 Oct. 1998	27.3	0.009	33 (36)	57 (64)	90	0.30	I
17 Oct. 1999	31.5	0.005	30 (37)	51 (63)	81	0.13	I
12–13 Nov. 1999	67.5	0.272	38069 (99)	134 (1)	38203	0.10	II
11 Apr. 2000	45.7	0.020	1689 (99)	17 (1)	1706	0.80	III
07 May 2000	53.0	0.156	977 (52)	889 (48)	1866	4.70	
10 May 2000	16.7	0.024	2484 (98)	39 (2)	2523	0.08	III
28 Sep. 2000	55.0	0.060	311 (51)	296 (49)	607	0.76	I
Canalda Basin							
26 Sep 1998	51.6	1.520	21 200 (45)	25 500 (55)	46 700	1.52	I
24 Oct. 1999	15.7	1.185	47 600 (59)	33 400 (41)	81 000	1.19	
12–13 Nov. 1999	67.5	3.845	117 100 (57)	89 400 (43)	206 500	2.46	II
16 Apr. 2000	15.9	1.595	65 200 (61)	41 400 (39)	106 600	1.57	III
07 May 2000	53.0	1.510	30 900 (57)	23 200 (43)	54 100	1.30	
28 Sep. 2000	55.0	1.307	16 300 (46)	19 300 (54)	35 600	1.25	I

[1] Total precipitation (P); [2] Peak discharge (Q_{peak}); [3] Percentage of given runoff within total runoff; [4] Maximum suspended sediment concentration (SSC_{max}); [5] Flood type (FT).

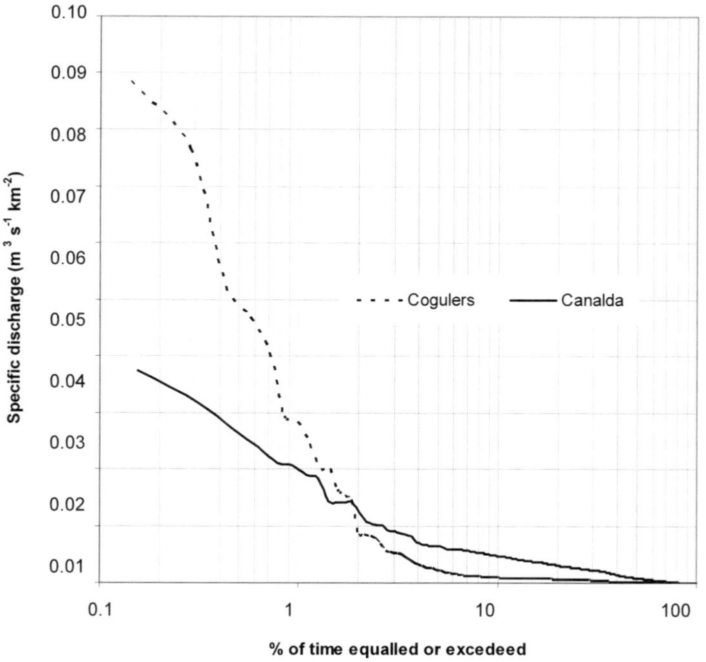

Fig. 3 Flow duration curves of the Cogulers and the Canalda river basins (1998–2000).

Overall, suspended sediment concentrations are lower than 0.5 g l^{-1} but, occasionally, concentrations up to 4 g l^{-1} have been measured linked to high rainfall intensities. Floods have been divided in relation to both the magnitude of the hydrological response and the suspended sediment dynamics as follows (Figs 4 and 5, Table 4):

(a) *Type I floods* (occurring during summer and early autumn) are characterized by relatively low peak discharges but relatively high concentrations of suspended sediment (e.g. up to 0.3 g l^{-1} in Cogulers), indicating a high sediment availability in the basins (e.g. Williams, 1989). Both, clockwise and anti-clockwise loops are found.

(b) *Type II floods* (mostly occurring during late autumn) showed the highest peak discharges, but low to moderate suspended sediment concentrations (e.g. 0.1 g l^{-1} in Cogulers), suggesting a dilution effect related to the dominant groundwater contribution and to the limited availability of sediment. Clockwise loops define the relation between discharge and concentration.

(c) *Type III floods* (occurring in winter and spring) show intermediate discharges and low sediment concentrations (e.g. <0.1 g l^{-1} in Cogulers), suggesting a sediment exhaustion effect after an initial maximum (L-trend). This fact may be attributed to the low sediment production during the winter season in the basins.

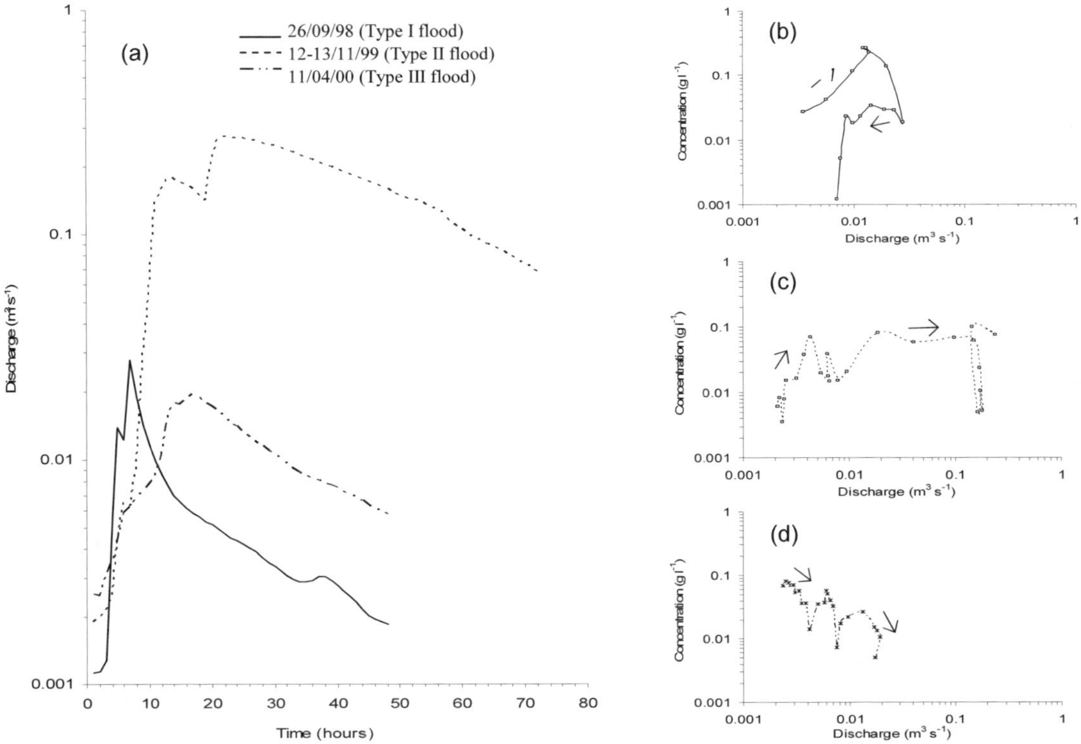

Fig. 4 Flood hydrographs (a) and discharge/concentration rating curves for different flood types in the Cogulers basin. (b) Type I flood (26 September 1998) with $P = 51.6$ mm/48 h. (c) Type II flood (12–13 November 1999) with $P = 67.5$ mm/48 h. (d) Type III flood (11 April 2000) with $P= 45.7$ mm/48 h.

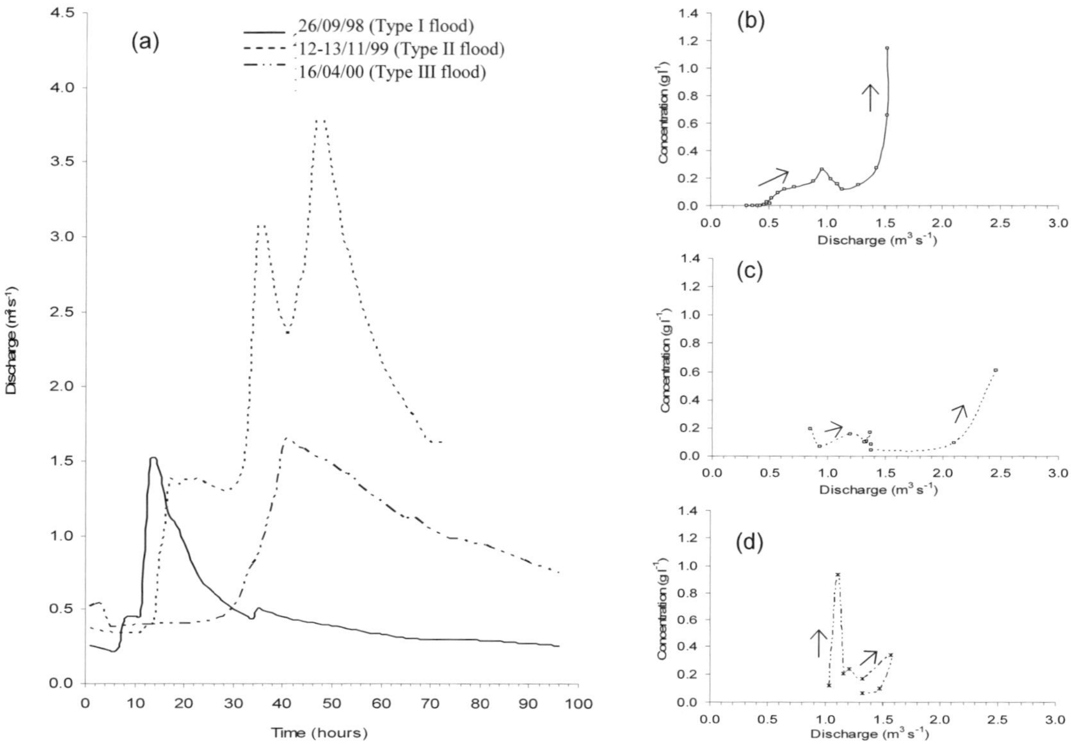

Fig. 5 Flood hydrographs (a) and discharge/concentration rating curves for different flood types in the Canalda basin. (b) Type I flood (26 September 1998) with P= 51.6 mm/48 h. (c) Type II flood (12–13 November 1999) with P = 67.5 mm/48 h. (d) Type III flood (16 April 2000) with P = 15.9 mm/24 h.

The Canalda basin shows a positive statistically significant relation between discharge and sediment concentration (r^2 = 0.50); in contrast, the relation in Cogulers is not significant (Fig. 6). This fact reflects the distinct nature of sediment transport between the two basins. Sediment sources and sediment availability in the Canalda basin is higher than in Cogulers. This fact may be related to the following factors: (a) lithology (exclusively conglomerates with poor and thin soils in Cogulers, and conglomerates but also evaporitic and clay Triassic and Cretaceous sediments in Canalda), (b) event runoff production, and (c) land use (there are bare crops and pastures in Canalda but not in Cogulers).

In the Cogulers basin, sediment availability is controlled by geomorphic controls, which ultimately may limit the sediment supply. The entire basin is on conglomerates. Fine sediment particles come almost exclusively from the stony and thin soils in the slopes. The narrow river channel with vegetated banks does not itself constitute a major sediment source. As a consequence, suspended sediment concentrations will be generally low, especially due to maximum runoff volume that occurs at the end of the rainy seasons (autumn and spring).

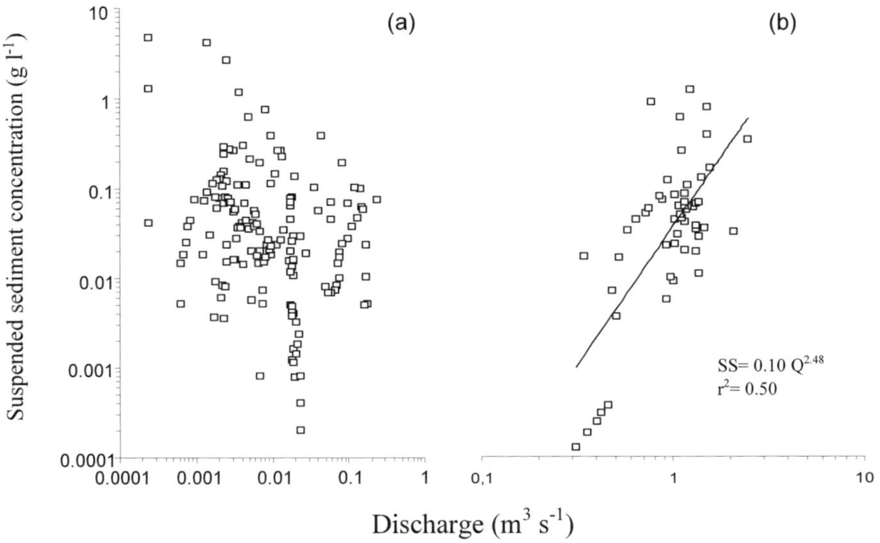

Fig. 6 Relation between discharge and suspended sediment concentrations for the (a) Cogulers and (b) Canalda basins (1998–2000).

In the case of the Canalda basin, river hydraulics (i.e. suspended sediment concentration is strongly sensitive to discharge, as indicated by the high exponent of the equation) acts as a primary factor controlling sediment transport, which is mostly eroded from alluvial deposits. In large channels, sediment is generally being entrained from the channel rather than by runoff derived from the slopes (Mossa, 1996; Hudson, 2003). Non-cohesive rock outcrops in the Canalda basin supply fine sediment to zero-order basins and small tributaries. Fine particles tend to settle once they reach the downstream main channel, due to the relatively lower gradient, which subsequently acts as the most important source of sediment during forthcoming floods.

Mean sediment yield is estimated at around 1 Mg km^{-2} year^{-1} in Cogulers and 3 Mg km^{-2} year^{-1} in Canalda, values in the low range of forested basins. Sediment yield in the study years are especially low if compared with average yields in Mediterranean counterparts. For instance, sediment yield in rivers in the Catalan Coastal Ranges reaches the order of 100 Mg km^{-2} year^{-1} (Batalla *et al.,* 1995; Rovira *et al.,* 2004). In other Pre-Pyrenean basins with badlands (Regüés *et al*., 2000) or incised gullies on flysch sediments (Seeger *et al*., 2004), specific yields can attain even higher values (>500 Mg km^{-2} year^{-1}).

The low runoff and the low magnitude of floods which occurred during 1998–2000 in the Ribera Salada drainage basin is seen as the main factor responsible for such differences in sediment yield. Mean precipitation for the period 1998–2000 was 590 mm (e.g. 390 mm in 1998), almost 25% less than the average in the region. Annual runoff (0.136 m^3 s^{-1} in Canalda) was, consequently, lower than the long-term average (1 m^3 s^{-1}), estimated from the Cardener River neighbouring drainage basin (period 1945–1995). The relatively small runoff and, especially, the low magnitude of floods in the watershed during the study period (e.g. maximum peak was 2.4 m^3 s^{-1},

0.040 $m^3\ s^{-1}\ km^{-2}$) are pointed out as factors to explain the low sediment yield in the Ribera Salada basins in comparison with average regional values. In addition, both the dense and extended vegetation cover in the slopes of the Ribera Salada and the absence of remarkable overland flow erosion processes (mass movements, gullies, badlands) limit the amount of sediment available to be transported by the river during low magnitude frequent floods.

CONCLUSIONS

Patterns of suspended sediment transport have been presented for two forested mountainous basins in the Catalan Pre-Pyrenees for the dry years 1998–2000. Conclusions are twofold:

(a) Three types of suspended sediment patterns have been identified according to flood magnitude and, especially, to the position of the flood in the hydrological year (season). In general, floods occurring at the end of summer and beginning of autumn (Type-I) tend to show the highest concentrations, although peaks of discharge are rather low. Large runoff events in autumn are the most effective in terms of the basin sediment yield (Type-II).

(b) Sediment yield in the study basins is of the order of 1 to 3 t km^{-2} $year^{-1}$, well below average values reported from other experimental basins in the region. Taking into account that suspended sediment concentrations can attain relatively high values (up to 4 g l^{-1}) during strong rainfall intensity events, the small floods and, subsequently, the low runoff during the study period are seen as the main factors for such low sediment yield values.

Acknowledgements This work was supported by a grant from the Spanish Ministry of Education and Science (CICYT, HID97-0397). Oscar Alisedo at the Forestry Institute of Catalonia assisted during fieldwork. Revisions by Paul F. Hudson and Wojciech Froehlich greatly improved this paper.

REFERENCES

Batalla, R. J., Sala, M. & Werritty, A. (1995) Sediment budget focused on solid material transport in a subhumid Mediterranean drainage basin. *Z. Geomorphol.* **29**(2), 249-264.

FEC (1985) Recursos geològics i sòl. In: *Història Natural dels Països Catalans* (ed. by Fundació Enciclopèdia Catalana), vol. 3, Barcelona, Spain.

Seeger, M., Errea, M. P., Beguería, S., Arañes, J., Martí, C. & García-Ruiz, J. M. (2004) Catchment soil moisture and rainfall characteristics as determinant factors for discharge / suspended sediment hysteretic loops in a small headwater catchment in the Spanish Pyrenees. *J. Hydrol.* **288**(3–4), 299–311.

Hudson, P. F. (2003) Event sequence and sediment exhaustion in the Lower Pánuco basin, eastern Mexico. *Catena* **52**, 57–76.

Mossa, J. (1996) Sediment dynamics in the lowermost Mississippi River. *Engng Geol.* **45**, 457–479.

Regüés, D., Balasch, J. C., Castelltort, X., Soler, M. & Gallart, F. (2000) Relación entre las tendencias temporales de producción y transporte de sedimentos y las condiciones climáticas en una pequeña cuenca de montaña mediterránea (Vallcebre, Pirineos Orientales). *Cuadernos de Investigación Geográfica* **26**, 41–65.

Rovira, A., Batalla, R. J. & Sala, M. (2004) Fluvial sediment budget of a Mediterranean River: The Lower Tordera (Catalan Coastal Ranges, NE Spain). *Catena* **60**, 19–42.

Walling, D. E. (1984) Dissolved loads and their measurement. In: *Erosion and Sediment Yield: Some Methods of Measurements and Modelling.* (ed. by F. Hadley & D. E. Walling), 111–117. GeoBooks, London, UK.

Williams, G.P. (1989) Sediment concentration versus water discharge during single hydrologic events in rivers. *J. Hydrol.* **111**, 89–106.

4 Bed load Transport and River–Channel Processes

Estimating the transport of bed material at low rate in gravel armoured channels

MICHAEL CHURCH & MARWAN HASSAN
Department of Geography, The University of British Columbia, Vancouver, British Columbia V6T 1Z2, Canada
mchurch@geog.ubc.ca

Abstract Classical sediment transport formulae represent high flux conditions. In contrast, bed material transport in most gravel-bed channels remains at very low rates most of the time, with only partial exchange of local bed sediments. Armouring and aggregated clast structures develop that are critical for the persistence of the low transport rate and for the stability of the channel. Classical formulae developed for fully mobile conditions grossly overpredict transport in this state. The transport cannot, in general, be calculated merely by an adjustment of bed material parameterization to reflect the surface armour, though that may work for specific histories of bed development. There is at present no general formulation available for bed material transport in channels with low transport and a structured bed.

Key words armoured bed; bed material transport; clast structures; gravel-bed rivers; sediment transport computation

INTRODUCTION

The transport of bed material in gravel bed channels occurs, most of the time, at very low rates, when it happens at all. Finer material is preferentially entrained and transported. Consequently, a distinctive surface gravel layer develops consisting of material that is, on average, coarser than that of the bulk material beneath (Parker & Klingeman, 1982; Wilcock & McArdell, 1993). In contrast, classical sediment transport formulae have been developed, mainly from flume studies, for high flux conditions in which there is little or no differentiation of the surface. It remains difficult to measure bed material flux, most of which occurs as bed load. Finding some means to estimate the transport in a relatively reliable way therefore remains an important undertaking.

Classical formulae index sediment character by a single reference size, usually D_{50} of the bulk sediment in the stream bed, and they invariably overestimate the actual sediment transport at low rates in gravel-bed channels. "Mixed size" formulae, capable of calculating transport for individual fractions of the bed material (Einstein, 1950; Egiazaroff, 1965; Parker *et al.*, 1982) and formulae calibrated to the bed surface grain size distribution (Parker, 1990; Wilcock & Crowe, 2003) appear not to be generally successful either. It is therefore important to examine further the character of gravel surfaces and their influence upon sediment transport. The objective of this paper is to attempt to rescue a more reasonable computational approximation for sediment transport than is afforded by straightforward application of available formulae.

THEORETICAL DEVELOPMENT

Dietrich *et al.* (1989) supposed that the degree of surface coarsening of a gravel bed reflects the degree to which the input sediment is "starving" the transport process—that is, the capacity of the stream, locally, to transport sediment, exceeds the input supply—so that the bed is winnowed to the point that additional sediment is difficult to entrain as the result of the increase in surface grain size. They explored this conjecture using a conceptual sediment transport model of the form:

$$q_b = k(\tau_b - \tau_c)^n \qquad (1)$$

in which q_b is the sediment transport per unit width, k is a constant of proportion (which covers things like sediment density), τ_b is the effective stress at the stream bed (the grain shear), and τ_c is the critical threshold stress to move the bed sediment. n is an exponent, usually 3/2. The well-known Meyer-Peter & Müller (1948) equation has exactly this form. The ratio of sediment transport under two different conditions, q_{b*}, can be written:

$$q_{b*} = [(\tau_b - \tau_{cs})/(\tau_b - \tau_{ct})]^n \qquad (2)$$

in which τ_{cs} and τ_{ct} represent the threshold stresses for the two conditions, in particular, for entrainment from the surface sediment, and for entrainment of the actually transported load. If the size distribution, hence threshold stress, of the load is not known, it might be possible to substitute τ_{cb}, the threshold stress for the bulk bed material. At least this should work under the assumption (implicit in classical formulations) that all sizes in the bed are equally likely to be transported.

Define a dimensionless value of the shear stress, $\tau_* = \tau/\gamma_s' D_{50}$, in which γ_s' is the submerged specific weight of the sediment and D_{50} is the median grain size of the relevant population. τ_* is the Shields number. Inverting it and introducing critical values for the threshold of material entrainment, we have $\tau_c = \tau_{*c}\gamma_s' D_{50}$. Using this expression, equation (2) can be rewritten (Dietrich *et al.*, 1989) as:

$$q_{b*} = [\{(\tau_b/\tau_{ct}) - (D_{50s}/ D_{50t})\}/\{(\tau_b/\tau_{ct}) - 1\}]^{3/2} \qquad (3)$$

where we now replace n with its customary value. This equation shows how the ratio of grain sizes affects the sediment transport. If we replace D_{50t} by D_{50b}, the median size of the bulk bed material, then the grain size ratio is the familiar armour ratio, D_{50s}/D_{50b}. Transport declines rapidly as the ratio becomes larger but it is also, of course, affected strongly by the ratio τ_b/τ_{ct}, which indicates how close to threshold the shear stress is.

Further, as $q_{b*} \rightarrow 0$, $D_{50s}/D_{50t} \rightarrow (\tau_b/\tau_{ct})$, from which we can obtain the expected ratio of surface to transported grain sizes (or the expected armour ratio, under the hypothesis of equal mobility). Inspection of equation (2) shows that q_{b*} properly approaches 0 for usual values of τ (or τ_*) only when τ_{cs} increases, so bed surface condition is seen to control the adjustment process. Considering that $\tau_{cs} = \tau_{*cs}\,\gamma_s' D_{50s}$, it is clear that τ_{cs} can increase by one of two mechanisms. Either surface grain size can increase (the commonly observed adjustment, creating the surface armour layer), or the Shields number can rise. For widely graded sediment mixtures (i.e. typical streambed sediments), τ_{*c} is conventionally assigned the value 0.045 (Miller *et al.*, 1977; Komar & Li, 1988; Wilcock & Southard, 1988). Meyer-Peter & Müller (1948) estimated 0.047. The possibility that it may be variable has not been widely considered.

We examine the possibility by rearranging equation (2) instead in the form:

$$\tau_{cs} = \tau_b (1 - q_{b*}^{2/3}) + \tau_{ct}\, q_{b*}^{2/3} \tag{4}$$

which gives $\tau_{cs|qb*}$ when q_{b*} is the observed transport ratio. In this equation, we infer by considering the limits $0 \leq q_{b*} \leq 1$ and recalling that τ_b is the grain shear, that the first term on the RHS represents the portion of the shear stress that is not associated with the sediment transport, while the second term is the portion that is. Given $\tau_{cs|qb*}$—that is, the actual threshold stress for the surface—we can obtain a predicted Shields number:

$$\tau_{*c|qb*} = \tau_{cs|qb*}/\gamma_s' D_{50s} \tag{5}$$

Alternately, we can determine what would be the value of D_{50s} if all of the adjustment for the reduced transport were taken up by surface armour:

$$D_{50s|qb*} = \tau_{cs|qb*}/0.045\, \gamma_s' \tag{6}$$

In computing this hypothetical quantity, it is understood that there is no adjustment to the conventional critical Shields number.

In the analyses that follow, we use ratios of observed transport, not transport computed from a bed load formula, so that our results are free of any systemic bias that a particular formula might introduce. We use equation (2) and the equations that we derive from it to study the effect of changing bed surface material grain size in the formulation of sediment transport.

EXPERIMENTAL STUDIES

We have examined the behaviour of the ratio q_{b*} using results from two series of experiments, one series conducted without sediment feed (Church *et al.*, 1998) and one with (Hassan & Church, 2000). In both cases, the actual transport was well below the computed capacity of the flow for the bulk sediment in the flume and the transported sediment was significantly finer than the total distribution of that in the flume. Details of the experiments, which were conducted in a 10 × 0.8 m flume, are reported in the above references. They were all run with low excess shear stress under which surface armour develops. Max[τ_b/τ_{cs}] = 2.62, but almost all values were <2 (data in Table 1). τ was determined from the depth-slope product once uniform flow was confirmed since we are interested in the average value applied to the bed. Sidewall corrections were performed, so the quoted values, τ_b, nominally refer to grain shear.

In both sets of experiments, individual runs continued for about 100 h. In the non-feed experiments, sediment transport declined dramatically—by three orders of magnitude—as the surface armoured and became structured. In the feed experiments, the flume was run without feed for about 16 h to develop an armoured bed, after which sediment feed, using the same mixture as the bulk sediment in the bed of the flume, was initiated and maintained at a rate that was some fraction (see Table 1) of the rate of evacuation of sediment out of the flume during the immediately preceding period (the "reference transport rate"). In these circumstances, the coarser sizes were abandoned in the flume. Although the transport stabilized after some hours for all but the lowest and highest feed rates (respectively 50% and 150% of the reference rate), the transport process was not, then, a true equilibrium process. No primary bedforms at larger than grain scale were observed. In both sets of experiments, degradation was of the order of 1 grain diameter. Experiments were repeated and shown to be repeatable.

We measured the grain size distribution on the bed surface and of the transported sediment at the end of each experiment and we use those figures (Table 1) to investigate the effect of the bed surface on the transport. We estimated q_{b*} by assuming that the observed transport during the first hour, q_{b1}, when flow was initiated over the unmodified bed material, represented full transport, and then formed ratios for later values of q_b. For example, $q_{b*96} = q_{b96}/\ q_{b1}$ is the transport ratio after 96 h based on the observed transport, q_{b96}, at that time. In addition to our own data, we show the summary results of Dietrich *et al.* (1989). Their runs, of 6 to 8 h length, were much shorter than ours.

Table 1 Data and computed quantities in the experiments[1].

Experiment	$\boldsymbol{D_{50s}}$ (mm)	$\boldsymbol{D_{50t}}$ (mm)	D_{50s}/D_{50t}	τ_b[5] (Pa)	τ_{cs}[6]	τ_{ct}[6]	τ_b/τ_{cs}	τ_b/τ_{ct}	$\boldsymbol{q_{b96}}$ (kg/h[8])
UBC non-feed experiments; Series I, D_{50b} = 2.2 mm; Series II, D_{50b} = 3.2 mm;									
I 3.1	4.0	1.3	3.00	4.21	2.91	0.95	1.45	4.43	0.11
4.1	3.9	1.0	3.84	4.91	2.84	0.73	1.73	6.73	0.097
5.1	3.0	0.85	3.52	3.53	2.19	0.62	1.61	5.69	0.049
6.1	4.7	1.3	3.59	5.98	3.42	0.95	1.75	6.29	0.10
7.1	4.1	1.7	2.47	7.26	2.99	1.24	*2.43*[7]	5.85	0.13
II 8.1	4.2	0.97	4.30	4.51	3.06	0.71	1.47	6.35	0.094
9.1	4.2	1.1	3.77	5.42	3.06	0.80	1.77	6.78	0.041
10.1	5.1	1.5	3.37	5.41	3.71	1.09	1.46	4.96	0.049
11.1	4.8	1.4	3.44	6.18	3.50	1.02	1.77	6.06	0.065
12.1	5.5	3.0	1.85	6.67	4.01	2.19	1.66	3.05	0.028
13.1	3.2	1.0	3.14	3.21	2.33	0.73	1.38	4.40	0.061
H 3.1	1.9	0.75	2.53	3.61	1.38	0.55	*2.62*	6.56	0.053
3.2	3.4	1.1	3.05	4.21	2.48	0.80	1.70	5.26	0.079
1	4.2	1.2	3.50	5.18	3.06	0.87	1.69	5.95	0.12
Average			3.11±0.20				1.62±0.14[7]	5.60±0.29	
UBC feed experiments (proportional feed rate in parentheses: see text); D_{50b} = 1.4 mm									
HM-1 (0%)	4.5	0.92	4.89	4.00	3.28	0.67	1.22	5.97	0.16
HM-3 (50%)	3.7	1.2	3.08	3.89	2.70	0.87	1.44	4.45	0.81
HM-5 (75%)	3.5	1.25	2.80	4.06	2.55	0.91	1.59	4.48	0.91
HM-4 (100%)	3.2	1.3	2.46	3.89	2.33	0.95	1.67	4.11	1.1
HM-6 (150%)	2.7	1.4	1.93	4.01	1.97	1.02	2.03	3.94	1.5
HM-7 (0%)	2.8	1.1	2.64	3.30	2.06	0.78	1.60	4.24	0.11
HM-8 (50%)	2.7	1.1	2.45	3.37	1.97	0.80	1.71	4.20	0.33
HM-9 (100%)	2.4	1.2	2.00	3.37	1.75	0.87	1.92	3.85	0.65
Average							1.65±0.26	4.41±0.67	
Dietrich *et al.* (1989) experiments[2]; D_{50b} = 3.7 mm, $< D_{50s}/R >$ = 0.041±0.006[4]									
17.4	3.7	3.7	1.00	5.2	2.70	2.70	1.93	1.93	
6.1	4.3	1.4[3]	3.10	4.7	3.13	1.02	1.50	4.61	
1.7	4.9	1.16[3]	4.22	3.9	3.57	0.84	1.09	4.64	

[1] Bold heading indicate measured quantities. Normal headings indicate computed ones. Measured quantities specified in the original papers are not repeated here.

[2] Experiment designation indicates sediment feed rate in g min^{-1} cm^{-1}.

[3] Calculated from D_{50s}/D_{50t}, read as the average of a range from their Fig. 3.

[4] R is hydraulic radius; D_{50s}/R is relative roughness.

Table 1 (cont.) Data and computed quantities in the experiments[1].

Experiment	q_{b^*obs96}	$q_{b^*(D50t)}$	$\tau_{cs\|qb^*}$ Pa	$D_{50s\|qb^*}$ mm	$D_{50s\|qb^*}/D_{50s}$	$\tau_{*c\|qb^*}$
Series H, D_{50b} = 1.4 mm, $<D_{50s}/R>$ = 0.080±0.017				UBC non-feed experiments		
I 3.1	0.0016	0.27	4.17	5.7	1.43	0.064
4.1	0.00085	0.36	4.87	6.7	1.72	0.077
5.1	0.0016	0.31	3.49	4.8	1.60	0.072
6.1	0.00077	0.36	5.94	8.2	1.74	0.078
7.1	0.00076	*0.58*	7.21	*9.9*	*2.41*	*0.109*
II 8.1	0.0031	0.24	4.43	6.1	1.45	0.065
9.1	0.00074	0.37	5.38	7.4	1.76	0.079
10.1	0.00057	0.25	5.38	7.4	1.45	0.065
11.1	0.00065	0.37	6.14	8.4	1.75	0.079
12.1	0.00024	0.45	6.65	9.1	1.65	0.075
13.1	0.00048	0.23	3.19	4.4	1.38	0.062
H 3.1	0.0016	*0.62*	3.57	4.9	*2.58*	*0.116*
3.2	0.0016	0.37	4.16	5.7	1.68	0.076
1	0.0020	0.35	5.11	7.0	1.67	0.075
Average	0.0012±0.0008	0.33±0.07			1.61±0.15	0.072± 0.006`
$<D_{50s}/R>$ = 0.053±0.010 (first ser.)/ 0.050±0.004 (second ser.)				UBC feed experiments		
HM-1 (0%)	0.0012	0.10	3.96	5.4	1.20	0.054
HM-3 (50%)	0.0058	0.25	3.79	5.2	1.41	0.063
HM-5 (75%)	0.0069	0.34	3.94	5.4	1.54	0.070
HM-4 (100%)	0.0085	0.39	3.76	5.2	1.63	0.073
HM-6 (150%)	0.012	0.57	3.85	5.3	1.46	0.088
HM-7 (0%)	0.0012	0.34	3.27	4.5	1.61	0.072
HM-8 (50%)	0.0037	0.40	3.31	4.5	1.67	0.076
HM-9 (100%)	0.0077	0.52	3.27	4.5	1.88	0.084
				Dietrich *et al.* (1989) experiments		
17.4	1.00	1.00	2.70	3.7	1.0	0.045
6.1	0.35	0.27	2.87	3.9	0.91	0.041
1.7	0.098	0.039	3.24	4.4	0.90	0.041

[5] Adjusted for sidewall effects from measurements of water surface slope and flow depth. We prefer this measure as it gives an average over the bed. Uniform flow was checked.
[6] From Shields' relation, using observed grain size
[7] Eccentric figures in italic. Average excludes those figures.
[8] In an 0.80 m wide flume; figures not adjusted. Reported in convenient units.

Our results, when correlated with the appropriate measure of effective grain stress, that is $\tau_{eff} = \tau_b - \tau_{cs|qb^*}$, scale appropriately (Fig. 1). $\tau_{cs|qb^*}$ is calculated from equation (4) and represents the actual surface threshold condition. It is, however, back-calculated from knowledge of the sediment transport and would not be available for a prediction. Our problem, then, is to investigate whether there is some *a priori* means by which this quantity, which depends on the entire surface state, can be estimated or adjusted for in sediment transport predictions.

We first studied the effect of substituting bed material bulk grain size for the transport size in equation (3) (Fig. 2). It is apparent that the bulk sediment does not index the transported sizes adequately, although in many cases a relatively major correction was achieved. The transport process was not a self-similar phenomenon

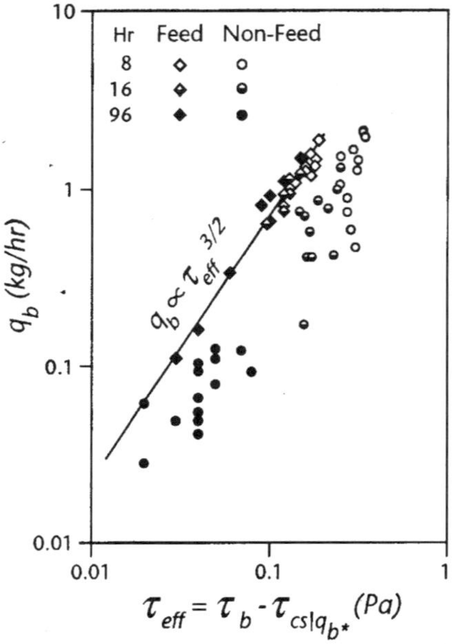

Fig. 1 Correlation of bed load transport observed in our experiments with the true grain stress. The non-feed experiments scatter below the correlation for the feed experiments. Reasons for the scatter include a probably variable history of surface development and sampling variability for the very low transport rates. Notice, in particular, that the surface development, as indexed by $\tau_{cs|qb*}$, correctly adjusts for the changing transport with time.

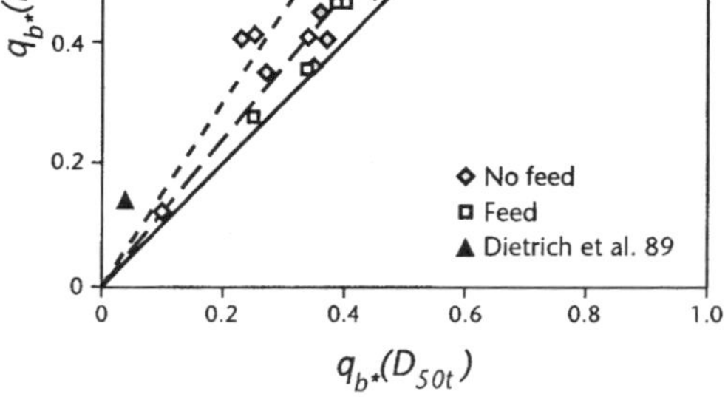

Fig. 2 Comparison of apparent reduction in transport, as calculated from equation (3), when D_{50b} is used to estimate transported sizes with that when D_{50t} is used.

with respect to grain size, even under conditions of sediment feed (since the feed was identical in size distribution to the bulk sediment in the bed), implying that sediment exchange between the transport and the bed continued throughout our runs. The result is disappointing, since the armour ratio is an accessible measure of bed state in a gravel-bed stream channel.

We next studied the effect of using D_{50} of the transported grain sizes in equation (3). Figure 3 illustrates the results. In our experiments, observed values, $q_{b\text{-}obs}$, hence $q_{b*\text{-}obs}$, remain far below $q_{b*}(D_{50t})$—the predicted transport using the observed transported grain size. In comparison, the results of Dietrich *et al.* (1989) fall close to the predicted values. Not surprisingly, the non-feed experiments generally fall even farther from the expected results than the feed experiments. We suppose that the evolution of the bed surface, and differences in evolution between non-feed and feed cases, account for the observations.

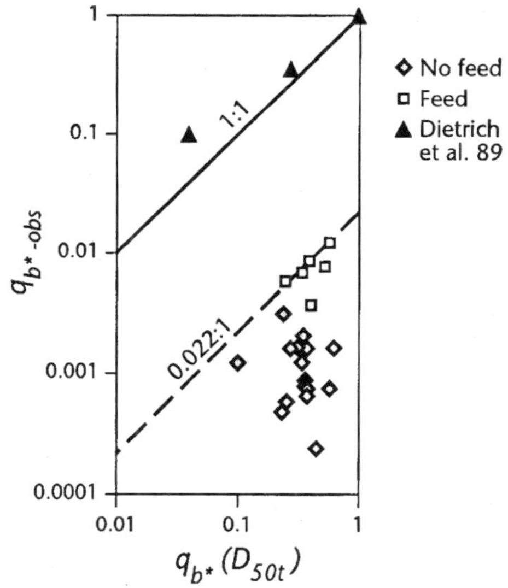

Fig. 3 Concordance diagram for q_{b*obs} *vs* $q_{b*pred}(D_t)$. The Dietrich *et al.* (1989) results slightly exceed the expected result. This is a consequence of the form of equation (2) that is used. Using a slightly different representation (arising from a different sediment transport function) Dietrich *et al.* (1989) obtained even closer agreement.

ANALYSIS

We first consider the specification of τ_{*c}. Although conventionally assigned the value 0.045, that specification hides wide variation in observed values. Buffington & Montgomery (1997) found that substantial variation in results is systematically associated with the means by which the value is determined.

There are two distinct approaches to determine τ_{*cs}. Estimates may be based on observed sediment transport or upon the observed conditions for mobilization of sediments from the bed. In the former case, estimates have been based either on extrapolating observed rates of sediment transport to zero, or on observing the largest grains trapped in transport (on the assumption that the largest grains indicate the competence of the flow). In the latter case, grains must, by some means, be confirmed to be entrained directly from the local bed. The two methods are not equivalent because it may take a lower force to sustain grains in transport than it does to entrain them. We are concerned with estimating τ_{*cs} for transport calculations, and we are tolerably certain, in most of our observations, that there is not a representative exchange process with the local bed, so it appears most appropriate to base estimates of τ_{*cs} on observed transport. This procedure also appears prudent for extension of results to rivers, where one cannot be sure that sediment being transported is derived from a local exchange process with the bed.

Buffington & Montgomery (1997) report a range $0.052 \leq \tau_{*cs}D_{50} \leq 0.086$ for reference transport data. This range covers most of our back-calculated observations of $\tau_{*c|qb*}$ (Table 1), but those of Dietrich *et al.* fall below that range and close to the classical value of 0.045. There are two possible interpretations of these results: either our data suffer from the range of inconsistencies identified by Buffington & Montgomery (1997) or various of their data are subject to the phenomena we are attempting to investigate. Regardless of the question of bias in our procedures, our methods were internally consistent. We conclude that the range of variation in τ_{*c} (or conversely, in q_{b*}) observed by us is real and that the phenomena underlying it may well explain a substantial part of the range detected by Buffington & Montgomery (1997).

We next examined the relation between $\tau_{*c|qb*}$ (as computed from equation (5)) and q_{b*} in order to determine whether q_{b*} has some effect on τ_{*c} (Fig. 4). There appears to be some systematic relation between q_{b*} and τ_{*c}, with *higher* values of relative transport being associated with higher values of τ_{*c}. Lower rates of feed are also associated with higher values of τ_{*c}. The second relation is expected. Presumably, at lower feed rates, there is a greater adjustment of the surface, by some means, to compensate the mobile sediment "deficit". However, the former result appears, at first sight, to be both counterintuitive and contrary to the second one. It must be recalled that we are dealing with ratios of actual sediment transport to the transport expected at full mobility. The ratio actually increases with feed (~ transport) rate (see Table 1) and, as that occurs, the surface becomes finer. D_{50s} is inversely associated with feed rate. Re-examining equation (5), we see that it is also inversely related to the estimate of $\tau_{*c|qb*}$, whereas $\tau_{cs|qb*}$ does not vary so systematically. Hence, there is an induced relation. Our experiments are insufficient in number, and not sufficiently well controlled, to pursue the association farther.

Figure 4 shows that, at the lower reference feed rate (0.64 kg h^{-1}), the relation with transport rate declines to reach the mean value of τ_{*c} for zero feed (i.e. zero incoming transport) around 0.072. For the higher reference feed rate (1.2 kg h^{-1}), however, the value declines into the range of classically quoted Shields numbers, reaching 0.054. The difference probably is related to the finer state of the bed surface (hence lower expected value of τ_{*c}) achieved with the higher reference feed rate.

We next examined the relation between D_{50s} and $D_{50s|qb*}$, using equation (6) to obtain the latter quantity. The result is portrayed in Fig. 5. D_{50s} in our experiments is finer than predicted. Furthermore, there appears to be a fairly consistent relation:

$$D_{50s} = 0.60\, D_{50s|qb*} \qquad (7)$$

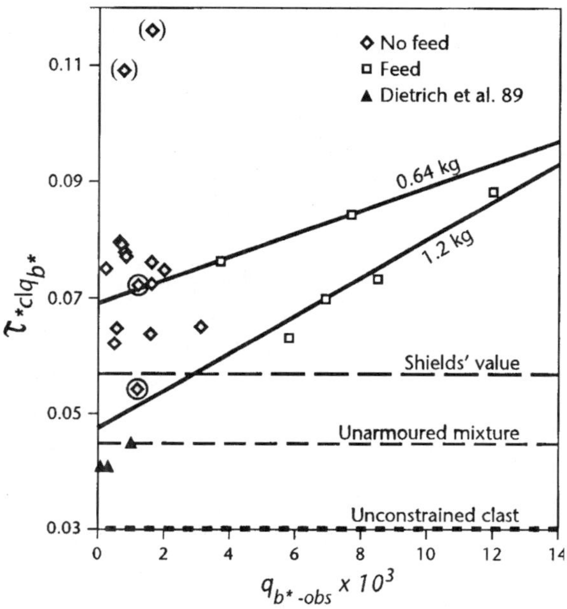

Fig. 4 Relation between $q_{b\text{-}*obs}$ and $\tau_{*c|qb*}$, with correlations for two reference rates of transport in our experiments. Sediment feed rates varied as proportions of the reference rate (see Table 1). The circled data represent the no-feed limit in each sequence of sediment feed experiments. Data in parentheses are anomalous.

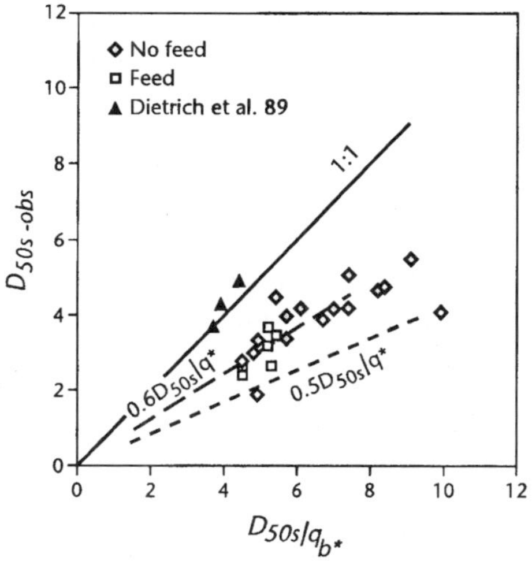

Fig. 5 Relation between D_{50s} and $D_{50s|qb*}$ from UBC experiments, with correlation and limit relations superimposed.

This result indicates that a portion of the adjustment to reduced sediment feed is taken up by some means other than surface armouring. We suppose that this is the reason why a simple adjustment of transport based on the ratio D_{50s}/D_{50t} is insufficient. The data of Dietrich *et al.* (1989) do not show a similar effect. Within the bounds of variation imposed by the choice of transport equation, hence the exact form of equation (2), their results show 1:1 agreement between D_{50s} and $D_{50s|qb*}$. A significant difference between their experiments and ours is that they began with established sediment feed onto the initial, unmodified sediment bed in the flume, so their bed immediately adapted to that imposed transport rate. The bed modification that occurred in the first 16 h of non-feed conditions in our experiments—conducted in order to establish something more akin to an actual gravel river bed—did not occur in their experiments.

We have previously essayed the question why τ_{*c} should vary systematically in conditions of low sediment transport (Church *et al.*, 1998; Hassan & Church, 2000), and why D_{50s} should depart from the calculated value for the incoming sediment load. We have ascribed the difference to the development of particle clusters, lines and nets amongst the larger clasts, all of which we have observed in the field and developed in our flume in both the non-feed and feed experiments analysed herein. At the higher rates of sediment feed, development was restricted beyond that which occurred in the first 16 h, but it was not destroyed. In general, we would not expect the structures to be broken until $\tau > \tau_c$ for the local bed structures, a level not reached in our experiments and not often reached in many gravel-bed channels, even though incoming finer sediment can be transported onward. Examined from the viewpoint of sediment mobilization, the rise in τ_{*c} can be interpreted in terms of structural strengthening of the bed by the imbricate and jammed nature of the stone structures. Smaller grains participate in such structural effects in only limited degree, but are effectively hidden by the large clast network. This is the reason why $\tau_{cs|qb*}$ differs from τ_{cs}.

Can the effect be predicted? We have obtained a roughly consistent relation between $D_{50s|qb*}$ and D_{50s} observed, a result that—if consistent—would permit a successful adjustment of sediment transport estimates. Our feed experiments cross the trend but show variance similar to that of the non-feed experiments (Fig. 5). Equation (7) then becomes a candidate adjustment. But the consistent way in which our experiments were initiated, and the non-conformity of the Dietrich *et al.* (1989) experiments, suggest that the result is by no means general. Our non-feed experiments represented fairly severe "starvation" of the channel and so it is likely that the correlation would not fall outside the envelope $D_{50s} = 0.50\ D_{50s|qb*}$ of our results, but there is no reason to suppose that results would be consistent from one flow history to another.

DISCUSSION

It is apparent from our investigation, and from related field work (Church *et al.*, 1998) that sediment transport in armoured gravel bed channels may fall well below what might be predicted from classical sediment transport relations. There appear to be circumstances (Dietrich *et al.*, 1989) in which an adjustment can be based on the grain size distribution in the load. The circumstances are those in which sediment transport arises from an equilibrium exchange of sediment between the transport and the modified bed surface. This remains a somewhat onerous constraint since the size distribution of the load usually is not known, but recent, "surface-based" sediment transport formulae might calculate the load in such circumstances.

We compared our results with predictions from the Meyer-Peter & Müller (1948) formula, and from the surface-based Parker (1990) formula (Fig. 6). The Meyer-Peter & Müller (1948) formula covers the condition of partial transport when grain size ratios in the range 2 to 3 are applied (that is, the reference grain size used in the formula is increased by a factor of 2 to 3 above the mixture size). These values are generally larger than the actual ratios D_{50s}/D_{50b} in the flume, which were 1.72 ± 0.52 for the non-feed experiments and 2.28 ± 0.49 for the feed experiments. The ratios D_{50s}/D_{50t}, on the other hand were 3.11 ± 0.20 and 2.78 ± 0.93, respectively, which are

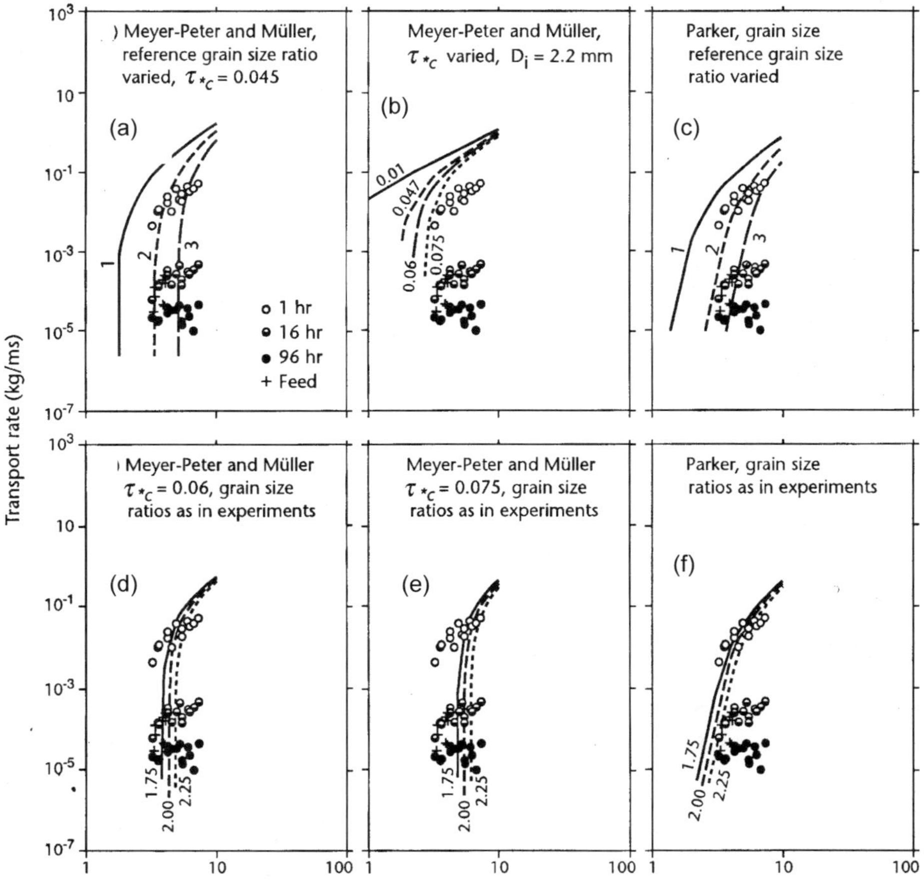

Fig. 6 Predictions from sediment transport formulae superimposed upon our experimental data. (a) Meyer-Peter & Müller (1948) formula with various reference grain size ratios; (b) same, critical Shields number varied; (c) Parker (1982) formula with reference grain size ratios; (d) Meyer-Peter & Müller (1948) formula with τ_{*c} = 0.06 and grain size specification adjusted to cover the range of surface conditions observed in the experiments; (e) same, except τ_{*c} = 0.075; (f) Parker formula with grain size specification adjusted to cover the range of surface conditions observed in the experiments. The reference grain size ratios are most appropriately considered as ratios D_m/D_{50t} for Meyer-Peter and Müller (1948), where D_m is mean grain size, and as D_{50s}/D_{50t} for Parker (1982), according to the original formulations. Note the different scaling of the ordinate and abscissa.

closer to the observed best fit values. Adjusting τ_{*c} (Fig. 6(b)) does not effect the desired adjustment: transport remains overestimated, even when τ_{*c} is raised to 0.075, although the threshold for motion appears, then, to be well estimated. The continuing influence of the intact surface structure is not successfully estimated by the formula. The Parker formula (Fig. 6(c)) estimates the 1 h transport tolerably well with a grain size ratio of 2 (i.e., the size employed in the formula is $2D_{50b}$), but is on the low side of the final (96 h) transport with a ratio of 3.

As a further test, we applied the formulae using critical tractive forces and ratios D_{50s}/D_{50t} actually observed in our experiments. Results are shown in Figure 6(d)–(f). The formulae cross the trend of the experimental data. In general, they predict a much more rapid rate of increase in transport, the threshold once crossed, than occurred in our experiments. All of our data cluster in the region near the initiation of motion indicated by the formulae. Again, the variation observed in the experiments is plausibly ascribed to the effect of surface structures in reducing entrainment and transport.

There appears to be no basis in the observations to predict the appropriate corrections for this condition. Adjustments based on grain size are insufficient when structural channel bed arrangements constrain grains and present additional resistance to flow. In a previous analysis of the experimental data presented in this paper, Hassan & Church (2000) showed that from 15 to 50% of the shear force exerted on the bed was dissipated by the structural arrangement of the bed. Before transport can be computed, such effects must be accounted for, as is done *ex post facto* in Fig. 1. There is, at present, no general means by which to effect the adjustment because surface structure effects are not formulated.

CONCLUSIONS

It appears from our investigations that, although bed material transport at low rate—absent structural arrangement of the surface—may be predicted by adjustments of bed load formulae to account for the surface armour (cf. the results of Dietrich *et al.*, 1989), when flow history has created a structurally reinforced surface the appropriate adjustment is not covered by extant formulations. Whether viewed as an adjustment of form resistance (so that the effective stress is smaller than calculated) or as an increase in the reluctance of sediments to be moved, transport remains smaller—possibly dramatically smaller—than formulated estimates. Our experiments cover too narrow a range of conditions to reveal a general correction, although an appropriate ratio adjustment of surface grain size would approximately work for our experiments. A more fundamental approach may be accessible through a structural description of the developed bed surface or through advances in estimating form resistance on gravel streambeds.

There remain complexities yet to be discovered in the transport of gravel in rivers.

Acknowledgments Research described in this paper was supported by the Natural Sciences and Engineering Research Council of Canada. The paper was written whilst the senior author was a Visiting Research Fellow at Hatfield College and the Department of Geography at the University of Durham, UK. The paper was a keynote presentation for the International Conference on River and Catchment Dynamics held in Solsona, Spain on 15–20 May, 2004. We appreciate the careful review by Tom Lisle, which led us to clarify many points.

REFERENCES

Buffington, J. M. & Montgomery, D. R. (1997) A systematic analysis of eight decades of incipient motion studies, with special reference to gravel-bedded rivers. *Water Resour. Res.* **33,** 1993–2029.

Church, M., Hassan, M. A. & Wolcott, J. F. (1998) Stabilizing self-organized structures in gravel-bed stream channels: field and experimental observations. *Water Resour. Res.* **34**, 3169–3179.

Dietrich, W. E., Kirchner, J. W., Ikeda, H. & Iseya, F. (1989) Sediment supply and the development of the coarse surface layer in gravel-bedded rivers. *Nature* **340**, 215–217.

Egiazaroff, I. V. (1965) Calculation of non-uniform sediment concentrations. *J. Hydraul. Div. ASCE* **91**(4), 225–247.

Einstein, H. A. (1950) The bedload function for sediment transport in open channel flows. *US Dept. Agric., Soil Cons. Serv, Tech. Bull. 1026.*

Hassan, M. A. & Church, M. (2000) Experiments on surface structure and partial sediment transport on a gravel bed. *Water Resour. Res.* **36**, 1885–1895.

Komar, P. D. & Li, Z. (1988) Applications of grain pivoting and sliding analyses to selective entrainment of gravel and to flow competence evaluations. *Sedimentology* **35**, 681–695.

Meyer-Peter, E. & Müller, R. (1948) Formulas for bedload transport. In: *Proc. IAHSR* (2nd Meeting, Stockholm, Sweden), 39–64.

Miller, M. C., McCave, I. N. & Komar, P. D. (1977) Threshold of sediment motion under unidirectional currents. *Sedimentol.* **24**, 507–527.

Parker, G. (1990) Surface-based bedload transport relation for gravel rivers. *J. Hydraul. Res.* **28**, 417–436.

Parker, G. & Klingeman, P. C. (1982) On why gravel bed streams are paved. *Water Resour. Res.* **18**, 1409–1423.

Parker, G., Klingeman, P. C. & McLean, D. G. (1982) Bedload and size distribution in paved gravel-bed streams. *J.Hydraul. Div. ASCE* **108**, 544–571.

Wilcock, P. R. & Southard, J. B. (1988) Experimental study of incipient motion of in mixed-size sediment. *Water Resour. Res.* **24**, 1137–1151.

Wilcock, P. R. & McArdell, B. W. (1993) Surface-based fractional transport rates: mobilization thresholds and partial transport of a sand-gravel sediment. *Water Resour. Res.* **29**, 1297–1312.

Wilcock, P. R. & Crowe, J. C. (2003) Surface-based transport model for mixed-size sediment. *J. Hydraul. Eng.* **129**, 120–128.

Geomorphological Processes and Human Impacts in River Basins (Proceedings of the International Conference held at Solsona, Catalonia, Spain, May 2004). IAHS Publ. 299, 2005.

Geomorphologic units and sediment transport in a very low relief basin: Rio Quequén Grande, Argentina

LILIANA B. TERUGGI[1], GUSTAVO A. MARTINEZ[2], PAOLO BILLI[3] & EMANUELE PRECISO[3]

1 *Dipartimento di Ingegneria Civile, Università degli Studi di Firenze, via Santa Marta 3, 50139, Firenze, Italy*
lteruggi@dicea.unifi.it

2 *Centro de Geologia de Costas y del Cuaternario, Universidad Nacional de Mar del Plata, Funes 3350, Mar del Plata 7600, Argentina*

3 *Dipartimento di Scienze della Terra, Università degli Studi di Ferrara, Corso Ercole I d'Este n.32, 44100 Ferrara, Italy*

Abstract The aim of this work was to identify the geomorphological units within the drainage network of the Rio Quequén Grande low relief basin, situated in the southeast section of the province of Buenos Aires, and to quantify the present sediment transport rate. For this purpose field surveys and satellite images were analysed. The former included measurement of bed material grain size and channel hydraulic geometry, in order to set up a bed and a suspended load measurement programme. Six main geomorphic units were identified: Ranges, Perirange Aeolian Hillocks, Hillocks with Lagoons, Relic Hills, a Poor Drainage Alluvial Plain and an Alluvial Plain. Bed load discharge and suspended sediment concentration measured during a bankfull flood are very low, given the relatively high stream power of 287 W m^{-1}. The measured bed load discharge was three to five orders of magnitude lower than six applied bed load equations, presumably due to low sediment availability.

Key words Argentinian Pampas; geomorphology; sediment transport

INTRODUCTION

Few studies have dealt with channel processes and sediment transport in very low gradient rivers in humid climate conditions. The majority of research to date has focused on large alluvial rivers such as the Mississippi (Belt, 1975; Winkley, 1982) and the Ganges–Brahmaputra (Coleman, 1969; Abbas & Subramanian, 1984), since their dynamics are directly connected to human activities that normally develop densely on lowland alluvial or coastal plains. The Argentinian Pampas are almost flat, and the drainage and river systems are poorly developed because of a very low regional gradient. Characteristics of the area include normally very low drainage density (0.2 km km^{-2}) and long resident time of runoff on the land surface. The aim of this study is to analyse the role of geomorphology in controlling the drainage network structure and to quantify the present sediment yield of a low relief basin.

STUDY AREA

The Rio Quequén Grande basin is located in the southeast section of the province of Buenos Aires, Argentina, and is a typical example of the rivers that drain the flat pampas of the region (Fig. 1). The basin has a surface of about 9940 km^2, but only 2.5% of it is taken up by low-relief Sierras (maximum elevation = 510 m a.m.s.l.), while the average slope gradient is about 0.03. Most of the basin consists of an almost flat plain (the pampas, generated from Quaternary wind-blown deposits) with a gradient of about 0.003. The river has cut its channel into the aeolian deposits (loess and reworked loess), that underlie the majority of the region, and has a slightly entrenched morphology extending almost as far downstream as its outflow into the Atlantic Ocean, near the city of Necochea. The main channel is carved into bedrock and does not have a clearly distinctive alluvial plain. Depositional bodies such as longitudinal or point bars are uncommon and in places the channel profile is stepped due to the occurrence of hard "caliche" layers in the bedrock. In the upstream reaches bed material consists of sand and gravel (D_{50} ranges from 0.09 to 9.92 mm), while in the downstream reaches, sand predominates (D_{50} ranges from 0.08 to 0.32 mm), though coarser particles stemming from erosion of the "caliche" layers may be present as well (Teruggi *et al.*, 2003).

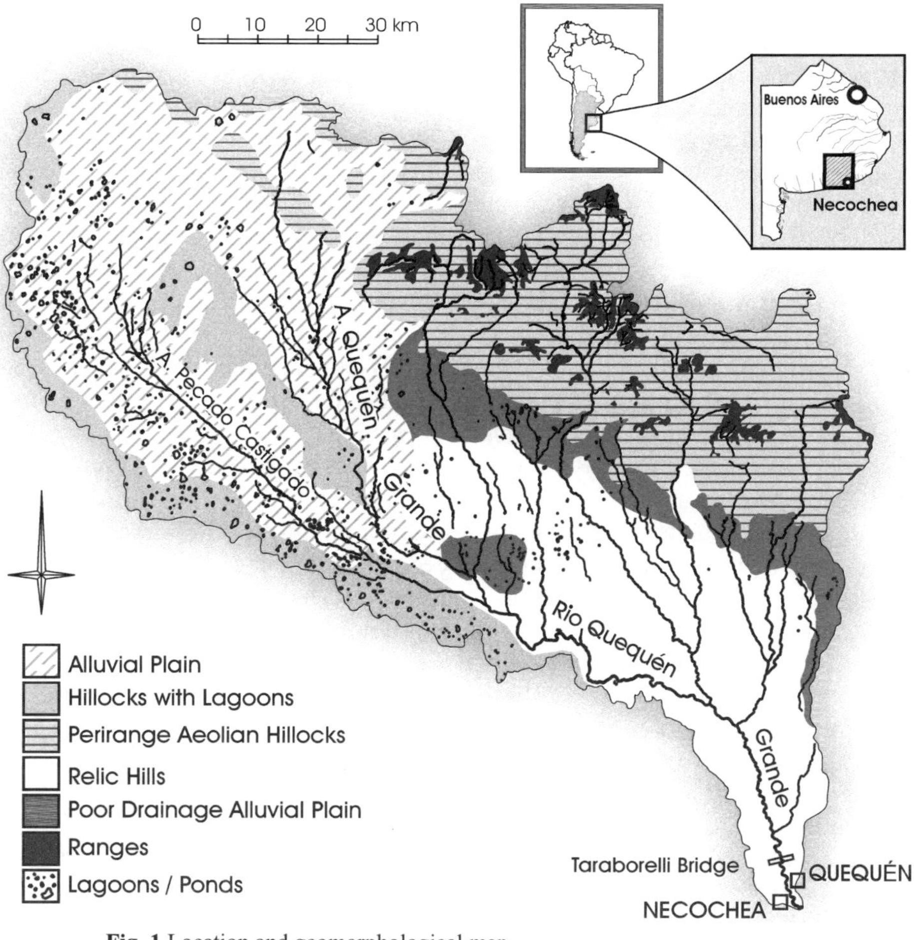

Fig. 1 Location and geomorphological map.

Mean annual precipitation in the basin is about 800 mm, with the highest monthly values occurring in the summer (Varela & Teruggi, 2002). Unfortunately there are many gaps in the time series of daily flow data, often due to damage caused to the flow gauge installation during peak discharge. Nevertheless, processing of the available data indicates a mean daily discharge of about 16 $m^3 s^{-1}$, while the maximum and minimum recorded discharges are 1700 and 1.6 $m^3 s^{-1}$, respectively.

BASIN GEOMORPHOLOGY

The geomorphology of the Pampian region is difficult to ascertain because of the very low relief and the superposition of erosional depositional forms of both aeolian and fluvial origin. These processes were governed by climatic oscillations that occurred during the Upper Cenozoic. Glacial and stadial events associated with cold and dry conditions in the region resulted in landscape re-activation and the development of aeolian forms. Interglacial and interstadial events, which are characterized by warmer and wetter conditions, led to landscape stabilization, with pedogenesis being the dominant process (Tricart, 1973).

The basin geomorphology was analysed by means of field surveys and satellite images. Geomorphic units were classified according to their morphology, lithology, genesis, slope and relief characteristics. Image analysis processing using Landsat TM 7+ images was carried out to identify different geomorphic environments. Statistical analysis between bands was performed and the coefficient of correlation calculated. A variety of false colour images were created by fusing different TM bands. These were used for preliminary visual analysis. The most useful integration technique in terms of the geomorphologic objectives was Principal Component Analysis, which enhances aeolian features and structural lineaments. Image analysis techniques are the best tools for geomorphological mapping in the area because low relief geomorphological features become readily visible. Complementary, topographic maps (scale 1:50 000) were used to extract physiographic data. Finally, field work allowed us to check the preliminary map and to add geomorphic information.

The following is a brief description of the main geomorphological units identified in the area (Fig. 1):

(a) Ranges. This unit includes the low-elevation relief of the Tandilia System, whose morphology is controlled by uplifted bedrock blocks consisting of ortoquartzitic or igneous–metamorphic crystalline basement. These hills make up the basin headwater and the most evident divide. Here, stream flow may be ephemeral and relatively steep and narrow valleys develop.

(b) Perirange Aeolian Hillocks. These hillocks have a complex and generally rounded morphology and are located around the ranges, forming an undulating landscape. They originated as a result of the bedrock ranges, having acted as an orographic barrier to west and southwestern palaeowinds, increasing local loess sedimentation during the cycles of major aridity in the Late Pleistocene and Early Holocene. In some hillocks close to the ranges, the crystalline basement is near the surface and is covered by a thin layer of loess.

(c) Hillocks with lagoons. This unit consists of three elongated reliefs running in a west–east direction and perpendicular to the regional gradient. These reliefs make up the divide of the Rio Quequén Grande tributaries of the west sector and are responsible for the eastward deviation of the upstream portion of A. (arroyo) Pecado Castigado and A. Quequén Grande. The divide is formed by a series of aligned lagoons, the product of an old deflation-accumulation relief (Martínez, 1997).

(d) Relic Hills. The relic hills represent an old fluvio-aeolian landscape preserved by calcareous crust, which is crossed by the middle and lower reaches of the Rio Quequén Grande. The remains of a planation level and two old terraces dissected by incision processes were identified near the main river course. These indicate a significant lowering in the channel base level.

(e) Poor Drainage Alluvial Plain. This landscape has a very low gradient (0.2%), with flat to smoothly undulating morphology. The drainage network consists of a series of minor ephemeral, subparallel streams and small subcircular ponds which vary in size from 0.014 to 1.10 km^2.

(f) Alluvial Plain. There is an alluvial plain in the middle and upper reaches of the A. Quequén Grande and A. Pecado Castigado and their tributaries.

(g) Parabolic palaeodunes of aeolian origin are superimposed onto the plain. Dating to the Late Holocene, they have a relative relief of 0.50 m and a west–east direction. This stabilized dune field is now completely vegetated and inactive. Only first and second order tributaries are controlled by this relic landscape.

SEDIMENT TRANSPORT

A measuring station for recording water and sediment discharge was established at Taraborelli Bridge (Fig. 1), 10 km upstream of the river outlet. Field surveys included the measurement of channel hydraulic geometry (Fig. 2) and bed material texture (Teruggi *et al.*, 2003). Bed load transport was measured with a Helley-Smith sampler, while suspended sediment samples were taken with a US DH-59 sampler.

The bed material D_{50} is 1.12 mm and the bed gradient is 0.0006. Bed load transport was measured at 11 verticals (Fig. 2). Table 1 shows some grain-size distribution data for bed material and bed load samples collected at each of the 11 verticals considered.

During the bankfull flood of 24 October 2002, with a measured discharge of 478 $m^3 s^{-1}$ and a stream power of 287 W m^{-1}, mean bed load discharge was 2.4 t day^{-1} (0.048 t $day^{-1} m^{-1}$) and mean concentration of suspended sediment was 0.24 g l^{-1}. These values indicate a very low bed load transport rate, respectively, one and two orders of magnitude lower than that measured in Italian high relief river systems with similar climatic conditions such as the Reno River (Billi & Salemi, 2004) and the Ombrone

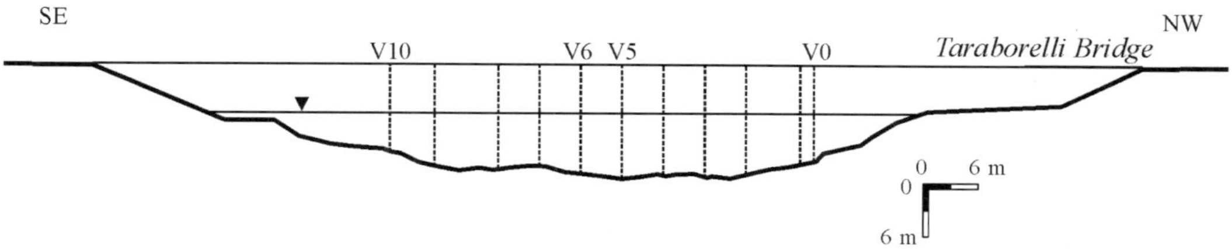

Fig. 2 Cross-section of Rio Quequén Grande at Taraborelli Bridge.

River (Billi & Paris, 2004), though these rivers are characterized by a very low bed load yield, as evidenced by the severe erosion of the beaches supplied by their mouths. The bed load rates of these rivers are in turn two orders of magnitude lower than that measured in the Tanana River in Alaska (Harrold & Burrows, 1983), another high relief river system with a comparable drainage basin surface, while the suspended sediment discharge of the latter is twice that measured for the Rio Quequén Grande for the same discharge.

In order to obtain some reference data, the sediment transport rate for the 24 October 2002 flood was also calculated by a few of the equations most commonly reported in the literature for a wide spectrum of river environments (Table 2). They are equations originally applied to rivers in equilibrium, where no supply limitation exists. In our case, equations gave a wide range of results (three orders of magnitude) and even the lower value predicted by Meyer-Peter & Muller's (1948) equation is three orders of magnitude higher than the values measured in the Rio Quequen Grande.

Stevens & Yang (1989) have tested several bed load formulas and found that for field data the formulas of Engelund & Hansen (1967) and Ackers & White (1973) overestimate bed load with mean ratios between computed and observed data (Rqb) of 1.51 and 1.07, respectively. According to the same authors, the equation of Laursen

Table 1 Grain-size characteristics of bed and bed load sediments.

		D_{10} (mm)	D_{16} (mm)	D_{50} (mm)	D_{84} (mm)	D_{90} (mm)	SD (φ)
Bed material		0.12	0.29	1.12	5.31	8.08	2.04
Bed load Sediment (verticals measured)	V_0	0.08	0.10	0.26	0.42	0.48	0.99
	V_1	0.08	0.11	0.24	0.39	0.45	0.86
	V_2	0.07	0.08	0.15	0.28	0.30	0.77
	V_3	0.11	0.17	0.30	0.48	0.61	0.98
	V_4	0.10	0.16	0.29	0.45	0.50	0.88
	V_5	0.16	0.20	0.33	0.64	1.03	1.01
	V_6	0.15	0.19	0.32	0.54	0.76	0.99
	V_7	0.08	0.09	0.30	0.48	0.59	1.07
	V_8	0.16	0.20	0.34	0.54	0.67	0.83
	V_9	0.25	0.28	0.45	0.97	1.18	1.07
	V_{10}	0.32	0.50	4.76	8.90	9.74	1.79

Table 2 Bed load equations applied and results.

Equations	Sediment Transport Potential (t day^{-1})
Ackers–White (1973)	141300
Engelund–Hansen (1967)	34710
Laursen	16630
Meyer-Peter–Muller (1948)	1353
Toffaleti (1968)	3717
Yang (1973)	16300
Measured data	2.4

(1958) (Copeland & Tomas, 1989) and Toffaleti (1968) underestimate bed load (Rqb = 0.70 and 0.88, respectively), while Meyer-Peter & Muller (1948) underestimated it to a great extent (Rqb = 0.16). Alonso (1980) obtained similar results for the equations of Ackers & White (1958), Engelund & Hansen (1967), Laursen (1958) and Meyer-Peter & Muller (1948). More recently Molinas & Wu (2001) tested a few sediment transport equations for sand-bed rivers and obtained even larger discrepancies between computed and measured bed load with higher ratios for Ackers & White (1958) (Rqb = 4.5) and Engelund & Hansen (1967) (Rqb =1.21), while for Toffaleti (1968) the ratio is a little lower.

CONCLUSIONS

Six main geomorphic units were identified: Ranges, Perirange Aeolian Hillocks, Hillocks with Ponds, Relic Hills, an Alluvial Plain and a Poor Drainage Alluvial Plain. The almost flat morphology gives way to poorly developed drainage systems and a very large number of temporary ponds and marshes of deflation origin.

During field work we observed the scarcity of sediment along the bed channels, with bed rock (incrustation of "caliche") outcropping in some reaches of the studied courses. This observation was confirmed by the recorded bed load discharge during a bankfull flood, which was very low given the relatively high stream power. The bed load rate measured was from three to five orders of magnitudes lower than that predicted by the most well-known bed load equations.

The geomorphological analysis indicates that the present basin morphology is controlled by a relict landscape that was active during the last full glacial period. The present landscape is highly vegetated, stabilized to the present climatic conditions, with river channels that have a high capacity of transport but, as a consequence of low relief and of such stabilization and a reduced supply of sediment.

This area of research requires further investigation, and sediment transport monitoring should be continued to study the response of rivers with low relief basins to floods of varying magnitude. Climatic oscillations during the Quaternary appear to be the main control on the drainage system, as well as palaeotopography and neotectonics.

Acknowledgements This work was funded by the Italian Ministry of Foreign Affairs, granted by the Office of Scientific and Technological Cooperation of the Italian Embassy in Buenos Aires. The authors would like to thank John Laronne for his invaluable comments on the manuscript, M. Farenga and A. Ferrante for their assistance with the drawings and field work, and J. Carden for editing.

REFERENCES

Abbas, N. & Subramanian, V. (1984) Erosion and sediment transportation in Ganges River Basin (India). *J. Hydrol.* **69**, 173–182.

Ackers, P. & White, W. R. (1973) Sediment transport: new approach and analysis. *J. Hydraul. Div. ASAE* **99**(1), 2041–2060.

Alonso, C.V. (1980) Selecting a formula to estimate sediment transport capacity in non-vegetated channels. In: *CREAMS, A Field Scale for Chemicals, Runoff and Erosion from Agricultural Management System* (ed. by W. G. Knisel), 426–439. US Dept. of Agriculture Conservation Research report 26.

Belt, C. B. (1975) The 1973 flood and man's constriction of the Mississippi River. *Science* **189**, 681–684.

Billi, P. & Paris, E. (2004) Bedload transport and channel dynamics of the Ombrone River (Tuscany, central Italy). In: *Acts of the 32nd International Geological Congress*, 1131. Florence, Italy.

Billi, P. & Salemi, E. (2004) Misura delle portate solide del F. Reno ed inquadramento geomorfologico e sedimentologico dell'alveo. In: *Abstracts of the 2° Giornata di studio: Il monitoraggio idrotorbidimetrico dei corsi d'acqua per la stima dei processi erosivi e il bilancio dei solidi sospesi*. Bologna, Italy.

Coleman, J. M. (1969) Brahmaputra River: channel processes and sedimentation. *Sed. Geol.* **3**, 129–239.

Copeland, R. R. & Tomas, W. A. (1989) Corte Madera Creek Sedimentation Study. *Numerical Model Investigation. US Army Engineer Waterway Experiment Station*, Vicksburg, MS. TR-HL-89-6.

Engelund, F. & Hansen, E. (1967) *A Monograph on Sediment Transport in Alluvial Streams*. Teknisk Forlag, Copenhagen, Denmark.

Harrold, P. E. & Burrows, R. L. (1983) Sediment transport in the Tanana River near Fairbanks, Alaska, 1982. *USGS Water Resources Investigations Report 83-4213*.

Laursen, E. (1958) The total sediment load of streams. *J. Hydraul. Div. ASCE*. **84**(HY1), 1530-1-1530-36.

Martínez, G. A. (1997) Geomorphology and Late Cenozoic geology of the southern side of Tandilia Range, Buenos Aires province, Argentina. In: *International Symposium Geomatics in the Era of Radarsat. Expanded Abstracts* (Ottawa, Canada), 482–483.

Meyer-Peter, E. & Muller, R. (1948) Formulas for bed-load transport. In: *Proc. II Congress IAHR*, 39–64. Stockholm, Sweden.

Molinas, A. & Wu, B. (2001) Transport of sediment in large sand-bed rivers. *J. Hydrol. Res.* **39**(2), 135–146.

Stevens, H. H. & Yang, C. T. (1989) Summary and use of selected fluvial sediment-discharge formulas. *USGS Water Resources Investigations Report 89-4026*.

Teruggi, L. B., Billi, P., Farenga, M., Martínez, G., Ferrante, A. & Trassens, M. (2003) Dinamica dei Sedimenti del Fiume Quequén Grande, Provincia di Buenos Aires, Argentina: Studio Preliminare. In: *Atti del Convegno GeoSed 2003*, 223–228.

Toffaleti, F. B. (1968) A procedure for computation of total river sand discharge and detailed distribution, bed to surface. *Technical Report No.5*. Committee on channel stabilization, US Army Corps of Engineers.

Tricart, J. L. (1973) Geomorfología de la pampa deprimida. INTA. Colección Científica no XII.

Varela, L. B. & Teruggi, L. B. (2002) Caracterización hidrológica de la cuenca del Río Quequén Grande. In: *Manejo Integral de Cuencas Hidrográficas y Planificación Territorial*. (ed. by L. B. Teruggi), 19–29. Mar del Plata, Argentina.

Winkley, B. (1982) Response of the Lower Mississippi to river training and realignment. In: *Gravel-Bed Rivers* (ed. by R. D. Hey, J. C. Bathurst & C. R. Thorne), 659–680. Wiley, Chichester, UK.

Yang, C. T. (1973) Incipient motion and sediment transport. *J. Hydraul. Div. ASCE* **99**(HY10), 1679–1704.

Determining the competence of mountainous Mediterranean streams using lichenometric techniques

FREDERIC GOB[1], NICOLAS JACOB[2], JEAN-PAUL BRAVARD[2] & FRANCOIS PETIT[1]

1 *Laboratoire d'Hydrographie et Géomorphologie fluviatile, University of Liège Dept. of Geography, Allée du 6 Août, 2. 4000 Liège, Belgium*
fgob@ulg.ac.be

2 *University of Lyon 2, Louis Lumière 5600, 5 Avenue Pierre Mendès-France, 69676 Bron cedex, France*

Abstract This article deals with the application of lichenometry to the study of mountainous Mediterranean boulder-bed rivers, characterized by strong floods. Lichens present on blocks are a means of dating the reworking of this material, and therefore determining the maximal competence of floods. By associating the diameters of the blocks dated using lichenometry to the specific streampower of corresponding floods, several diameter/critical specific streampower relations for elements between 0.5 and 2.7 m were proposed. In this way several different relations were determined for the rivers studied, highlighting the uneven dissipation of energy due to the roughness of bed forms in these rivers.

Key words boulder bed rivers; flood events; lichenometry; Mediterranean streams; specific stream power

INTRODUCTION

Still relatively little is known about the fluvial dynamics of gorge sectors of mountainous Mediterranean streams, particularly as far as the mobility of gravel bed load is concerned. Yet, these environments offer an opportunity to calculate the specific streampowers necessary to mobilize generally coarse material (b axis > 1000 mm). However, these values are rarely available in the literature dedicated to this subject (Costa, 1983; Petit *et al.*, 2000). Moreover, it is interesting to be able to calculate the competence of flow in terms of specific streampower in order to understand the conditions in which riverbed features are formed. But processes remain difficult to study for a number of reasons: the absence of ancient descriptions or archives, the impossibility of observing the processes during flood conditions and the size of the alluvial material itself.

In an effort to overcome these difficulties, lichenometric techniques are employed for this study. These techniques are not widely applied to the study of watercourses, despite the fact that many authors have demonstrated their value in measuring hydraulic conditions (Gregory, 1976) or dating features, as has been shown in periglacial studies (Carling & Glaister, 1987; Macklin *et al.*, 1992; Maas *et al.*, 1998; Jacob, 2003). In a recent article, Gob *et al.* (2003) showed that lichenometry is a reliable instrument in determining the conditions for mobilizing coarse alluvial bed load. This technique is useful as it is easy to execute and especially because it provides information about past events that is not revealed by using radioactive, coloured or radio-emitting markers (Mussot, 1969; Ergenzinger *et al.*, 1989). Indeed, the study demonstrated that dating alluvial units (deposits, re-organization or stabilization) is possible over a period of at least 200 years (Jacob *et al.*, 2002).

STUDY AREA

Four rivers were selected; three on the eastern border of the Massif Central, the Chassezac, the Ardèche and the Loire, and the fourth in Corsica, the Figarella (Fig. 1).

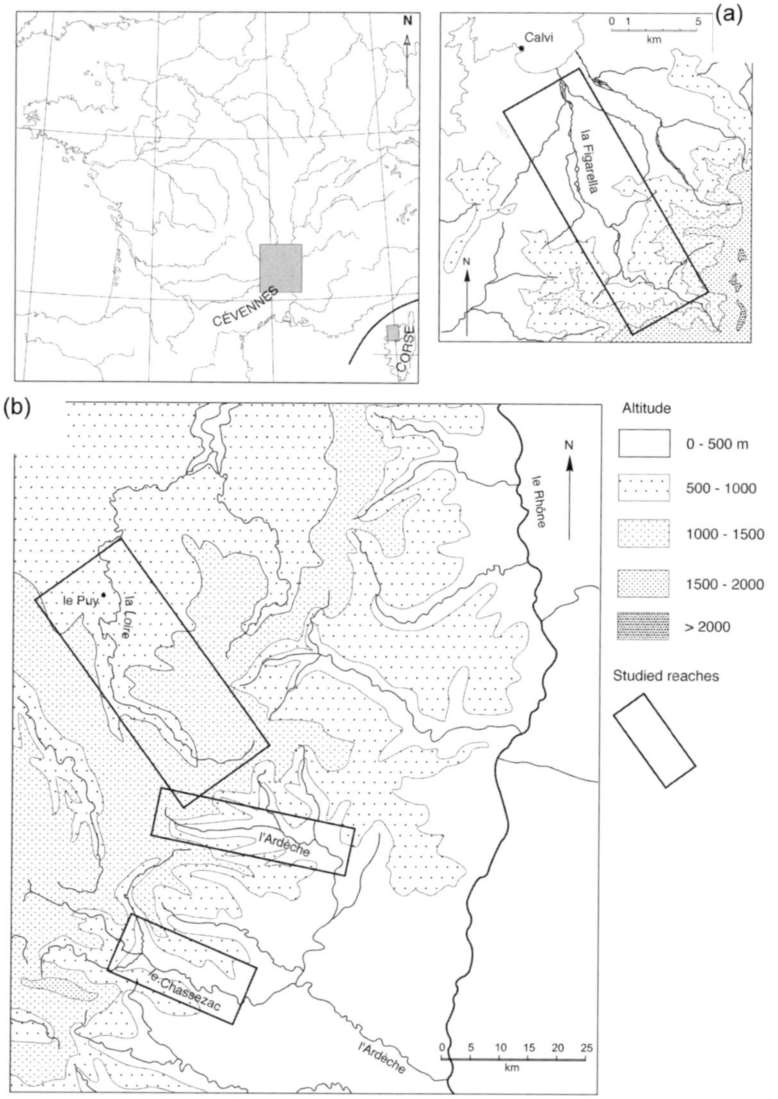

Fig. 1 Map showing (a) the Figarella River basin; (b) the Upper Loire, the Ardèche and the Chassezac basins.

Table 1 Characteristics of the river basins studied.

	Catchment area (km^2)	10 largest blocks max. diameter (m)	Mean slope ($m\ m^{-1}$)
Upper Ardèche R.	291	1.5	0.02
Upper Chassezac R.	507	2.9	0.014
Upper Loire R.	880	0.95	0.012
Figarella R.	50	3.2	0.05

The sectors chosen meet numerous conditions necessary for the successful application of lichenometry. Gravel bed sectors or sectors where the bed is composed of mobile blocks or coarse bed load were selected in order to refine mobility equations for elements greater than 1000 mm. The drainage basins of the rivers chosen are exposed to a Mediterranean or Peri-Mediterranean climate. Furthermore, the areas studied are found in mountainous or mid-mountainous regions. Generally, the substratum is crystalline and metamorphic with a predominance of granite and gneiss in the upstream part of the watercourse. In the selected areas, the slopes are steep and reach between 50 and 70% and the thalwegs have average longitudinal gradients of 0.012 to 0.05 $m\ m^{-1}$, but these values may reach 0.08 and 0.1 $m\ m^{-1}$ in the headwater basin (Table 1).

Hydrological characteristics and strength of streamflow

Episodes of violent autumn rainfall explain the strength of streamflow. In the Cèvennes, heavy rain may last for days. Rainfall totals of over 400 mm may be recorded, as occurred in September–October 1958, which witnessed 600 mm of precipitation (Fig. 2). Likewise, in October 1827, 972 mm of rain was recorded in 21 h at Joyeuse (Ardèche).

These conditions account for the formation of powerful flood waves, whose discharge often reaches or exceeds 1000 m^3, for example 2200 $m^3\ s^{-1}$ for the Chassezac in 1958. Steep longitudinal slopes allow very rapid propagation of the flood peak downstream, at a speed of 12–15 $km\ h^{-1}$ for the Ardèche and the Chassezac. Interestingly, surface speeds of 10 $m\ s^{-1}$ were measured on the Chassezac in 1958 (Guilhot, 1959). In Corsica, episodes of heavy rain produce flashfloods whose discharges exceed 150 $m^3\ s^{-1}$. Such streamflows are characterized by very high specific streampowers. The average values are around 3000–5000 $W\ m^{-2}$, while in the narrowest sectors, these rivers may reach values of up to 1000 $W\ m^{-2}$.

METHODOLOGY

The techniques traditionally used to evaluate bed load transport are difficult to undertake and are not generally effective in the type of river selected for this study. Based on natural markers, lichenometry allows the mobility of coarse material to be estimated and facilitates the study of past morphogenic floods and exceptional flood events.

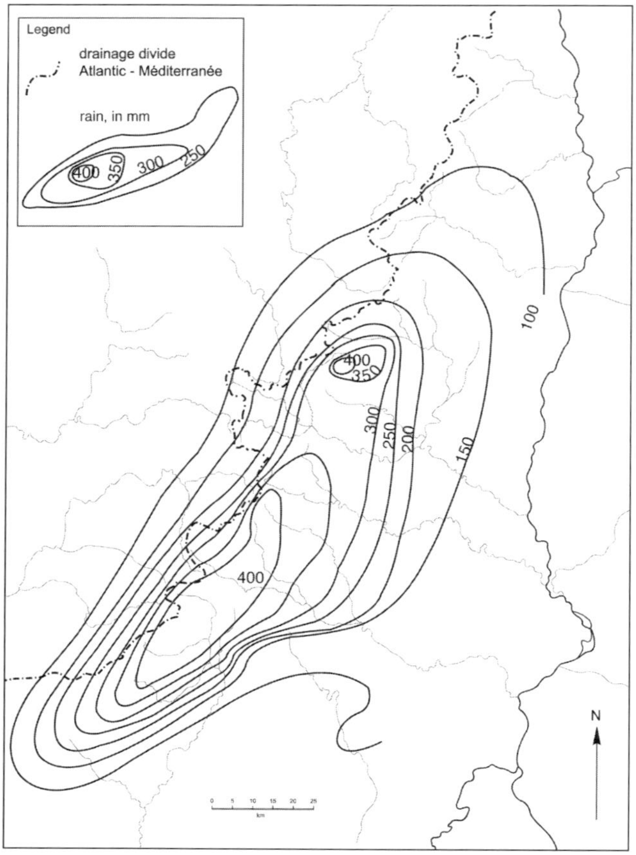

Fig. 2 Precipitation recorded between the 29th and the 30th of September 1958 on the Cévennes. This type of very heavy rainfalls creates very powerful floods

Definition of lichenometry and working hypothesis

Grove (1988) defined lichenometry as "a method of dating based on the assumption that the largest lichen growing on a given substrate is the oldest individual and that, if the growth rate for a given species is known, the maximum lichen size will provide a minimum age for the substrate". In fluvial environments, one considers that lichen present on blocks is destroyed during floods when the elements are mobilized. Following the flood, a new generation of lichen may begin to grow and, subsequently, by identifying the age of the lichen, the date of the mobilizing flood may be established.

Sampling in the selected riverbeds involves measuring the diameter of the largest thalli present on each block and also recording the intermediate axis of the blocks. Then, using a lichen growth curve, the last time that the blocks were moved may be dated and the hydrological parameters (specific streampower) of the morphogenic flood may be related to the diameters of the mobilized elements. An exhaustive chronical of hydrological events must be available so that the flood events and the mobilizations indicated by the lichen may be correctly corresponded.

Growth curve

Rhizocarpon geographicum l.s. was chosen for this study. This species of lichen is widely used by geomorphologists as it is common and easy to identify. The age of the thalli is determined by measuring their diameter and comparing this to the growth curve of the species being considered. Lichen growth curves are generally obtained indirectly, by measuring a large number of lichen on artificial substrates, whose age is already known: tomb stones, bridges, monuments, standing stones. Growth is determined by the environment (humidity, exposure, etc.) and therefore varies according to the regions studied (Fig. 3). The choice of envelope curve and the principles of building and applying a growth curve were presented and discussed in a recent publication (Jacob *et al.*, 2002).

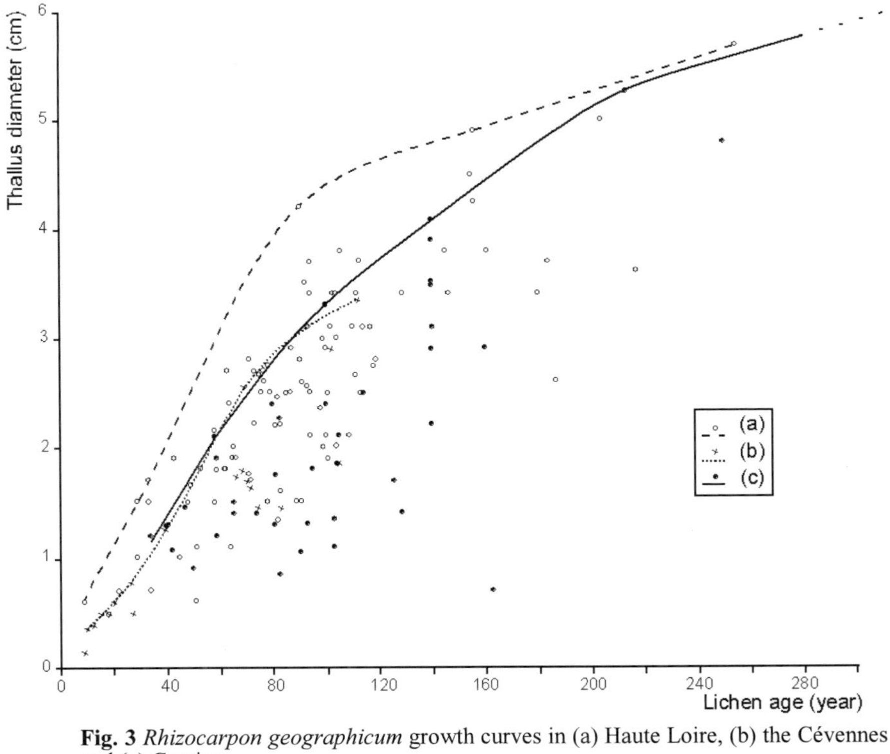

Fig. 3 *Rhizocarpon geographicum* growth curves in (a) Haute Loire, (b) the Cévennes and (c) Corsica.

Lichen populations in river beds may be closely correlated with major hydrological events in each basin. Size-frequency diagrams (Fig. 4) highlight these different populations, indicating their age and associating them with hydrological events. In the case of the Upper Loire, the elements making up the bed are regularly mobilized. The narrow width of the valley means that ancient deposits have not been stored: the traces of the oldest floods date to 1943.

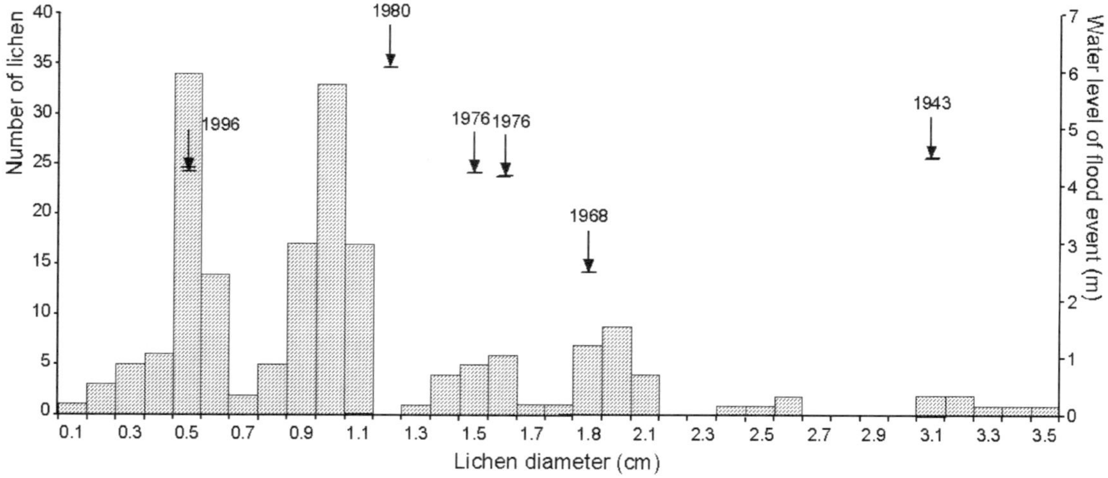

Fig. 4 Size-frequency diagram of *Rhizocarpon geographicum* population in the Upper Loire. Water levels have been recorded at the gauging station of Brives (close to le Puy-en-Velay). One can see a good correlation between the peaks of thalli and flood events.

Characteristics of the bed load studied and the critical specific streampower

The aim of this study is to determine the maximal competences of mountainous Mediterranean torrential rivers. This involved investigating the discharge required to mobilize the largest elements. To these ends, in each case we only considered the average of the ten largest particles mobilized (which corresponds to the D_{95} at least) or the largest element mobilized.

The researchers worked with specific streampowers which are increasingly used to study the dynamics of hydrosystems (Fergusson, 1981; Vanden Berght, 1995; Gintz *et al.*, 1996). Specific streampower is relatively easy to use as only the widths, slopes and discharges are considered in the equation:

$$\omega = (\rho\, g\, Q\, s)/w \tag{1}$$

where ω is the specific streampower (W m^{-2}), Q the discharge (m^3 s^{-1}), s the slope of the water surface (m m^{-1}) and w the width (m). The widths were measured on site by taking the width at the bankfull discharge or, in the case of gorge sectors, using the width constrained by rock surfaces. The slopes were calculated using maps or large scale plans when these were available.

The final elements necessary to calculate specific streampower are discharges. These were provided by la Direction Régionale de l'Environnement (the regional directorate of the environment) for the regions of Corsica, Rhône-Alpes and Auvergne. The discharges measured at the gauging station were extrapolated for the study sites using the equation for modest sized basins, based on the area of the valley sides (Bravard & Petit, 1997):

$$q = Q\, (a/A)^{0.8} \tag{2}$$

where q is the discharge sought, Q the discharge measured at the station (m^3 s^{-1}), a the

Table 2 Critical flood discharges used.

	Year	Discharge at the gauging station ($m^3 s^{-1}$)
Upper Ardèche River	1977	1465
(290 km²)	1982	1315
	1992	1900
Chassezac River	1890	3250
(510 km²)	1899	2200
	1907	2750
	1958	2250
	1980	2960
	1995	1460
Upper Loire River	1980	2000
(880 km²)	1996	1300
Figarella River	1973	153
(28 km²)	Q_3	50

area of the basin upstream from the sector studied and *A* the area at the station (km^{-2}). The discharge values for past flood events were identified through historical research.

Table 2 gives the list of floods that were highlighted using the size-frequency diagram for the different rivers studied. It should be added that due to the rarity of hydrological data on the Figarella, for the purposes of this study, blocks without lichen were considered to be regularly reworked. To calculate specific streampowers, the discharge of the 3-year flood (i.e. the amount of time necessary for blocks to be colonized by lichen) was taken. With regard to blocks without lichen in the Chassezac, account was taken of the fact that an 18-year flood had occurred 5 years before the measuring campaign was undertaken.

RESULTS

The blocks studied range in size from 0.5 to 3.2 m inclusive. Observations show that the coarsest particles may be mobilized at maximal specific streampowers. In the case of the Chassezac, 7000 W m^{-2} is necessary to remobilize particles with a b-axis of 2 m. Figure 5 shows the relations between the size of elements and the streampower necessary to mobilize them for each river. The dots correspond to the largest elements mobilized by a given flood at each site. The size-critical streampower relation is represented by the envelope curve which allows the dots corresponding to blocks smaller than that which the river may mobilize, to be excluded. In this way, the maximal competences of floods may be defined for the four rivers considered.

Table 3 summarizes the results, indicating the range of diameters considered and the specific streampowers developed, as well as the relation that may be determined for each river studied.

Two groups of curves may be distinguished (Fig. 5). Those for the upper Ardèche and the Haute Loire are quite close and seem to follow on from the relations put forward by Costa (1983) and Williams (1983). It should be noted that Costa also uses an envelope curve, while Williams uses straight line regression. The other relations determined for the Figarella and the Chassezac are notably different: their position on

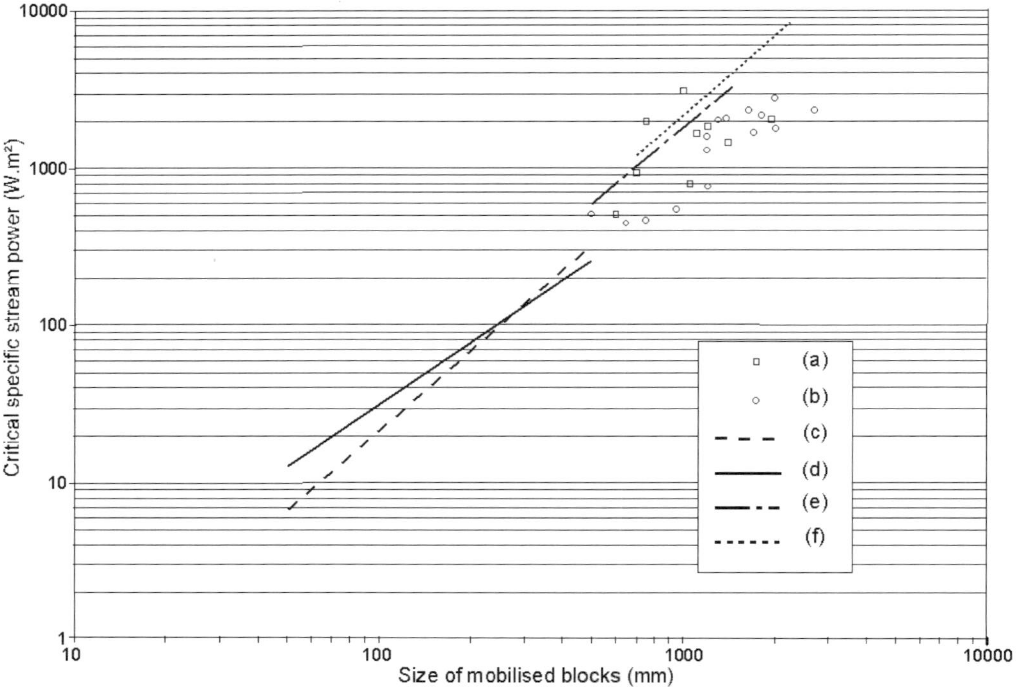

Fig. 5 Relation linking the Critical stream power and the diameter of mobilized blocks in several mountainous boulder bed rivers: (a) Upper Loire, (b) Upper Ardèche, (c) rivers of Colorado (Costa, 1983), (d) observations from literature (Williams, 1983), (e) Figarella River (Gob *et al.*, 2003), (f) Chassezac River (Jacob, 2003).

Table 3 Table showing the equation $\omega_0 = a\, d_i^b$ where ω_0 represents specific stream power (W m^{-2}) and d_i the size of the mobilized elements (mm).

	a	*b*	Block size range (m)
Upper Ardèche	0.1238	1.29	0.65–2.7
Chassezac (Jacob, 2003)	0.025	1.647	0.7–2.3
Upper Loire	0.6069	1.09	0.6–1.9
Figarella (Gob *et al.*, 2003)	0.0253	1.62	0.9–2
Costa (1983)	0.009	1.69	0.5–1
Williams (1983)	0.079	1.3	0.1–1.5

the graph shows that for blocks with equivalent diameters, greater critical specific discharges are required for mobilization than in the Ardèche and the Loire.

DISCUSSION

The rivers may be classed into two distinct groups, even though their hydrological regimes and the material being considered are comparable. As the energy necessary for mobilization to occur is higher in the gorges of the Figarella and the Chassezac, one may interpret that a greater amount of energy is dissipated by the roughness of the bed resulting from skin friction and form drag from individual bed particles (Wohl, 2000). The bed morphology may account for this roughness of features. In the Chassezac, meanders are numerous, which is not the case in the other rivers. Added to this are irregularities and depressions cut into the bedrock. In the Figarella as in the Chassezac (see Table 1), there are very coarse blocks. In each case, this may be explained by a combination of factors: (a) The structural conditions: the joints are widely spaced in the granite outcrops of these river basins, facilitating the liberation of large blocks (Valadas, 1984). In Corsica, freeze-thaw action at the headwater, which lies at a very high altitude, helps account for the large number of coarse blocks. Moreover (b), the evolution of the beds since the late-glacial period has been marked by the evacuation of the inherited pebble sheet, which has resulted in the formation of paving. Indeed, with time sediment sorting has rendered bed load reworking more and more difficult. In both valleys, the largest blocks were mobilized by exceptional floods at the end of the 19th century, but ordinarily, are stable.

This is not the case in the Ardèche and the Haute Loire, where the material is of more modest size, for different structural reasons. In the Ardèche, the inherited pebble sheet has not been completely evacuated due to the width of the alluvial plain. The Loire and the Ardèche have more rectilinear beds and a smoother and better sorted bed load. One may hypothesize that a smaller proportion of energy is taken up by the roughness of the bedforms (Petit *et al.*, 2000).

CONCLUSIONS

This paper underlines the value of lichenometry in the study of hydrology, and its advantages in relation to traditional techniques. Indeed, lichenometry allows a rapid evaluation of the mobility of deposits. The long lifespan of the thalli allow former floods and certain exceptional events to be recorded. In this way, the time period available for study is considerably wider than that provided by the use of colour or radio-emitting markers. Finally, it is possible to work with a large number of blocks as lichen is a natural marker. Lichenometry is an efficient instrument for the study of boulder-bed rivers and the morphogenic effects of extreme stream flows. However, precautions are necessary in employing this tool: for organisms sensitive to their surroundings, the field of application is limited to boulder beds in open torrential environments (without a canopy of trees).

Through lichenometry, size-streampower relations may be developed for blocks whose diameter is greater than that presented in the literature so far. Indeed, this study has allowed the sedimentary dynamics of very active mountainous Mediterranean streams to be better understood and quantified. Finally, the results suggest that in these rivers, the roughness of the bed consumes a large amount of energy. The presence of lichen, or in other words the presence of markers indicating ancient flood episodes, is linked to the preservation of river features. When the material remains relatively stable, as in the Chassezac, or when the large alluvial plain allows for changes in the course of the channel, as in the Ardèche, features that were laid down during the first half of the 19th century remain conserved.

Acknowledgements This study has benefited from the support of the "Programme Tournesol" and bilateral agreements between the Communauté Française de Belgique and the Ministère francais des Affaires Etrangères. Frédéric Gob is financed through a grant offered by the F.R.I.A. of the Communauté Française de Belgique. Comments made by Paul Carling improved this paper.

REFERENCES

Bravard, J. P. & Petit, F. (1997) *Les cours d'eau: dynamique du système fluvial*. A. Colin, Paris, France.

Carling, P. A. & Glaister, M. S. (1987) Reconstruction of a flood resulting from a moraine dam failure using geomorphological evidence and dam break modelling. In: *Catastrophic Flooding* (ed. by L. Mayer & D. Nash) (18th Annual Geomorphological Symposium, Miami University, Ohio), 181–200.

Costa, J. E. (1983) Palaeohydraulic reconstruction of flashflood peaks from boulder deposits in the Colorado Front Range. *Geol. Soc. Am. Bull.* **94**, 986–1004.

Ergenzinger, P., Schmidt, K .& Busskamp, R. (1989) The pebble transmitter system (PETS): first results of a technique for studying coarse material erosion, transport and deposition. *Z. Geomorphol.* **33**, 503–508.

Ferguson, R. I. (1981) Channel forms and channel changes. In: *British Rivers*. (ed. by J. Lewin), 90–125. G. Allen & Unwin, London, UK.

Gintz, D., Hassan, M. A. & Schmidt, K. H. (1996) Frequency and magnitude of bedload transport in a mountain river. *Earth Surf. Processes Landf.* **21**, 433–455.

Gob, F., Petit, F., Bravard, J. P., Ozer, A. & Gob, A. (2003) Lichenometric application to historical and subrecent dynamics and sediment transport of a Corsican stream. *Quatern. Sci. Rev.* **22**, 2111–2124.

Gregory, K. J. (1976) Lichens and the determination of river channel capacity. *Earth Surf. Processes Landf.* **1**, 273–285.

Grove, J. M. (1988) *The Little Ice Age*. Methuen, London, UK.

Guilhot, R. (1959) Note sur les crues de septembre-octobre 1958 dans le bassin du Chassezac. *Mémoires et Travaux de la Société Hydrotechnique de France* **1**, 60–65.

Guillot, P. (1959) Aspects hydrométéorologiques des crues cévenoles des 30 septembre et 4 octobre 1958. *Mémoires et Travaux de la Société Hydrotechnique de France* **1**, 50–59.

Jacob, N. (2003). Les Vallées en gorges de la Cévenne vivaraise, Montagne de sable et château d'eau. PhD Thesis, Univ. Paris IV-Sorbonne, France.

Jacob, N., Gob, F., Bravard, J. P. & Petit, F. (2002) Croissance du lichen *Rhizocarpon geographicum l.s.* sur le pourtour nord-occidental de la Méditerranée: observations en vue d'une application à l'étude des lits fluviaux rocheux et caillouteux. *Géomorphologie. Relief, Processus, Environnement* **4**, 283–296.

Maas, G. S., Macklin, M. G., Warburton, J., Woodward, J. C. & Meldrum, E. (1998) A 300 years history of flooding in an Andean mountain river system: the Rio Alizos, southern Bolivia. In: *River Basin Sediment Systems: Archives of Environmental Change* (ed. by D. Maddy, M. G. Macklin & J. C. Woodward), 297–324. Balkema, Rotterdam, The Netherlands.

Mackling, M. G., Rumsby, B. T. & Heap T. (1992) Flood alluviation and entrenchement: Holocene valley-floor development and transformation in the British uplands. *Geol. Soc. Am. Bull.* **104**, 631–643.

Mussot, R. (1969) Intérêt de l'utilisation des traceurs radioactifs dans la recherche en morphologie fluviale. *Bulletin de l'Association de Géographes Français,* **373/374**, 459–466.

Petit, F., Perpinien, G. & Deroanne, C. (2000) Détermination des puissances spécifiques critiques dans des rivières à charge de fond caillouteuse. *Revue géographique de l'Est* **40**, 59–65.

Valadas, B. (1984) Les hautes terres du Massif Central français, contribution à l'étude des morphodynamiques récentes sur versants cristallins et volcaniques. PhD Thesis, Univ. Paris-I, France.

Van den Berght, J. H. (1995) Prediction of alluvial pattern of perennial streams. *Geomorphology* **12**, 259–279.

Williams, G. P. (1983) Paleohydrological methods and some examples from Swedish fluvial environments. *Geogr. Ann. A* **65**(3-4), 227–243.

Wohl, E. E. (2000) *Mountain Rivers*. Water Resources Monograph 14. AGU, Washington, USA.

Bed load under low sediment transport in a large regulated river: the lower Ebro, NE Spain

DAMIA VERICAT & RAMON J. BATALLA

Department of Environment and Soil Sciences, University of Lleida,
Av. Alcalde Rovira Roure 177, E-25198 Lleida, Spain
dvericat@macs.udl.es

Abstract This study describes the bed load transport regime observed under low sediment transport conditions and constant discharge in the gravel-bedded lower Ebro River (NE Spain). The data show a tendency toward size-selective transport during bed load sampling. The maximum bed load transport rate was lower (35%) than average rates previously obtained under similar discharges. This decline is related to the development of an armour layer during winter floods, which has reduced sediment availability and, consequently, influenced the bed load transport rate and grain-size during the sampling period. Bed load transport observed at individual sampling verticals varied up to 3.7 times the mean rate computed at the same vertical. The highest transport rates and highest variance around the sampling mean occurred in mid-channel, with both the variance and rate diminishing towards the channel margins.

Key words bed load; Ebro River; grain-size; gravel-bed river

INTRODUCTION

Sediment transport is a key component of catchment and river processes. Information on the sediment load of a river can be used for a variety of purposes. It can be used to estimate erosion rates, to assess effects of land use changes on the sediment yield and river dynamics, and to evaluate downstream geomorphic effects of instream human activities, such as dam construction and removal, and gravel mining. Bed load is an essential element of a river's sediment transport regime but accurate assessment of bed load transport rates is difficult because of the inherent variability of the transport process (Einstein, 1937). This variability poses important problems for the calibration and use of bed load samplers (Emmett, 1980; Hubbell, 1987), and must be taken into account when estimating the precision of bed load measurements (e.g. McLean & Tassone, 1987 in Hubbell, 1987; Gomez *et al.*, 1989). The temporal variability has been linked to the movement of dunes in sandy rivers (Gomez *et al.*, 1990) and with cyclic slugs of sediment in coarser beds (Ehrenberger, 1931; Hamamori, 1962). Bed load rates vary across a given river section too. All or most of the bed load transport can occur in a narrow part of the channel width, which does not always correspond with the area with the highest velocity or shear stress. Moreover, distribution of bed load transport in a cross section can shift with changes in hydraulic conditions or sediment characteristics (Emmett, 1980). The objective of this study is to report the variance of bed load transport rates observed during sampling at almost constant discharge in the Ebro River. The samples analysed are part of a sediment transport monitoring programme carried out by a research group from the University of Lleida since 2002 in order to obtain direct field data on river processes, both to assess the magnitude of the reservoir-induced sediment deficit and to inform on-going fluvial restoration projects (Vericat & Batalla, 2005a).

STUDY AREA

The Ebro watershed (85 530 km^2) is located in the northeastern part of Spain (Fig. 1). Mean annual runoff at Tortosa is about 14 300 hm^3 (i.e. 170 mm, $Q_{mean} = 450$ m^3 s^{-1}), of which almost 60% is stored in reservoirs (Batalla *et al.*, 2004). Maximum peak flow at Tortosa, near the watershed outlet, was estimated at 12 000 m^3 s^{-1} in 1907 (Novoa, 1984). Dams have reduced the magnitude of frequent floods (Q_2–Q_{25}) by 25% (Batalla *et al.*, 2004) and trap most sediment transported from upstream reaches (Batalla, 2003), and have caused morphological and ecological adjustments to the lowermost reaches of the river (Vericat & Batalla, 2004, 2005a). The lower Ebro River is a gravel-bed river with a mean slope of 8.5×10^{-4}. It flows as a single thread meandering channel, but is constrained by bedrock outcrops when the river crosses the Coastal Range (Fig. 1). The bed material consists mainly of gravel and sand. The median grain-size of the surface is 58 mm at the Móra d'Ebre monitoring section, where bed load measurements are taken (Fig. 1). The median subsurface grain-size is 21 mm. An armour layer tends to develop under partly competent flows such as the small floods occurred during winter 2004, just before bed load samples reported here were taken (Vericat & Batalla, 2005a).

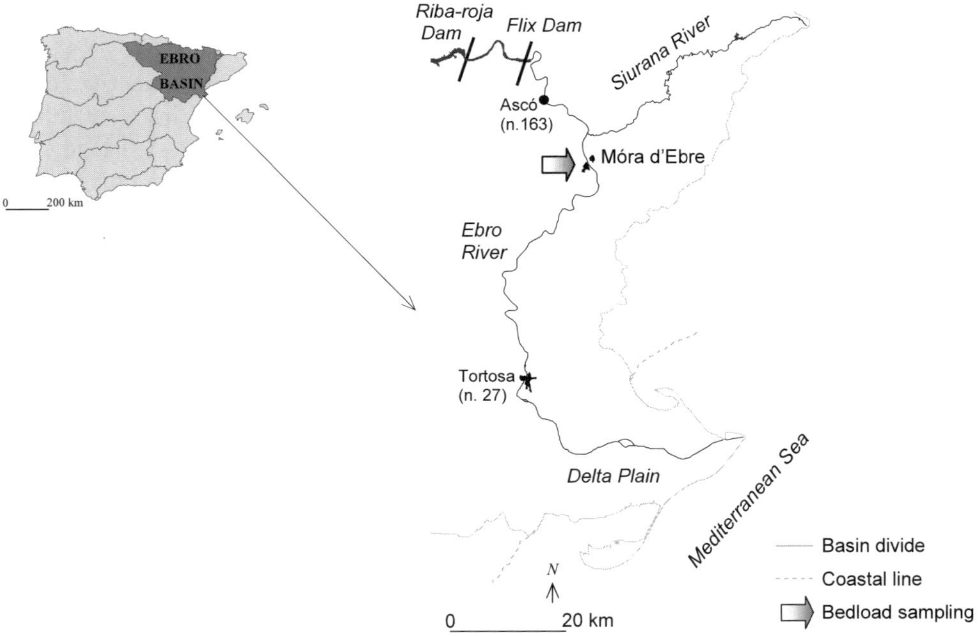

Fig. 1 Location of the monitoring section in the lower Ebro River (NE Spain).

METHODOLOGY

Bed load transport was measured on two consecutive days at the Móra d'Ebre monitoring section, 27 km downstream from the lowermost dam (Fig. 1). Bed load was sampled at seven equally spaced verticals (Fig. 2) using a 29 kg cable-suspended Helley-Smith sampler with a 76-mm intake (Helley & Smith, 1971). Bed load was sampled for a nearly constant discharge of 900 $m^3 s^{-1}$. Median bed load sizes collected ranged from 8 to 16 mm, implying that a 76-mm opening Helley-Smith sampler is adequate for most grain-size classes that may be encountered during sampling (Emmett, 1980). The bed load sampler was deployed using a manual crane. The sampling time was 5 min which prevented the sampler bag (0.25 mm mesh) from becoming more than 50% full, thereby maintaining adequate sampling efficiency (Vericat & Batalla, 2005a,b). The sampling interval between verticals was 15 min. A total of 35 measurements were taken, but only 18 samples collected bed load. They were taken to the laboratory where they were dried, sieved and weighed to obtain the total mass and the grain-size distribution.

Discharge was estimated by routing hydrographs from the upstream gauging station in Ascó (n.163) using the Muskingum method (Shaw, 1983), and further compared with discharges recorded at the downstream station in Tortosa (n.27) (Fig. 1). The actual discharges tended to be systematically overestimated by around 3% (Vericat & Batalla, 2005a).

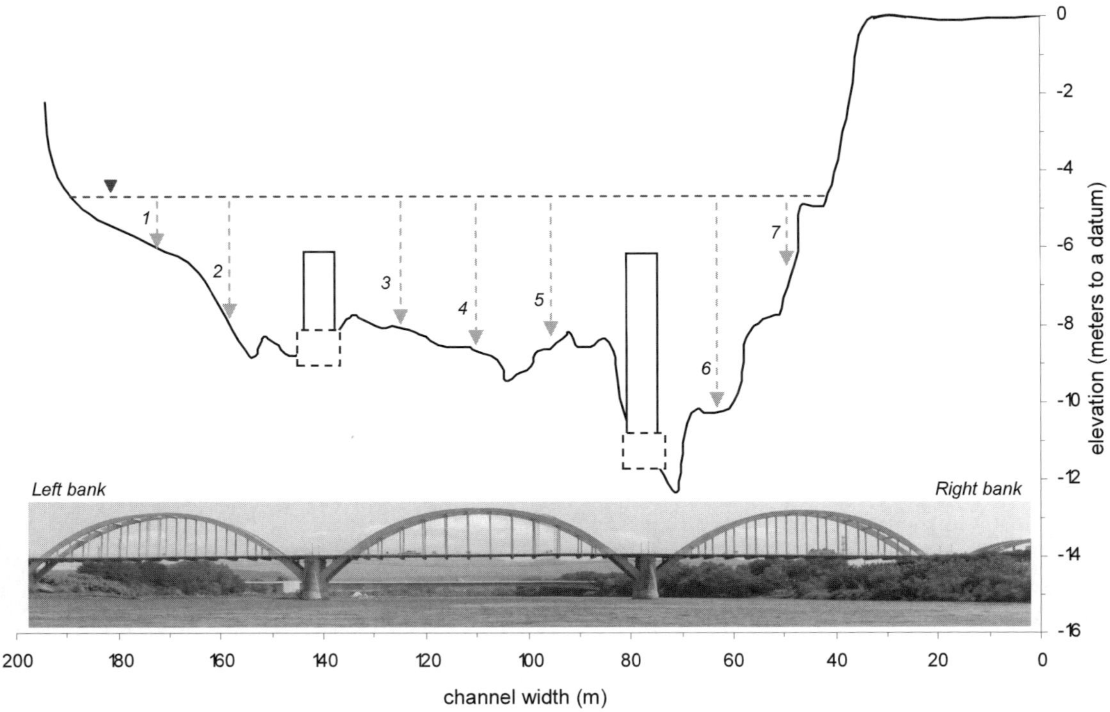

Fig. 2 Bed load sampling section in Móra d'Ebre (Fig. 1). Sampling verticals and water stage at 900 $m^3 s^{-1}$ are indicated.

RESULTS AND DISCUSSION

The relatively high flows during 1–2 June 2004 were the consequence of unusually water releases from the Riba-roja Dam (Fig. 1), due to dam maintenance operations. Maximum discharge at Móra d'Ebre was 925 $m^3 s^{-1}$ and the mean discharge during the period when bed load sampling was undertaken was 910 $m^3 s^{-1}$, ranging from 890 to 925 $m^3 s^{-1}$ (i.e. according to the flow duration curve, discharges equalled or exceeded 13% of time). Maximum water depth during sampling was close to 8 m and the maximum variation was 0.1 m. According to Shields' (1936) equation, the mean discharge during the sampling period was competent (26.7 N m^{-2}) to entrain at most the D_{20} (35 mm) of the bed surface distribution. Discharge dropped to 200 $m^3 s^{-1}$ (and transport ceased) immediately after the sampling period ended, and the bed-material was sampled at this point in time.

Bed load transport

The mean bed load transport rate on 1 June was 1.2 g $m^{-1} s^{-1}$, with a maximum of 10.7 g $m^{-1} s^{-1}$ observed in vertical n.3 (V_3) near to the mid-section of the channel (Fig. 2). On 2 June the mean bed load transport rate was 4.5 g $m^{-1} s^{-1}$. Again V_3 showed the highest rate (33.4 g $m^{-1} s^{-1}$), which was much lower than the mean bed load transport rate obtained under same discharge at the same vertical during 2003 and 2004 (51.4 g $m^{-1} s^{-1}$, N = 30). Median bed load size varied from 1 to 39 mm, and the maximum sampled particle size was 45 mm. The coarsest median bed load size (39 mm) was sampled at V_4 (mid width), while the lowest median (1 mm) was sampled at verticals close to channel margins (Fig. 2). Largest grain-sizes were close to the maximum theoretically movable size (35 mm) at the available discharge.

The mean fractional bed load transport rate (i_{bi}) scaled by the proportion of each size fraction in the surface bed material (f_{i-s}) was calculated (Fig. 3). Relative size was calculated by scaling each size (D_i) by the median bed surface size (D_{50-s}) (e.g. Gomez, 1995). Figure 3 suggests a tendency towards size-selective transport during bed load sampling. On one hand, particles around 16 mm were the main contributors to bed load, indicating that an active phase of winnowing of such fine surface size-fractions was still taking place during sampling. On the other hand, the low mobility of the finest particle sizes (<4 mm) might be attributed to the lack of movement of the coarse particles (>16 mm) that shelter them (e.g. Kirchner *et al.*, 1990). The coarse riverbed surface (i.e. armour layer) created during small winter floods likely reduced surface sediment availability and, consequently, determined bed load rate and grain-size during the sampling period.

Variance of bed load transport

The bed load samples collected in this study were not numerous enough to accurately assess the temporal variability around the true bed load mean (e.g. Carey & Hubbell, 1986; Gomez *et al.*, 1990; Gomez & Troutman, 1997). However, the standard deviation for each sampling vertical provides an indication of bed load temporal variance,

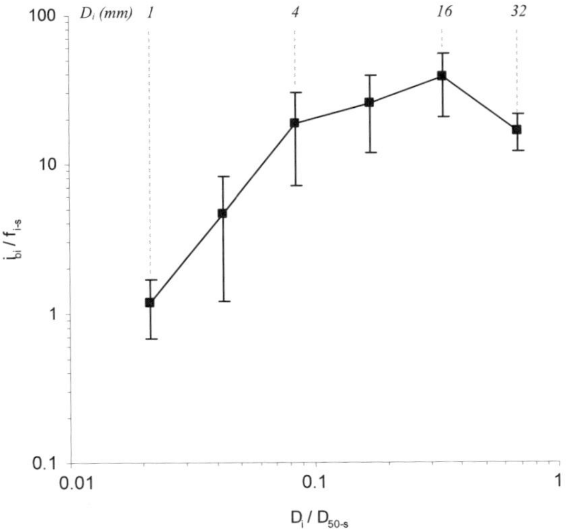

Fig. 3 Mean fractional bed load transport rate as a function of relative size. The mean fractional transport rate (i_{bi} in g m^{-1}s^{-1}) has been scaled by the proportion of each size fraction in the surface bed material ($f_{i\text{-}s}$). Relative size has been calculated scaling each size (D_i) by the median bed surface size ($D_{50\text{-}s}$) in 2004. Vertical lines represent the standard deviation around the mean ($\varepsilon = \sigma/\sqrt{N}$, where σ is the standard deviation and N the number of data).

and we use it to compare sampled values at given locations with the cross-section mean for the entire sampling period. Using this approach, it is clear that the larger the bed load transport the larger is the variance around the sampling mean (Fig. 4). Similar results were obtained from the Fall River, Colorado (Pitlick, 1987, in Hubbell, 1987). In the lower Ebro River, the maximum variation occurred in vertical located situated around the mid channel width (V_3 and V_4), where the coarsest bed load was collected. The maximum bed load transport rate at each vertical exceeded almost 2.4 times the mean rate at the same point, on average, while the maximum discrepancy was 3.7 at V_4 (Table 1), suggesting that temporal variability is also present in this large gravel-bed river. Comparable results were obtained in the Fraser River, Canada (McLean *et al.*, 1999).

Table 1 Bed load transport rates collected during the five transects at each vertical.

Vertical[1]	Bed load rate (i_b in g m^{-1} s^{-1}) t_1[2]	t_2	t_3	t_4	t_5	Mean bed load rate (i_{bm} in g m^{-1} s^{-1})	Maximum bed load rate (i_{bmx} in g m^{-1} s^{-1})	Dimensionless bed load rate (i_{bmx} / i_{bm})
1	0.00	0.00	0.00	0.00	0.00	0.00	0.00	–
2	0.00	0.07	0.05	0.11	0.00	0.05	0.11	2.40
3	0.45	3.65	10.65	33.45	0.10	9.66	33.43	3.50
4	3.25	0.00	5.40	26.00	0.65	7.06	25.98	3.70
5	0.55	0.00	0.60	2.52	0.11	0.76	2.52	3.30
6	0.02	0.01	0.00	0.00	0.00	0.01	0.02	2.90
7	0.00	0.00	0.00	0.00	0.00	0.00	0.00	–

[1] Sampling vertical along the cross-section (Fig. 2).
[2] Transect number.

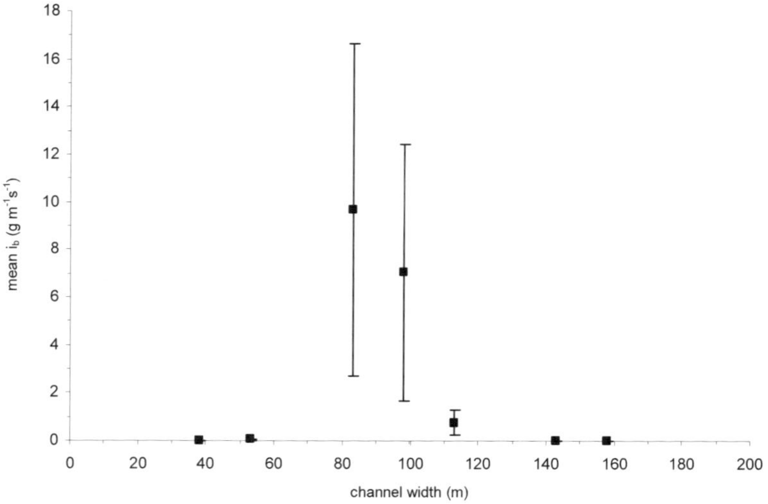

Fig. 4 Mean bed load transport rate at each vertical during the five transects. Vertical lines represent the standard deviation around the mean.

Table 2 Vertical variance in each transect (v_n^*).

t* n†	1	2	3	4	5	Mean v_n
1	0.0	0.0	0.0	0.0	0.0	0.0
2	0.0	0.1	0.0	0.0	0.0	0.03
3	0.9	6.1	4.4	3.9	1.1	3.3
4	6.4	0.0	2.2	3.0	6.2	3.6
5	1.1	0.0	0.3	0.3	1.0	0.5
6	0.0	0.0	0.0	0.0	0.0	0.0
7	0.0	0.0	0.0	0.0	0.0	0.0

* Transect number.
† Vertical along the cross-section (Fig. 2).
v_n is the variance of each vertical n at a given transect t around the sampling mean.

The mean bed load transport rate also varied from vertical to vertical. Maximum bed load transport rates were recorded in the channel centre (Table 2), as were the largest variances in transport rates (Fig. 4). In all the transects the maximum bed load transport rate was observed in the section between the two piles of the bridge (V_3, V_4, V_5, Fig. 2), a 50 m long sub-section that is not the deepest in the river channel. In contrast, V_6 is deeper but, there the bed load transport rate did not exceed 0.02 g m^{-1} s^{-1}, that is, three orders of magnitude lower than those obtained in the centre of the channel. Marginal sections V_1 and V_7 did not move bed load at any time during the sampling period (Table 2, Figs 2 and 4).

The variance of bed load transport rates at a single vertical around the mean (v_n) was been calculated as follows:

$$v_n = G_n/G_m$$

$$G_m = \Sigma\, i_{b,n} \times w_n$$

$w_n=(w_{n,\,n-1} / 2) + (w_{n,\,n+1} /2)$

where, G_n is the bed load transport rate (g s^{-1}) at a given vertical, G_m is the mean bed load transport rate (g s^{-1}) for the entire cross-section, w_n is the interpolated width, and $w_{n,\ n\pm1}$ is the distance between verticals. The variance of each vertical (v_n) was calculated for each of the five transects (Table 2). Once again, V_3 and, particularly, V_4 showed the highest variance around the sampling mean. These results confirm that during the sampling period bed load transport was concentrated in the centre of the channel, within section V_5. The transport rate in sections V_3 and V_5 during transects 1 and 5 showed the best overall agreement with the mean bed load transport rate ($v_n \approx 1$). The high discrepancy between verticals and cross-sectional measurements (v_n) prevents us from deriving a factor to correct for spatial variability when measuring bed load at a single point.

FINAL REMARKS

This paper reports on bed load transport observed during five transects (each consisting of seven equally-spaced sample verticals), of the lower Ebro River (a large gravel-bed river located in Spain) made under low sediment transport conditions and constant discharge. Results indicate that:

(a) There was a tendency towards size-selective transport during bed load sampling. The armour layer created during winter floods reduced sediment availability and, consequently, determined the bed load transport rate and grain-size observed during the sampling period.

(b) The bed load transport rate at individual sampling verticals varied up to 3.7 times the mean rate computed for the same vertical.

(c) The highest bed load transport rates and highest variance around the sampling mean were observed in mid-channel and the transport rate and variance both diminished toward the channel margins.

Our results provide a first glimpse of the variability of bed load transport in the lower Ebro River under near-steady flow and low sediment transport conditions. The data will help refine the measurement programme that is underway in this river. However, the characteristics of bed load transport must be investigated over a wider range of flow conditions, utilizing a greater number of transects to obtain an accurate assessment of the bed load transport regime of the River Ebro.

Acknowledgements This experimental study was carried out within the framework of the research project REN2001-0840-C02-01/HID, funded by the Spanish Ministry of Science and Technology. The first author received a grant from the Spanish Ministry of Education. Hydrological data was supplied by the Ebro Water Authorities. Albert Rovira at the University of Lleida assisted during fieldwork. Michael Church and Brett Eaton reviewed initial versions of the manuscript. Basil Gomez did the paper review and provided constructive critique that helped us to substantially improve the paper.

REFERENCES

Batalla, R. J. (2003) Sediment deficit in rivers caused by dams and instream gravel mining. A review with examples from NE Spain. *Cuaternario y Geomorfologia* **17**(3-4), 79–91.

Batalla, R. J., Kondolf, G. M., Gomez, C. M. (2004) Reservoir-induced hydrological changes in the Ebro River basin, NE Spain. *J. Hydrol.* **290**, 117–136.

Carey, W. P. & Hubbell, D. W. (1986) Probability distributions for bedload transport. In: *Proc. 4th Fed. Inter-Agency Sediment. Conf.* **2**, 131–140.

Ehrenberger, R. (1931) Direkte geschiebemessungen an de Donau bei Wien und deren bisherige ergebnisse. *Die Wasserwirtschaft* **34**, 1–9.

Einstein, H. A. (1937) Bedload transport as a probability problem. PhD dissertation published in *Sedimentation* (ed. by H.W. Shen), Colorado State University, Colorado, USA, 1972.

Emmett, W. W. (1980) A field calibration of the sediment trapping characteristics of the helley-smith bedload sampler. *US Geol. Survey Professional Paper 1139.*

Gomez, B. (1995) Bedload transport and changing grain-size distributions. In: *Changing Rivers Channels* (ed. by A. Gurnell & G. Petts), 177–199. John Wiley, Chichester, UK.

Gomez, B., Naff, R. L. & Hubbell, D. W. (1989) Temporal variations in bedload transport rates associated with the migration of bedforms. *Earth Surf. Processes Landf.* **14**, 135–156.

Gomez, B., Hubbell, D. W. & Stevens, H. H. (1990) At-a-point bedload sampling in the presence of dunes. *Water Resour. Res.* **26**, 2717–2731.

Gomez, B. & Troutman, B. M. (1997) Evaluation of process errors in bed load sampling using a dune model. *Water Resour. Res.* **33**(10), 2387–2398.

Hamamori, A. (1962) A theoretical investigation on the fluctuations of bedload transport. *Report R4, 14 p., Delft Hydraul. Lab., Delft, The Netherlands.*

Helley, E. J. & Smith, W. (1971) Development and calibration of a pressure-difference bedload sampler. *US Geol. Survey Open-file Report, Water Resources Division, Menlo Park, California, USA.*

Hubbell, D. W. (1987) Bed load sampling and analysis. In: *Sediment Transport in Gravel-Bed Rivers* (ed. by C. R. Thorne, J. C. Barthurst & R. D. Hey), 89–118. John Wiley, Chichester, UK.

Kirchner, J. W., Dietrich, W. E., Iseya, F. & Ikeda, H. (1990) The variability of critical shear stress, friction angle, and grain protrusion in water-worked sediments. *Sedimentology* **37**, 647–672.

McLean, D. G., Church, M. & Tassone, B. (1999) Sediment transport along lower Fraser River 1. Measurements and hydraulic computations. *Water Resour. Res.* **35**(8), 2533–2548.

Novoa, M. (1984) Precipitaciones y avenidas extraordinarias en Catalunya. In: *Proc. Jornadas de Trabajo sobre Inestabilidad de laderas en el Pirineo*, Barcelona, Spain; 1–15.

Shaw, E. M. (1983) *Hydrology in Practice*. Van Nostrand Reinhold, London, UK.

Shields, A. (1936) *Answendung Ahnlichkeitsmechanik und der Turbulenzforschung auf die Geschiebebewegung*. Mitt. Preussichen Versuchsanstalt Wasserbau Schiffbau, Berlin, Germany.

Vericat, D. & Batalla, R. J. (2004) Efectos de las presas en la dinámica fluvial del curso bajo del río Ebro. *Cuaternario y Gemorfologia* **18**(1–2), 37–50.

Vericat, D. & Batalla, R. J. (2005a) Sediment transport in the lower Ebro River (NE Iberian Peninsula). *Geomorphology* (in press).

Vericat, D. & Batalla, R. J. (2005b) Sediment transport in a highly regulated fluvial system during two consecutive floods (Lower Ebro River, NE Spain). *Earth Surf. Processes Landf.* **30**, 385–402.

A particle-size sediment budget for a Mediterranean river (the lower Tordera, northeastern Spain)

ALBERT ROVIRA
Landscape Architecture and Environmental Planning, University of California, Berkeley, 202 Wurster Hall #2000, California 94720-2000, USA
arovira@macs.udl.es

Abstract The sediment budget for the Lower Tordera has been evaluated for the dry period 1997–1999 by measuring the bed load transport in two cross-sections. The mobilization of most of the river bed fractions in both cross-sections is produced during floods of moderate magnitude and frequency. However, results show that coarser fractions (>16 mm) mobilized in the upper parts of the study channel do not reach the lower ones. This can be associated with some combination of changes in channel geometry (e.g. slope reduction) and the effects of gravel extractions. In contrast, finer particles (<4 mm) moved during low discharges and small floods in the upper cross-section are evacuated during floods in outlet basin. Overall, the sediment budget between the cross-sections exhibits a tendency to accumulate sediment along the channel during dry periods when a hydrological bimodality in the fluvial channel is observed.

Key words bed load transport; hydrological bimodality; particle size; sediment budget

INTRODUCTION

A sediment budget describes the input, transport, storage, and export of sediment from a geomorphic system (Reid & Dunne, 2003). The basic equation (sediment input = sediment output + change in sediment storage) can be refined in many ways and used to compare the likely outcomes of different land-management options or climatic change scenarios, or evaluate the significance and implications of the impacts that have occurred (e.g. Moore & Newson, 1986; Kesel *et al.*, 1992; McLean & Church, 1999). It plays a key role in basic and applied geomorphological studies over a wide range of scales and levels of complexity (Reid & Dunne, 2003). The main purpose of this paper is to present the bed load sediment budget of the Lower Tordera River basin. The Tordera River is a representative example of the fluvial water and sediment uses in a Mediterranean coastal basin of the Iberian Peninsula. From the 1950s to the 1980s, the Tordera River was severely affected by instream gravel mining where more than 3 million m^{-3} were extracted from the riverbed channel. Since the mid 1980s gravel extraction has been prohibited; however, the effects of gravel extraction are still evident (Rovira *et al.*, 2005b). The sediment budget has been evaluated by different grain size fractions (particles transported as bed load) in two cross-sections (one perennial and one ephemeral) in a single fluvial channel reach.

STUDY AREA

The Tordera River basin (894 km^2) is located in the northern part of the Catalan Coastal Ranges (Fig. 1). The bedrock is mainly composed of plutonic and metamorphic rocks. The climate is classified as subhumid Mediterranean with maritime influences. Mean annual rainfall ranges from 1000 mm in the summit areas to 600 mm in the coastal areas. The study site is located in the Lower Tordera basin, covering an 11-km fluvial channel reach from Fogars de Tordera (upper part) to Blanes Bridge, close to the Mediterranean Sea (Fig. 1). At the upper parts, the river exhibits a well-defined single thread channel with a mean channel width of 70 m. The mean channel slope is measured at 0.005. At the outlet basin, the river shows a multi-thread channel with a mean channel width of 100 m. The mean river bed slope is 0.002. There, part of the year (later spring to autumn) the river is ephemeral, while in the upper sections there is continuous water flow more than 84% of time (1967–2000). The mean annual water volume for the study period (1997–1999) is estimated at 74 hm^3, well below the long-term average (149 hm^3 year^{-1}). The period is thus classified as dry.

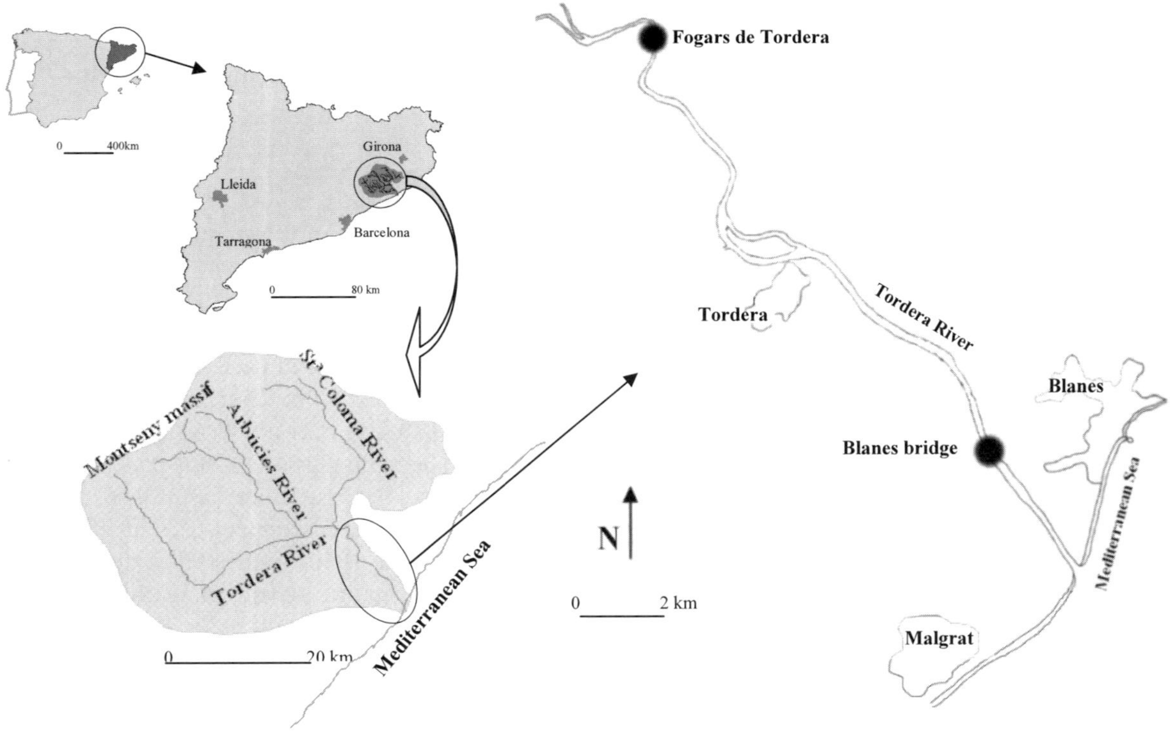

Fig. 1 Location of the study area.

METHODS

The sediment budget was constructed by measuring the bed load transport at two different cross-sections (Fig. 1): Fogars de Tordera (starting point of the study reach channel, 745 km^2) and Blanes Bridge (ending point of the study reach channel, 894 km^2). Bed load samples were collected during baseflows and floods using a handheld and a cable-suspended Helley-Smith sampler (76 mm intake, 15 kg bag capacity). At the Fogars de Tordera cross-section, 113 samples were collected on a weekly basis for discharges ranging from 0.2 to 4.3 $m^3 s^{-1}$ (equalled or exceeded 75 and 20% of time, respectively). Samples were always collected at the same point. The sampling time was established at 5 min. At the Blanes Bridge cross-section, 384 bed load samples were collected for discharges ranging from 4.5 to 39 $m^3 s^{-1}$. Samples were collected during floods. Fifteen sampling points were established along the cross-channel section in order to integrate the transversal variability of the bed load transport into the channel width. The sampling interval and the sampling time of the samples were established at 7 and 5 min, respectively. Technical limitations (e.g. duplication of material, displacement time and human resources) impeded simultaneous sampling in both cross-sections. The sampling period was from January 1997 to June 1999. At the laboratory, the bed load samples were dried, sieved and weighed. The total mass and the grain-size distribution were computed. The grain-size distribution for each bed load sample; the mean grain size distribution (for all samples), and the grain-size distribution of the river bed were obtained. The grain-size distribution of the river bed (Fig. 2) was characterized from bulk samples taken in the Fogars de Tordera and Blanes Bridge cross-sections. At the Fogars de Tordera cross-section, 48 kg of bed material was collected during summer at three different points (lateral bars and thalweg channel). There, no clear difference was observed between surface and subsurface material. At the Blanes bridge section, a total of 13 kg of surface and subsurface material was taken from across the channel. Ephemeral bars and recently active bed load paths in the riverbed were included in the sampling. In both cross-sections, the D_{max} of the river bed material did not exceed 1% of the sample (Church *et al.*, 1987). In addition, the D_{95} of the distribution was well represented (Ferguson & Paola, 1997).

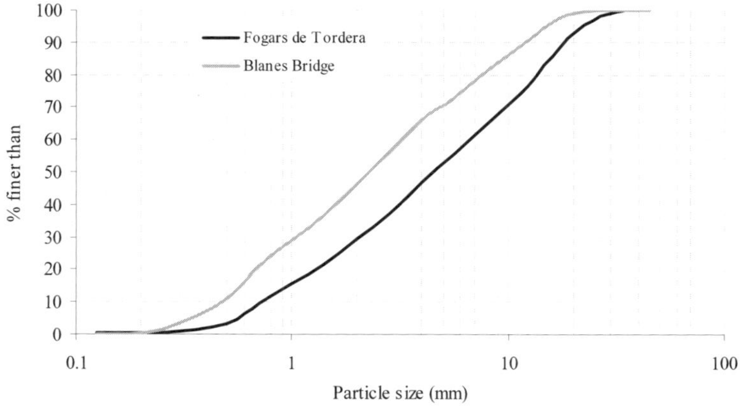

Fig. 2 Grain-size distribution of the river bed at the Fogars de Tordera and Blanes Bridge cross-section.

In the upper reaches (Fogars de Tordera), the sediment yield was calculated using the bed load rating relation method. The error associated with the sediment-rating curve was estimated and applied. At the lower section, no significant statistical relation between discharge and bed load transport was obtained. The sediment yield was computed by applying the interpolation load algorithm (Walling, 1984). This method is based on the assumption that the average transport rate obtained from two consecutive samples is representative of a given time period (mean interval approach), or similarly, an instantaneous sample is representative of the time period extending half way towards the preceding and succeeding samples (mid-interval approach). For calibration, bed load yield was also computed at the Fogars de Tordera cross-section using both approaches. Differences between the two methods (8%) were subsequently used to correct results yielded at the Blanes Bridge cross-section. The good agreement between both methods (bed load rating relation and interpolation load algorithm) shows the potential application of the latter, despite this method apparently having been developed for dissolved load. The total annual sediment bed load by grain size fraction was estimated for the median of each grain size class by computing the total annual sediment transported in each cross-section, and by transforming the mean grain size distribution percentages obtained for all samples to weight (Table 1).

RESULTS AND DISCUSSION

At the Fogars de Tordera cross-section, the distribution of the river bed particles (Fig. 2) ranges from very fine sand (0.25 mm) to very coarse gravels (64 mm). The mean bulk particle size (D_{50}) is 4.5 mm with a standard deviation of 7.1. Bed material is “poorly sorted” with a sorting index value of 1.8. River bed distribution shows a positive skewness (0.23) indicating the major domain of the coarse material (71% of gravel sizes). In that cross-section the relationship between discharge (Q) and bed load rate (i_b) is explained by the equation $i_b=0.18Q^{0.85}$ (r^2 = 0.67; $p < 0.05$; N = 75). In comparison with the upper parts of the basin, where the river shows a gravel bed channel and both the patch dynamics and the breaking of the armour layer control the bed load transport (Garcia *et al.*, 1999), the slope coefficient (0.85) in that equation suggests lower hydraulic dependency of the bed load transport; because in the Lower Tordera the bed load yield is highly determined by the continuous movement of the sandy particles on the channel bed (Batalla *et al.*, 2005). Field observations indicate that only 10 cm of water depth (0.14 $m^3 s^{-1}$) are needed to initiate the movement of sand particles. The error of the coefficient is estimated in 0.07 and the error of the estimation is assessed at 39%. However, since bed load transport is highly nonlinear, massive errors in estimates for higher discharges could likely occur.

In 1997, the total sediment computed to be passing through this cross-section was estimated at 65 600 t (±25 584); 37 000 t (±14 430) in 1998, and around 22 000 t (±8580) in 1999. The variability of the water and sediment transported annually is mainly explained by the number and magnitude of the floods recorded yearly (Rovira *et al.*, 2005b). For successive years, results (Table 1) show no significant differences on the distribution of the percentage of the riverbed particles transported during discharges equalled or exceeded 20% of time. Altogether, around 50% of the total

Table 1. Annual sediment budget by particle size at the Fogars de Tordera section for the study period (1997–1999).

Year 1997	GSD	Fogars de Tordera	GSD	Blanes Bridge	Balance
Fraction	% finer than [(1)]	(t)	% finer than	(t)	(t)
0.125–0.25	0.01	7	0.02	15	–8 [(4)]
0.25–0.5	0.03	17	0.35	243	–226
0.5–1	5.41	3547 [(2)]	9.22	6481	–2934
1–2	31.38	20586	26.13	18381	+2205 [(5)]
2–4	41.21	27035	35.87	25228	+1807
4–8	17.58	11533	21.01	14779	–3246
8–16	3.84	2518	5.89	4145	–1627
16–32	0.50	327	1.30	916	–589
32–60	0.06	38	0.2	142	–104
Total sediment		65609 [(3)]		70330	–4721
Total water yield (hm^3)		100		117	+17
Year 1998	% finer than	(t)	% finer than	(t)	(t)
0.125–0.25	0.02	6	0.02	2	+4
0.25–0.5	0.01	5	0.25	25	–20
0.5–1	1.58	582	10.22	1042	–460
1–2	20.71	7650	28.07	2862	+4 787
2–4	49.79	18391	35.19	3588	+14 802
4–8	23.79	8785	19.72	2011	+6 774
8–16	3.77	1392	5.20	530	+862
16–32	0.33	122	1.13	115	+7
32–60	0.01	3	0.2	20	–16
Total sediment		36936		10197	+26 739
Total water yield (hm^3)		50		23	–27
Year 1999[(6)]	% finer than	(t)	% finer than	(t)	(t)
0.125–0.25	0.02	5	0.02	1	+4
0:25–0.5	0.02	3	0.16	13	–9
0.5–1	0.56	126	6.14	475	–350
1–2	18.08	4041	25.32	1961	+2080
2–4	55.01	12294	37.82	2929	+9365
4–8	24.40	5453	23.97	1856	+3597
8–16	1.84	411	5.75	445	–34
16–32	0.07	16	0.74	57	–42
32–60	0	0	0.08	6	–6
Total sediment		22350		7744	+14606
Total water yield (hm^3)		32		16	–16

[(1)] Mean grain size distribution obtained from all bed load samples collected in that year; [(2)] Total tons estimated by different grain size fractions; [(3)] Total sediment transported estimated from the bed load rating relation method (Walling; 1984); [(4)] Negative values indicate that inputs are lower than outputs; [(5)] Positive values indicate that inputs are higher than outputs; [(6)] Only comprise the first 6 months.

material transported during those discharges ($\leq 4.5\ m^3\ s^{-1}$) ranges between 1 and 2 mm and more than 90% of the particle sizes ranges between 0.5 and 4 mm, on average.

Those particle sizes (0.5–4 mm) represent the D_5 and D_{46} of the river bed distribution. However, it is observed that particles finer than 0.5 mm (D_3 of the river bed distribution) and coarser than 16 mm (D_{84} of the river bed distribution) are not transported during those discharges (Table 1). The transport of the finer particles is expected to be in suspension when the water column can support the particle weight (Richards, 1982). Larger particles (>D_{84}) are mainly mobilized during higher discharges (>4.5 m^3 s^{-1}). The calculation of the critical shear stress (Shields, 1936) for the coarser particle sizes (e.g. >D_{84}) indicates that a minimum peak water discharge of 25 m^3 s^{-1} (equalled or exceeded less than 10% of the time) is needed to initiate the motion of those fractions. Andrews (1983) and Misri *et al.* (1984) suggested that critical dimensionless shear stress values for the entrainment of any particle are not constant, but depend strongly on the ratio of this size to the median size of the bed material. To explain that relation, they proposed an equation of the type $\tau_{ci}^* = a(D_i/D_{50})^b$, where τ_{ci} is the critical dimensionless shear stress; D_i (mm) is the largest sampled particle collected with the Helley–Smith sampler, and D_{50} is the mean diameter of the subsurface bed material. In accordance with Andrews (1983), the largest particles collected in the bed load samples were grouped into 18 size classes. The average τ_{ci}^* for the samples within each class was computed and it was assigned to the mid-point of each size class. Finally, a log–log least square regression between τ_{ci}^* and the ratio D_i/D_{50} was fitted and transformed to obtain the equation: $\tau_{ci}^* = 0.138(D_i/D_{50})^{-0.63}$ ($r^2 = 0.76$, $N = 18$; $p < 0.05$), with standard errors of 0.04 for the intercept and 0.08 for the exponent. The intercept value (0.138) in this relation is similar to that reported by Batalla & Martín–Vide (2001) in the Arbúcies River, one of the main upstream tributaries of the Tordera River; but very high compared with other studies (Parker *et al.*, 1982; Andrews, 1983; Ashworth & Ferguson, 1989). Batalla & Martín–Vide (2001) suggest that the deviation of the intercept *a* may be a consequence not only of an overestimation of the shear stress needed for the particle entrainment, but also due to its low sensitivity to variations in depth and velocity. The exponent *b* in this equation is not very close to the value –1, suggesting that critical shear stresses are directly proportional to particle size and that particle weight is the principal force restricting movement. In other words, there is no clear tendency towards equal mobility, at least down to a critical shear stress <18 N m^{-2}. Nevertheless, the results have to carefully read since the range of sampled discharges is not representative of the whole conditions (e.g. >4.5 m^3 s^{-1}). Furthermore, sampling efficiency problems (Wilcock, 1988) and shear stress computation (Batalla & Martín-Vide, 2001) may also distort the results. Altogether, the results show that approximately 80% of time, the partial mobilization of the river bed fractions (D_5–D_{50}) is reached; whereas the general entrainment of the bed material probably arises during flood events (discharges >4.5 m^3 s^{-1}).

At the Blanes Bridge cross-section, the distribution of the river bed (Fig. 2) ranges from very fine sand (0.25 mm) to coarse gravels (32 mm). The percentage of gravels is estimated at 54% against 46% for sand particles. The mean bulk particle size (D_{50}) is 2.3 mm with a standard deviation of 4.3. The bed material is "poorly sorted" with a sorting index value of 1.8. The grain size distribution is symmetric with a slight positive skewness (0.09). There, water circulation is only produced during the flood season (late autumn to spring), especially during dry years (such as 1998 and 1999). In 1997 the total annual sediment transported through this cross-section was estimated at

70 000 t (±27 300); 10 000 t (±3900) in 1998 and 8000 t (±3120) in 1999. More than 90% of the sediment is exported during flood events (Rovira *et al.*, 2005a). The distribution of the particle sizes obtained for different floods shows no significant differences in the mobilization of the bed material despite the magnitude of maximum estimated water peak discharge (55, 23 and 13 m^3 $^{-1}$, respectively) (Fig. 3). In consequence, the mobilization of the river bed particles is similar, although a change in the hydraulic conditions is produced. In that case, the relation between τ_{ci} and the ratio D_i/D_{50} was constructed by grouping the largest particles collected in the bed load samples into 19 size classes. Then, a log–log least square regression was obtained: $\tau_{ci}^* = 0.102(D_i/D_{50})^{-0.80}$ ($r^2 = 0.71$, $N = 19$; $p < 0.05$), with standard errors of 0.1 for the intercept and the exponent. Similarly to the Fogars de Tordera cross-section, the intercept value (0.102) in this relation is higher than that from other studies. The sampling errors (e.g. probability of capturing the largest particle in the Helley-Smith sampler); the overestimation of the critical shear stress (e.g. bedform shear stress is not taken into account) and the use of local measurements (e.g. water depth) that are related to D_i, whereas the conditions associated with the entrainment are originated upstream, can explain this fact. The exponent *b* in this equation (–0.80) is closer to the value –1, suggesting a slight tendency towards equal mobility during those discharges. Overall, 36% of the sediment transported (as bed load) ranges between 2 and 4 mm and more than 80% of particles are estimated to be between 1 and 8 mm (Table 1). Those particle sizes (1–8 mm) represent the D_{30} and D_{80} of the grain size distribution of the river bed. The calculation of the critical shear stress for the coarser particle sizes (e.g. $>D_{84}$) shows that the motion of those fractions is initiated by peak discharges up to 44 m^3 s^{-1}.

Altogether, the results indicate that in both cross-sections, the entrainment of the major part of the river bed particles is produced during flood events. However, coarser fractions (>16 m) observed in the upper reaches are not reaching the outlet basin. In contrast, particles transported during low discharges and small floods at the Fogars de Tordera are evacuated at the Blanes Bridge cross-section during floods (Fig. 4). This

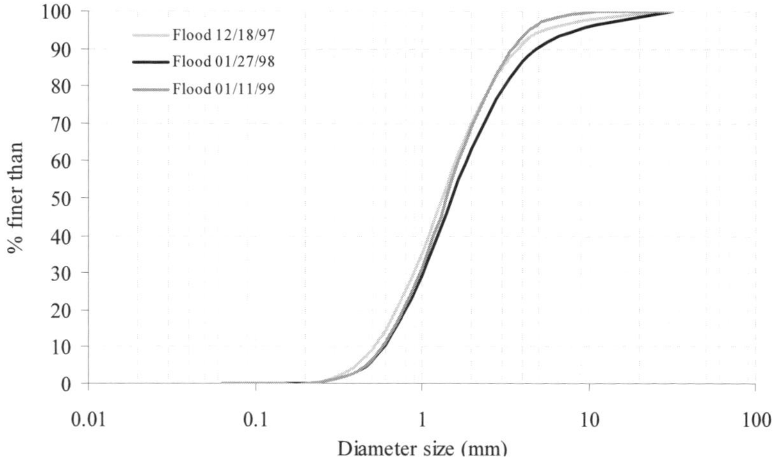

Fig. 3 Mean grain-size distribution of the bed load samples collected at the Blanes Bridge cross-section during the sampled floods.

fact can be associated with some combination of changes in both channel geometry (e.g. channel widening, slope reduction, etc.) that lead the reduction of the river's capacity to transport the largest grains; and the effects of the gravel extractions, which trap the coarser fractions in the large gravel pits. For instance, the discharge to entrain the coarser particles of the Fogars de Tordera riverbed in the lower cross-section is estimated at >100 $m^3 s^{-1}$. Field observations indicate that the aggradation of the bed is mainly produced where gravel mining was more intensive. In contrast, the water still tends to erode the bed channel in the lowermost sections where the channel geometry was not modified (Rovira *et al.*, 2005b). Similar results where observed by Gaillot & Piégay (1999) in the Fium Seccu Creek (Corsica). The existence of a large gravel pit in the mainstem trapped all of the gravels and cobbles and most of the fine sediment, thus interrupting the supply of bed load. This fact can be reinforced by a selective deposition through lateral or vertical sorting (e.g. Paola & Seal, 1995; Ferguson *et al.*, 1998; Seal *et al.*, 1998). Hassan *et al.* (1999) observed that large proportions of the coarse particles placed in a sandy river were found in the vicinity of the islands and in low point bars where changes in the hydraulic geometry lead to a reduction of the specific stream power.

During summer and early autumn, a tendency to accumulate material along the channel is produced (Table 1). Low flows and the small floods recorded at Fogars de Tordera are not reaching the basin outlet. Sand can accumulate to aggrade stream beds if sediment loads are particularly high or transport capacity is perennially or seasonally low (Knighton, 1999). In contrast, during periods when continuous water circulation is recorded along the study channel there is a re-entrainment of the sediment accumulated along the channel. Similarly in the Fraser River (Canada), McLean & Church (1999) observed that over the long-term, the suspended sand load estimated to be transported in the upper sections was approximately equal to that of the lower ones and over the course of individual freshet events, some of the sand becomes temporarily stored.

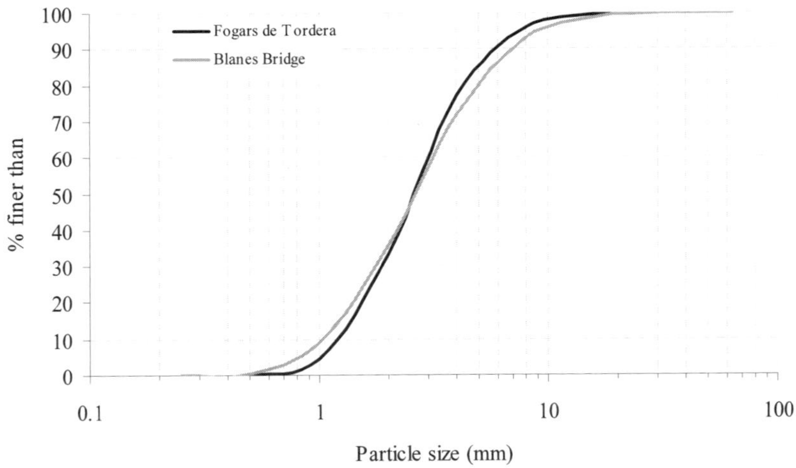

Fig. 4 Mean grain-size distribution of all bed load samples collected during the study period (1997-1999) at the Fogars de Tordera cross-section (low discharges and small floods) and Blanes Bridge cross-section (flood events).

CONCLUSIONS

In the upper reaches, approximately 80% of the time a partial mobilization of the river bed (D_5–D_{50} of the grain size distribution) is produced; whereas the general entrainment of the bed material probably arises during moderate and high discharges. The relation between τ_{ci}^* and the ratio D_i/D_{50} suggest a clearly size-selective entrainment, at least for discharges ≤ 4.5 m^3 s^{-1}. At the outlet basin, the mobilization of the major part of the river bed fractions (D_{30} and D_{80} of river bed distribution) is produced during floods. The relation between τ_{ci}^* and the ratio D_i/D_{50} suggest a tendency to equal mobility during those discharges. Coarser fractions (>16 mm) are not reaching the outlet basin. Both, a reduction of the capacity of the river to transport the largest grains and the effect of the gravel extractions, which trap the coarser fractions into the large gravel pits, mainly explain this fact.

A general tendency towards river bed aggradation is produced when the hydrological bimodality into the channel (perennial to ephemeral) is observed. But when there is water circulation along the channel, there is sediment transfer and the balance between sections is reached.

Acknowledgements Spanish Ministry of Education and Science funded this investigation (Grant HID96-0971). The Mediterranean Environmental Research Group (GRAM) at the University of Barcelona assisted during field-work. The author is highly grateful to Ramon Batalla for the comments on an early manuscript. Review by Michael Bliss Singer greatly improved the final version of the manuscript. The author is indebted to all of them. The paper was written when the author was a Postdoctoral fellow at the University of California at Berkeley.

REFERENCES

Andrews, E. D. (1983) Entrainment of gravel from naturally sorted riverbed material. *Geol. Soc. Am. Bull.* **94**, 1225–1231.

Ashworth, P. J. & Ferguson, R. I. (1989) Size-selective entrainment of bed load in gravel bed streams. *Water Resour. Res.* **25**, 627–634.

Batalla, R. J. & Martín-Vide, J. P. (2001) Thresholds of particle entrainment in a poorly sorted sandy gravel-bed river. *Catena* **44**, 223–243.

Batalla, R. J., Garcia, C. & Rovira, A. (2005) Sediment yield over a decade in a large Mediterranean river (The Tordera, Catalan Ranges, NE Spain). In: *Catchment Dynamics and River Processes. Mediterranean and other Climate Regions* (ed. by C. Garcia & R. J. Batalla,). Elsevier, Amsterdam, The Netherlands. (in press).

Church, M., McLean, D. G. & Wolcott, J. F. (1987) River bed gravels: sampling and analysis. In: *Sediment Transport in Gravel-bed Rivers* (ed. by C. R. Thorne, J. C. Bathurst & R. D. Hey), 43–88. John Wiley, Chichester, UK.

Collins, B. & Dunne, T. (1989) Gravel transport, gravel harvesting and channel-bed degradation in rivers draining the Southern Olympic Mountains, Washington, USA. *Environ. Geol. Water Sciences* **13**(3), 213–224.

Davis, J., Bird, J., Finlayson, B. & Scott, R. (2000) The management of gravel extraction in alluvial rivers: a case study from the Avon River, south-eastern Australia. *J. Phys. Geogr.* **21**(2), 133–154.

Ferguson, R. I. & Paola, C. (1997) Bias and precision of percentiles of bulk grain size distributions. *Earth Surf. Processes Landf.* **22**, 1061–1077.

Ferguson, R. I., Hoey, T. B., Wathen, S. J., Werritty, A., Hardwick, R. I. & Sambrook Smith, G. H. (1998) Downstream fining of river gravels: integrated field, laboratory and modeling study. In: *Gravel-Bed Rivers in the Environment.* (ed. by P. C. Klingeman, R. L. Beschta, P. D. Komar & J. B. Bradley), 85–114. Water Resour. Publ., Highland Ranch, Colorado, USA.

Garcia, C, Laronne, J. B. & Sala, M. (1999) Variable source areas of bedload in a gravel-bed stream. *J. Sed. Res.* **69**(1) 27–31.

Gaillot, S. & Piégay, H. (1999) Impact of gravel-mining on stream channel and coastal sediment supply: example of the Calvi Bay in Corsica (France). *J. Coast. Res.* **15**(3), 774–788.

Gilbert, G. K. (1917) Hydraulic-mining Debris in the Sierra Nevada. *US Geol. Survey Professional Paper* **105**.

Hassan, M.; Shick, A. & Shaw, P. (1999) The transport of gravel in an ephemeral sandbed river. *Earth Surf. Processes Landf.* **24**, 623–640.

Hill, H. (1896) Denudation as a factor of geologic time. *Trans. and Proc. New Zealand Inst.* **28**(11th of New Series), 666–679.

Kesel, R., Yodis, E. & McCraw, D. (1992) An approximation of the sediment budget of the lower case Mississippi River prior to major human modification. *Earth Surf. Processes Landf.* **17**, 711–722.

Knighton, D. (1999) Downstream variation in stream power. *Geomorphology* **29**, 293–306.

Kondolf, G. M. & Swanson, M. L. (1993) Channel adjustments to reservoir construction and instream gravel mining, Stony Creek, California. *Environ. Geol. Water Sci.* **21**, 256–269.

McLean, D. G. & Church, M. (1999) Sediment transport along lower fraser river: 2. Estimates based on the long-term gravel budget. *Water Resour. Res.* **35**(8), 2549–2559.

Ministerio de Economía (2002) Guía de las Buenas Prácticas Medioambientales de la Industria Extractiva Europea. Aplicación al caso español. *Eds. Dirección General de la Política Energética y Minas. Ministerio de Economía.* November 2002.

Misri, R. L., Garde, R. J. & Ranga Raju, K. J. (1984) Bed load transport of coarse non-uniform sediments. *ASCE J. Hydraul. Div.* **110**, 312–328.

Moore, R. J. & Newson, M. D. (1986) Production, storage and output of coarse upland sediments: natural and artificial influences as revealed by research catchments studies. *J. Geol. Soc.* **143,** 921–926.

Paola, C. & Seal, R. (1995) Grain size patchiness as a cause of selective deposition and downstream fining. *Water Resour. Res.* **31**(5), 1395–1407.

Parker, G., Klingeman, P. C. & McLean, D. G. (1982) Bedload and size distribution in gravel-bed streams. *ASCE J. Hydraul. Div.* **108**, 544–571.

Rapp, A. (1960) Recent developments of mountain slopes in Kärkevagge and surroundings, northern Scandinavia. *Geogr. Ann.* **42**(2/3), 65–200.

Reid, I. & Dunne, T. (2003) Sediment budgets as an organizing framework in fluvial geomorphology. In: *Tools in Fluvial Geomorphology* (ed. by G. M. Kondolf & H. Piegay), 463–500. John Wiley, Chichester, UK.

Richards, K. (1982) *Rivers. Forms and Process in Alluvial Channels*. Methuen & Co. Ltd, London, UK.

Rovira, A. (2001) Balanç de sediment i dinàmica fluvial en un riu de règim hidrològic transitori (Tram final de la Tordera). MSc Thesis, University of Barcelona, Barcelona, Spain.

Rovira, A., Batalla, R. J. & Sala, M. (2005a) Fluvial sediment budget of a Mediterranean river: the Lower Tordera (Catalan Coastal Ranges, NE Spain). *Catena* **60**, 19–42.

Rovira, A., Batalla, R. J. & Sala, M. (2005b) Response of a river sediment budget after historical gravel mining (the Lower Tordera, NE Spain). *Reg. Rivers Manage.* (in press)

Seal, R., Toro-Escobar, C., Cui, Y., Paola, C., Parker, G., Southard, J. B. & Wilcock, P. R. (1998) Downstream fining by selective deposition: theory, laboratory, and field observations. In: *Gravel-Bed Rivers in the Environment*. (ed. by P. C. Klingeman, R. L. Beschta, P. D. Komar & J. B. Bradley), 61–84. Water Resour. Publ., Highland Ranch, Colorado, USA.

Shields, I. A. (1936) Application of Similarity Principles and Turbulence Research to Bed-load Movement. Translate by W. P. Ott and J. C. Van Vchelin, US Soil Conservation Service Coop. Lab., California Inst. of Tech., Pasadena, California, USA.

Trimble, S. W. (1977) The fallacy of stream equilibrium in contemporary denudation studies. *American J. Sci.* **277**, 876–887.

Trimble, S. W. (1983) A sediment budget for Coon Creek basin in the driftless area, Wisconsin, 1853–1977. *American J. Sci.* **283**, 454–474.

Walling, D. E. (1984) Dissolved loads and their measurements. In: *Erosion and Sediment Yield: Some Methods of Measurements and Modelling* (ed. by R. F. Hadley & D. E. Walling), 111–177. Geo Books, London, UK.

Wilcock, P. R. (1988) Methods for estimating the critical shear stress of individual fractions in mixed-size sediment. *Water Resour. Res.* **24**, 1137–1151.

5 River Dynamics and Impacts

Accumulation of sediments within the channelized reach of the Upper Odra River, Poland

CZAJKA AGNIESZKA
Faculty of Earth Sciences, University of Silesia, Bedzinska 60, 41-200 Sosnowiec, Poland
czajka@wnoz.us.edu.pl

Abstract Construction of regular arrays of groynes along the Upper Odra channel created a new environment for sediment deposition. The inter-groyne infill is a significant sink for material transported in the Upper Odra channel. Sediment accretion rates were derived using artefacts as the alluvia age indicator. In the last 40 years the average rate of sediments accretion has been 5 cm year^{-1} while shortly after channel stabilization the rates ranged from 1.2 to 2.8 cm year^{-1}. During large floods up to 30 cm per event have been deposited.

Key words artefacts; overbank sedimentation; regulated river; stratigraphic markers

INTRODUCTION

It is widely recognized that the overbank storage of flood sediments generates the vertical accretion of a flood plain surface (Macklin, 1985; Magilligan, 1985; Macklin *et al*., 1992; Zwolinski, 1992; Moody & Troutman, 2000; Page *et al.*, 2003). Newly constructed groynes intensify the within-channel sedimentation creating an additional significant sink for the suspended material transported in the river channel. Accelerated storage of alluvia is both an anthropogenic and climatic effect on sediment budget equilibrium. The anthropogenic influences on fluvial sedimentation within regulated river reaches are so far not completely recognized. Such anthropogenically induced disturbances as land cover changes (Knox, 1999) or reservoir construction (Lajczak, 1995) are often mentioned by authors. Mean annual rates of deposition of suspended material are also controlled by factors such as human induced changes of the flood plain and river channel (Babinski & Klimek 1990; Macklin & Klimek, 1992; Ferguson & Brierley 1999; Ciszewski, 2002; Czajka-Kaczka, 2003). A number of recent studies have highlighted the possibility of detailed stratigraphic descriptions of sediments, thanks to relatively newly recognized, easily identified components of sediments such as heavy metals (Macklin & Klimek, 1992; Ciszewski, 2002), artefacts (Trimble, 1998; Czajka-Kaczka, 2003; Page *et al*, 2003), coal dust (French, 1998) or radionuclides e.g. ^{137}Cs (Froehlich & Walling, 1997; Walling & He, 1997). The above elements are human-made and authors have successfully used them to investigate accumulation rates of flood plains. European rivers have been regulated since the 16th century (in The Netherlands since 1000 AD) and the aim of this activity was to take control over the river processes (Hesselink, 2002). Regulated river channels are used for navigation purposes, and the construction of dykes concentrates flood flows within the narrow zone along the river channel, protecting arable land and urban areas against flooding. The Upper Odra channel regulation started in the 18th century (Born, 1948). The river flood plain is delimited by dykes and the channel width and depth are controlled by a regular array of groynes. As a result, the morphological activity of the river has been confined to the main river channel and during floods the sediments are deposited on the embankment flood plain. Construction of groynes trains the main flow away from the river banks but also results in deposition of material adjacent to the groyne structures (Ciszewski, 2002). This results in a decrease of the channel width from over 150 m before the rectification to approximately 50–70 m after the training works. This paper describes the rate of sediment accumulation within the channelized reach of Upper Odra River in the lee-side of the groynes (Fig. 1).

STUDY SITE AND METHODS

The reach of Upper Odra in question is a sand bed river with a catchment area of 7500 km^2. The river drains northward through the Raciborz Basin which is a

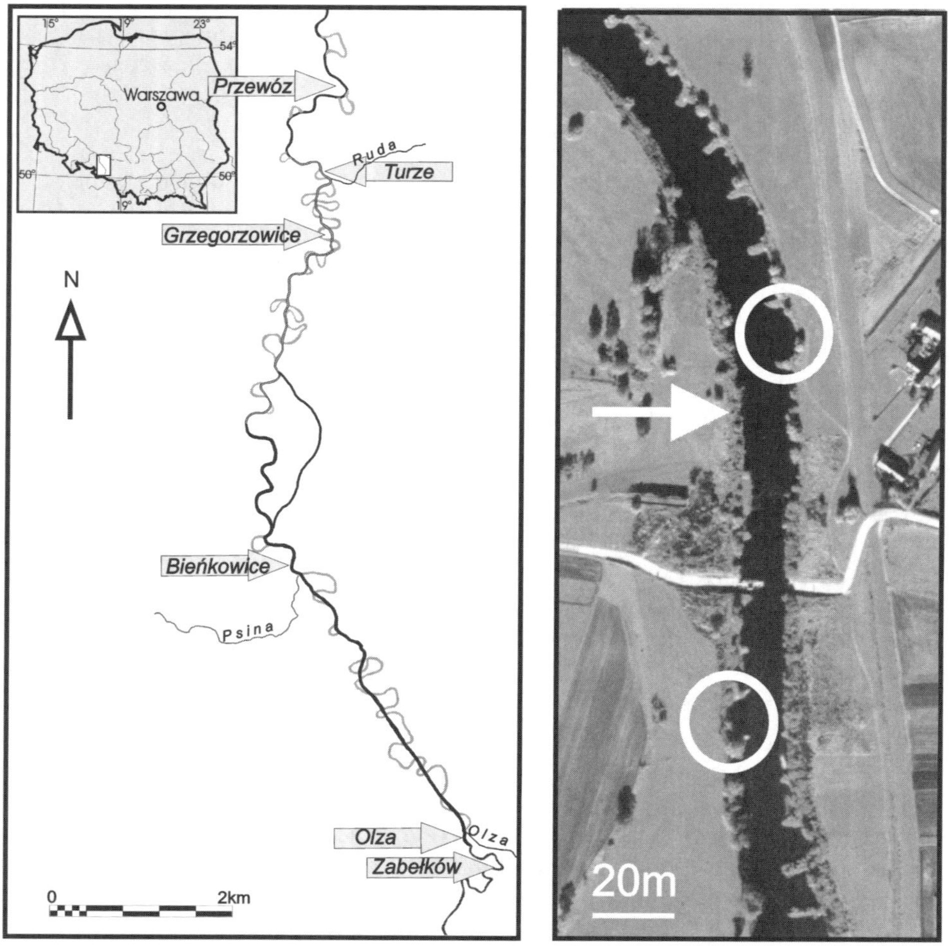

Fig. 1 Location of the study sites. The grey line indicates the channel course before the regulation. In the aerial photo the groynes are shown. White arrow points the Grzegorzowice site, the inter-groyne infill. Partly infilled basins are marked with white circles.

submontane depression. The Upper Odra catchment is underlain by Upper and Middle Miocene clay covered by glaciofluvial sediments of Pleistocene age. The headwater tributaries of the Upper Odra drain the most eastern part of the Sudety Mountains (in the Czech Republic) and the major tributary of the reach studied, the Olza River drains the western part of Carpathians which results in a large sand concentration in Upper Odra flood sediments. The mean discharge of the reach studied is approximately 65 m^3 s^{-1}. The coal-mining area around the river resulted in the need of river rectification for navigation purposes. The training works included the decrease of channel sinuosity by cutting off the meander bends, the decrease of channel width by constructing the groynes and reduction of the active flood plain by embankment. The works started in 1786. As an effect of channel shortening, the sinuosity of the Upper Odra channel decreased by 36%. The channel gradient increased by 37% to 0.4 m km^{-1}.The average bankfull width ranges from 30 to 50 m. After preliminary studies, the rate of overbank sedimentation along the channelized Upper Odra channel seems to be relatively high (Czajka-Kaczka, 2003). The stratigraphy of the Upper Odra overbank deposits was investigated in six sites adjacent to the river channel (Fig. 1). Sedimentary structures of the alluvia were described from exposures excavated in the river bank. Detailed stratigraphic description and sediment sampling for textural analyses were completed in each of the exposures. Sediments from each stratum of the horizontally accreted alluvia were collected and analysed for grain size, coal dust content and the presence of various artefacts. Flood sediments within the alluvia of Upper Odra were dated using two different age indicators: coal dust and artefacts. The river's proximity to the coal mine area resulted in the presence of coal dust in the flood sediments. The first coal mines within the Upper Odra catchment were developed in the Ostrava region in 1780. It is likely that the presence of coal mines in the Upper Odra catchment would be registered in alluvia if flood sediments of that age were present. Various artefacts within the horizontally accreted layers offered an opportunity to establish the contemporary rates of overbank sedimentation.

RESULTS AND DISCUSSION

Excavations were dug down until remnants of the 19th century bank revetments were found. Depth of the excavations ranged from 3 m in Bienkowice to 4.2 m in Grzegorzowice site. At all sites, alluvia consisted of flat- and cross-bedded, well-sorted alternating sands and silty sands with mean grain size between 4.2 and –1.08 phi. Thickness of a single stratum ranged from 0.5 to 70 cm. An examination of the alluvia age indicators allowed age estimates to be made and chronostratgraphies to be constructed. The coal dust was present in alluvia of each of the studied sites but only in one case, in Olza, coal dust was absent in the lowest part of the unit indicating that this material was deposited before 1780. Various artefacts within the alluvia studied gave an opportunity to estimate the sedimentation rates. These artefacts included e.g. a plastic bag in Grzegorzowice at a depth of 2 m, a sweet wrapper in Bienkowice at a depth of 110 cm, pieces of linoleum in Bienkowice, at a depth of 2.6 m, aluminium foil in Przewoz at a depth of 2.37 m and remains of groyne construction at the bottom of the excavations in Bienkowice, Grzegorzowice and Przewoz. The production of plastic in Poland started after the Second World War so it is highly probable that the deposition of alluvia containing artefacts occurred within the last 50 years. By determining the depth of the strata containing artefacts, more detailed age estimates were obtained. Rates of sediment accumulation per year were defined in thickness (cm year^{-1}) and calculated by dividing the thickness of sediments deposited above the stratum containing the dated artefact by the number of years since the artefact deposition. The rate of sedimentation can be expressed as:

$$\overline{R}_s = S_A / (Y_P - Y_{S_A})$$

where R_S = average rate of sedimentation (cm year^{-1}), S_A = depth of stratum containing artefact of known age (cm), Y_P = year of deposition of top layer of the unit and Y_{SA} = year of artefact deposition. Depths of artefacts deposition demonstrate high rates of vertical accretion within the studied reach over the past 200 years, peaking within the last 50 years. Table 1 presents the sediment accumulation rates in cm year^{-1} for dated periods. The accretion rates indicate differences among the reach studied (Fig. 2). The highest rate was noted in Bienkowice site for the period 1997–2001. During that time two large floods took place. Since the river banks in Bienkowice are the lowest among all sites studied the deposition at the time of the latest flood events was the most significant. The relatively high rate noticed in Bienkowice may be also influenced by the Psina tributary which drains the loess area west from the studied site. The lowest accretion rate was determined for Grzegorzowice site (Table 1). Variations in

Table 1 Average annual sediment accumulation rates for dated periods

Zabelkow		Olza		Bienkowice		Grzegorzowice		Przewoz	
period	rate (cm year^{-1})	period	rate (cm year^{-1})	period	rate (cm year^{-1})	period	rate (cm year^{-1})	period	rate (cm year^{-1})
2001–1997	1.5–8.5	2000–1954	0.1–3.5	2001–1994	1.5–30	2001–1987	1.1–4.5	2001–1997	0.1–10
1997–1985	3–7	1954–1882	0.5–3.9	1994–1980	1.5–7	1987–1954	1–12.5	1997–1960	1–10.5
1985–1905	1.1–4.5	–	–	1980–1950	1–5	1954–1850	1–2	1960–1789	0.1–3.3

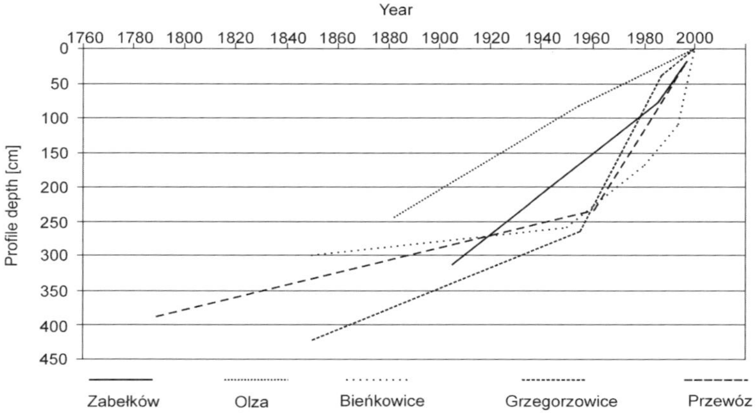

Fig. 2 Cumulative sedimentation curves for the study sites.

sediments rates between the study sites are associated with the differential height of the river banks. A number of factors influence the rate of deposition. The human-influenced changes of channel geometry created a new environment for sediment deposition. The accretion takes place within the reach with limited flood plain area where construction of the regular arrays of groynes stopped the lateral migration, forcing continuity of vertical accretion at the stabilized river banks. Channel rectification caused an increase of the rate of water energy per unit channel length and, as a consequence, the magnitude of sediment transport increased. As a result, during floods, significant amounts of sediment are deposited.

CONCLUSIONS

Calculated accretion rates were derived using artefacts as the alluvia age indicator. From the methodological point of view it confirms the utility of artefacts for dating young alluvia deposited adjacent to the groynes, which limit the channel width and prevent significant erosion of the sediments. Measurements represent time-averaged results. A progressive increase in annual accretion rate occurred in the most recent 40 years and reached an average over 5 cm year^{-1} while shortly after channel stabilization the rates raged from 1.2 to 2.8 cm year^{-1}. Calculating the accretion in cm per year shows only the average deposition, yet single flood events may deposit thick layers of material up to 0.3 m in the studied reach of Upper Odra. The results suggest that, after channel stabilization, deposition becomes the dominating process forming the river banks. At present the banks are a significant sink for the suspended material transported in the Upper Odra channel.

Acknowledgements The author would like to thank Michael Church for thoughtful comments and conscientious critique of the manuscript.

REFERENCES

Babiński, Z., & Klimek, K. (1990) The present day channel and floodplain of the Vistula River. In: *Evolution of Vistula River Valley during the last 15000 years, Part III, Geographical Studies* (ed. by L. Starkel), Special Issue no. 5, 62–75.

Born, A. (1948) *Monography of Odra*. Instytut Zachodni, Poznan, Poland (in Polish).

Ciszewski, D. (2002) Heavy metals in vertical profiles of the middle Odra River overbank sediments: evidence for pollution changes. *Water, Air, and Soil Pollut.* **143**, 81–98.

Czajka-Kaczka A. (2003) The rate of sedimentation on the regulated rivers floodplains, the upper Vistula and the upper Odra Rivers, Southern Poland. In: *Palaeofloods, Historical Data and Climatic Variability: Applications in Flood Risk Assessment* (ed. by V. R. Thorndycraft, G. Benito, M. Barriendos & M. C. Llasat). PHEFRA International Workshop.

Ferguson, R. J. & Brierley, G. J. (1999) Levee morphology and sedimentology along the lower Tuross River, south-eastern Australia. *Sedimentology* **46**, 627–648.

French, P. W. (1998) The impact of coal production on the sediment record of the Severn Estuary. *Environ. Pollut.* **103**, 37–43.

Froehlich, W., & Walling, D. E. (1997) The role of unmetalled roads as a sediment source in the fluvial systems of the Polish Flysch Carpathians. In: *Human Impact on Erosion and Sedimentation* (ed. by D. E. Walling & J. L. Probst) (Proc. Rabat Symp., April 1997), 159–168. IAHS Publ. 245. IAHS Press, Wallingford, UK.

Hesselink, A. W. (2002) History makes a river. Morphological changes and human interference in the river Rhine. PhD Thesis, University of Utrecht, The Netherlands.

Knox, J. C. (1999) Long-term episodic changes in magnitudes and frequences of floods in the Upper Mississippi River Valley. In: *Fluvial Processes and Environmental Change* (ed. by A. G. Brown & T. A. Quine), 255–282. John Willey & Sons Ltd, Chichester, UK.

Lajczak, A. (1995) The impact of river regulation, 1850–1990, on the channel and floodplain of the Upper Vistula River, Southern Poland. In: *River Geomorphology* (ed. by E. J. Hickin), 209–233. Wiley, Chichester, UK.

Macklin, M. & Klimek, K. (1992) Dispersal, storage and transformation of metal-contaminated alluvium in the upper Vistula basin, southwest Poland. *Appl. Geogr.* **12**, 7–30.

Macklin, M. G., Rumsby, B. T. & Newson, M. D. (1992) Historical floods and vertical accretion of fine-grained alluvium in the Lower Tyne Valley, Northeast England. In: *Dynamics of Gravel-bed Rivers* (ed. by P. Billi, C. R. Thorne & P. Tacconi), 573–589. John Wiley & Sons Ltd, Chichester, UK.

Macklin, M. (1985) Flood-plain sedimentation in the upper Axe Valley, Mendip, England. *Trans. Inst. Br. Geogr.* **10**, 235–244.

Magilligan, F. J. (1985) Historical floodplain sedimentation in the Galena River basin, Wisconsin, Illinois. *Ann. Assoc. Am. Geogr.* **75**(4), 583–594.

Moody, J. A. & Troutman, B. M. (2000) Quantitative model of the growth of floodplain by vertical accretion. *Earth Surf. Processes Landf.* **25**, 115–133.

Page, K. J., Nanson, G. C. & Frazier, P. S. (2003) Floodplain formation and sediment stratigraphy resulting from oblique accretion on the Murrumbidgee River, Australia. *Sedimentary Geology* **13**(1), 5–14.

Trimble, S. W. (1998) Dating fluvial processes from historical data and artefacts. *Catena* **31**, 283–304.

Walling, D. E. & He, Q. (1997) Use of 137Cs fallout in investigations of overbank sediment deposition on river floodplains. *Catena* **29**, 263–282.

Zwoliński, Z. (1992) Sedimentology and geomorphology of overbank flows on meandering river floodplain. *Geomorphology* **4**, 367–379.

Anthropogenic impacts on sediment discharge in the dry-hot valleys of SW China—the example of Longchuanjiang

X. X. LU[1], Y. M. ZHU[1,2] & Y. ZHOU[2]

1 *Department of Geography, National University of Singapore, 10 Kent Ridge Crescent, 119260 Singapore*
geoluxx@nus.edu.sg

2 *Department of Environmental Science, Kunming University of Science & Technology, Kunming, Yunnan, 650093, China*

Abstract The sediment discharge data measured in a dry-hot valley, Longchuanjiang basin in the Lower Jinsha tributary, were analysed with Mann-Kendall, Seasonal Mann-Kendall and Sen's tests. In the upper reach, water discharge and sediment yield increased by 2.0 and 2.48% per year, respectively, during the period from 1970 to 2001. In the lower reach they increased by 1.95 and 3.94%, respectively. The sediment discharge showed a significant increase (0.05 at Xiaohekou and 0.01 at Xiaohuangguayuan), despite numerous reservoir constructions and insignificant changes in water discharge. This could be attributed to the intensification of human activities, especially those related to land surface disturbance such as deforestation and afforestation, expansion of agriculture land, and road construction. The increase was more significant in the lower reach of the river due to the dry-hot climate. The profound increase observed in sediment discharge has significant implications for effective management of the on-going Three Gorges reservoir project.

Key words afforestation; dams; deforestation; dry-hot valley; sediment discharge

INTRODUCTION

Sediment discharge in the river is influenced by human activities such as reservoir construction and land use alteration, including afforestation and deforestation, intensification of agriculture, draining of wetlands and urbanization. Land use change can alter the evapotranspiration regime, the initiation of surface flow, water quality and soil erosion process on a basin scale or even influence the climate at local or global scale (Calder, 1993). Land use changes such as land conversion, road and reservoir construction, and urban area expansion have been observed in most parts of China during the past 50 years as a result of population pressure and socio-economic policies. This is particularly true in the dry-hot valleys where large populations are located. The dry-hot valleys, characterized by lower rainfall and higher temperature, are often found in the mountainous areas in southwest China (Zhang *et al.*, 2001). Despite lower rainfall, these valleys are heavily cultivated due to the abundance of solar energy, and form the agricultural basis in local areas.

Longchuanjiang basin is located in the Lower Jingsha region of the Upper Yangtze basin which was identified as one of the areas with the highest sediment yield in the Upper Yangtze basin (Dai & Tan, 1996; Lu & Higgitt, 1998). The lower part of the basin is a typical dry-hot valley in southwest China. Compared to other parts of the basin, the dry-hot valley has a lower rainfall (620 mm) and a higher temperature (21.9°C), resulting in a higher evaporation and high arid index. Because of the harsh climate in the dry-hot valley, the areas have a lower threshold "capacity" to human disturbance and become more susceptible to soil erosion. This paper aims to examine the time series of sediment discharge from 1970 to 2001 in the Longchuangjiang River, and to investigate the possible impact of human activities such as land use alteration, road and reservoir construction on sediment discharge in the dry-hot valley.

STUDY AREA

The research site, Longchuanjiang River, is located between 24°45′N~26°15′N and 100°56′E~102°02′E, in southwest China. It is a tributary of Jinsha River (the upper reach of Yangtze River). The length of the main channel is 231.2 km. Location of the river is shown in Fig. 1. There are two gauging stations along the river, Xiaohekou in the upper stream and Xiaohuangguayuan in the lower stream. The drainage sizes of Xiaohekou and Xiaohuangguayuan are 1788 and 5560 km^2, respectively. This basin has experienced a wide range of common human activities such as deforestation from the 1950s to the 1980s, afforestation since the late 1980s, construction of roads, reservoirs and canals, and the expansion of irrigated farmland.

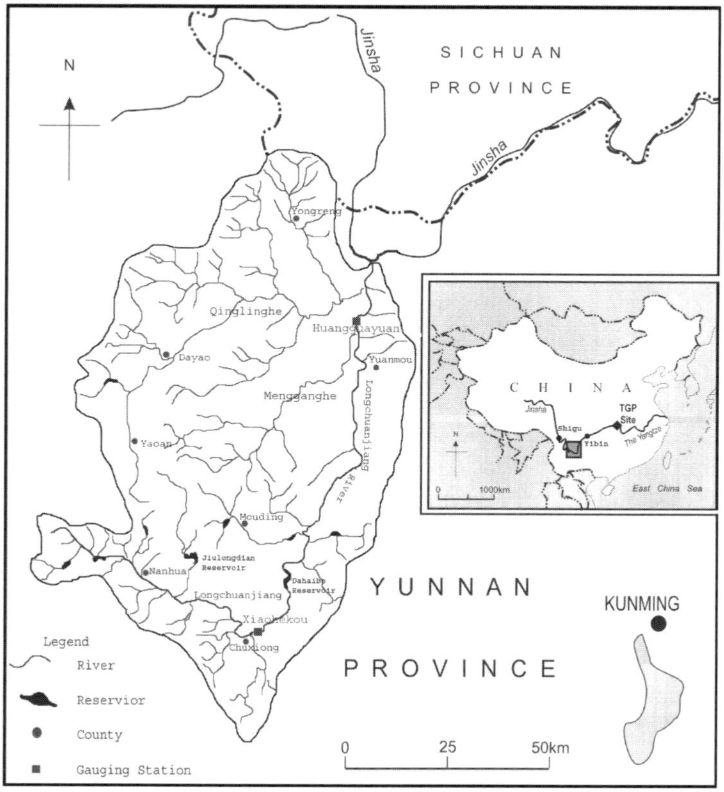

Fig. 1 Location and river system of the Longchuanjiang basin.

DATA AND METHODS

Monthly sediment discharge data from 1970 to 2001 were collected. The trend of three variables: annual sediment discharge, annual maximum monthly sediment discharge, and annual maximum daily sediment discharge, were tested with the Mann-Kendall test. Kendall's correlation of coefficient (Kendall's tau), an effective and general measure of correlation between two variables (Kendall, 1938; Mann, 1945), is extensively used for testing the trend in hydrological data (Helsel & Hirsch, 1992; Burn & Hag-Elnur, 2002; Xiong & Guo, 2004; Yue & Pilon, 2004). The trends of monthly sediment discharge were tested with the Seasonal Kendall test. Seasonal Kendall Trend Test (SKTT) is a modified version of the standard Kendall test to assess the trends in data which have a significant seasonal component (Hirsch *et al.*, 1993). The slope of the changing trend was tested by Sen's method. Sen's method is a nonparametric method used to estimate the value and confidence interval for the slope of an existing trend. It is widely used in hydrological data analysis because of its characteristics such as allowing missing data, gross data errors and outliers, and making no assumptions on distribution of the data.

Annual sediment discharge changes

Time series of annual mean sediment discharges, their five year moving average, the long-term means and the trend lines (regression) at the two stations are plotted in Fig. 2. The long-term annual mean sediment discharges at Xiaohekou and Xiaohuangguayuan are 0.47 and 5.55 million tons, respectively. At Xiaohekou, the annual sediment discharge has increased since 1982. For Xiaohuangguayuan, the sediment discharge has been experiencing sharp increase since 1990. During the 12 years from 1990 to 2001, only two years, 1991 and 1992, have sediment discharge lower than the long-term mean. In terms of overall trend, both stations show a trend of increase by linear regression. The results of MK test are listed in Table 1. All the three variables have a positive trend at both stations. The percentage of Sen's slope estimator for the long-term annual mean is 2.49% at Xiaohekou and 3.95% at Xiaohuangguayuan.

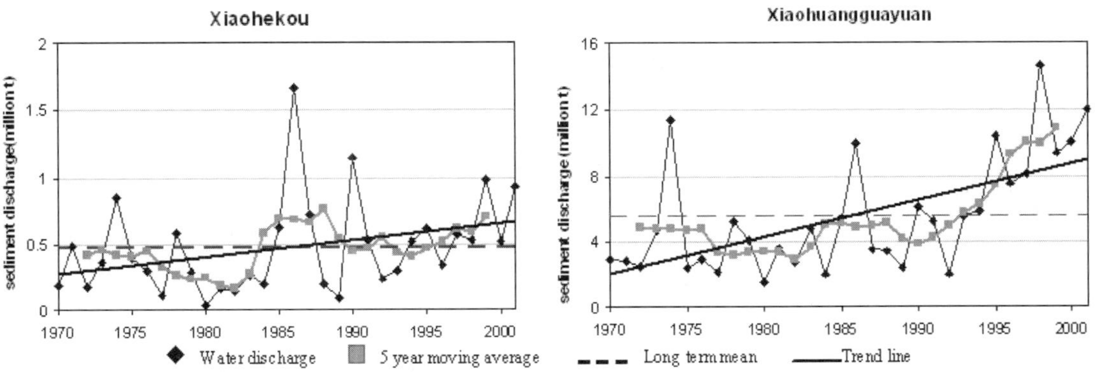

Fig. 2 Annual sediment discharge at Xiaohekou and Xiaohuangugayuan.

Table 1 Results of sediment discharge trend analysis with Mann-Kendall test.

Station	Dataset	n*	Mann-Kendall Test Statistics			Sen's Slope Estimator	
			Z	Trend	α**	Q	Q(%)
Xiaohekou	annual total discharge (10^6 t)	32	1.99	+	0.05	0.012	2.49
	maximum monthly discharge(kg s^{-1})	32	2.04	+	0.05	1.95	2.59
	maximum daily discharge(kg s^{-1})	32	1.95	+	0.1	14.97	2.12
Xiaohuang-guayuan	annual total discharge(10^6 t)	32	3.45	+	0.01	0.22	3.95
	maximum monthly discharge(kg s^{-1})	32	3.32	+	0.01	31.16	3.69
	maximum daily discharge(kg s^{-1})	32	1.87	+	0.1	139.72	1.95

* number of years from 1970 to 2001, ** level of significance.

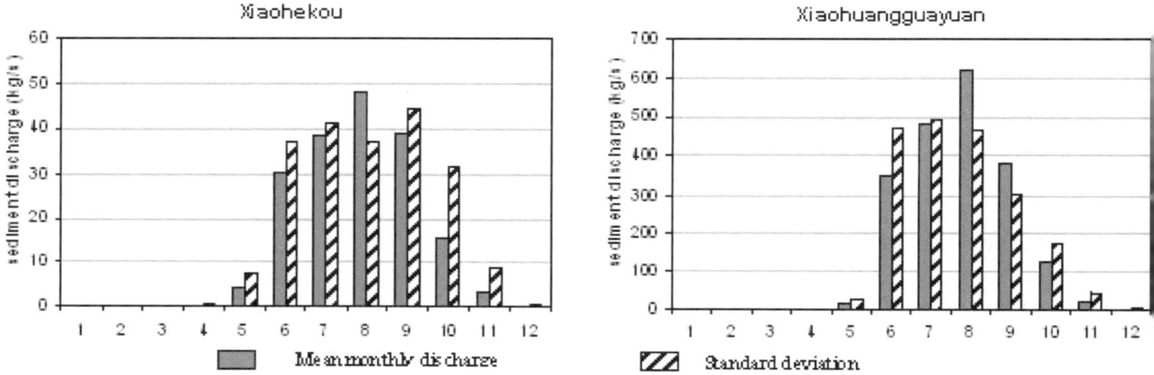

Fig. 3 Long-term monthly mean sediment discharge and standard deviation.

Seasonal sediment discharge changes

The mean monthly sediment discharge during wet season and its deviation are plotted in Fig. 3. The sediment discharge is highest in August and the amount of sediment load in three months, July, August and September, accounts for more than 80% of the total annual load for both stations. The standard deviation is very high compared to the long-term mean. Apart from August, it is higher than the long-term mean for the rest of the months.

The results of SMK test are listed in Table 2. Except for November, the monthly sediment discharge in the wet season shows a positive trend. At Xiaohekou, the sediment discharge has a significant increase (at 0.05 level) in September. At Xiaohuangguayuan, the increasing trend is also significant (at 0.05 level) in July, August and October, and is significant at 0.01 level in September. Xiaohuangguayuan has a higher percentage of Sen's slope estimator for most of the months than Xiaohekou.

Possible reasons for the sediment changes

The data from Chuxiong weather station, located in the upper part of Longchuanjiang basin, show an increase in temperature and a slight increase in precipitation since the 1950s. However, in the lower part (mainly Yuanmou dry-hot valley), the temperature

Table 2 Results of sediment discharge trend analysis with Seasonal Mann-Kendal test.

Station	Dataset	n*	Mann–Kendall Test Statistics			Sen's Slope Estimator	
			Z	Trend	α**	Q	Q(%)
Xiaohekou	May	32	0.03	+		0.012	0.29
	Jun	32	1.09	+		0.347	1.14
	Jul	32	1.54	+		0.706	1.84
	Aug	32	1.64	+		1.135	2.35
	Sep	32	2.11	+	0.05	1.088	2.79
	Oct	32	1.18	+		0.171	1.12
	Nov	32	–0.62	–		–0.001	–0.02
Xiaohuang–guayuan	May	32	0.46	+		0.010	0.06
	Jun	32	1.51	+		7.993	2.08
	Jul	32	2.59	+	0.01	16.130	3.06
	Aug	32	2.5	+	0.05	14.867	2.32
	Sep	32	2.92	+	0.01	15.341	3.85
	Oct	32	2.24	+	0.05	3.904	3.19
	Nov	32	–1.17	–		–0.040	–0.19

* number of years from 1970 to 2001, ** level of significance.

has decreased by 1.11°C and rainfall has increased by 131.8 mm from the 1950s to the 1990s (Zhang *et al.*, 2001), showing an opposite trend to the global climate change. Zhang *et al.* (2001) suggested that the climate change in the Yuanmou dry-hot valley was chiefly the result of the expansion of irrigated land, which led to the increase in evaporation and humidity of the air and ultimately resulted in the decreased temperature and increased precipitation.

Sen's slope (%) of temperature, precipitation, water discharge and sediment discharge changes are plotted in Fig. 4. The annual water discharge, annual sediment discharge and annual precipitation all show a positive trend, but the trends for water discharge are not significant for both stations. Annual increase in precipitation (0.84 and 0.63%) and water discharge (2.0 and 1.95%) are much slower than annual increase in sediment (2.48 and 3.94%). This suggests that the climate variation has a very limited impact on sediment discharge.

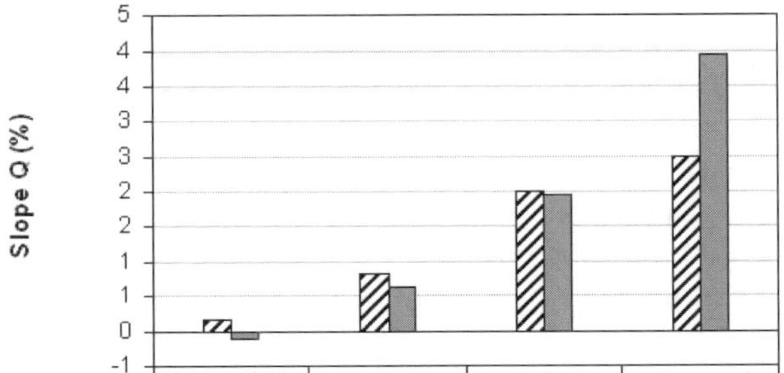

	temperature	precipitation	water	sediment
xiaohekou	0.179	0.838	2.005	2.487
xiaohuangugayuan	-0.115	0.627	1.957	3.947

Fig. 4 Sens's slope (%) of temperature, precipitation, water discharge, sediment discharge changes (1970~2001).

The forest cover in Longchuanjiang basin has changed significantly since the 1950s. It decreased from 55 to 24.1% from the 1950s to the 1980s due to population pressure and unsustainable decision making on socio-economic development. Since the late 1980s, many ecological engineering projects aimed at coordinating the economic development and environmental protection have been carried out in the basin by national to local level government. Most of these ecological engineering projects took reforestation as a measure and thus led to the increase of the forest area. In 1999, the overall forest cover percentage of the basin is 39.5%. But in some places under harsh environmental conditions, such as the dry-hot valley, the recovery of vegetation is very difficult once the vegetation has been destroyed. It has been observed that after deforestation, the transpiration and the water storage capacity of the soil will decrease, thus leading to an increase of overland water input into the river (Calder, 1993; Chen, 1999). Besides, the destruction of plant cover, together with other developing activities, will usually intensify the problem of soil erosion and finally result in the increasing sediment discharge in the river.

Many engineering projects such as road construction, mining and urban construction also contribute to soil erosion in the basin. During road building, the natural surface of the slope is cut and the vegetation on the land is cleared. The excavated materials comprising earth and stone are usually thrown into the river or heaped up along the road. When storms occur, most of the earth flows into the river, increasing the suspended sediment concentration of the water. Furthermore, in the area between Xiaohekou and Xiaohuangguayuan which is susceptible to frequent debris flows, the debris flows are conducted into Longchuajiang River to ensure safety of the roads. The length of road in Chuxiong has increased from 506 km in 1952 to 14 081 km in 2003. It is estimated that the sediment yield from the road construction is 100 t km^{-2} $year^{-1}$, and the increase of the sediment discharge in Xiaohuangguayuan gauging station since 1989 is partially due to road construction (Chuxiong Hydraulics Bureau, 2003).

In contrast, construction of reservoirs and dams could have reduced sediment discharge. Since the 1950s, 14 medium size reservoirs (storage capacity between 10 to 100 million m^3), 67 small size I (storage capacity between 1 to 10 million m^3) reservoirs, 418 small size II (storage capacity between 0.1 to 1 million m^3) reservoirs and 30 060 ponds, have been built. These reservoirs and dams with a total capacity of more than 300 million m^3 can trap considerable amounts of sediment. The information of deposition of nine medium size reservoirs in this basin is listed in Table 3. Dahaibo was built on the main channel of Longchuanjiang River between Xiaokekou and Xiaohuangyuan in 1959. It is estimated that 90% of the sediment passed through Xiaohekou is trapped by the Dahaibo reservoir (Chuxiong Hydraulics Bureau, 2003).

Sediment yields measured in seven plots (Table 4), in 10 small size II reservoirs and in the two gauging stations in the river are plotted in Fig. 5. The average sediment yield from the plots, reservoirs and river are 19 700, 3537, and 601 t km^{-2} $year^{-1}$, respectively. The sediment yield from the plots is much higher than yields from reservoirs and rivers, suggesting that these reservoirs might be an important control on sediment delivery to the main river (Lu & Higgitt, 2001). However, the construction of these reservoirs did not change the increasing trend of sediment discharge in the river. This contrasts reports in other tributaries of the Upper Yangtze basin (Dai & Tan, 1996; Lu & Higgitt, 1998). For example, Lu & Higgitt (1998) reported that the observed decreases in sediment yield were due to the large reservoir schemes on the tributary rivers.

Table 3 Sediment deposition in reservoirs.

Reservoir	Construction year	Reservoir storage (10^6 m^3)	Total deposition (10^6 m^3)	(%)
Maobanqiao	1957	19.4	6.8	35.1
Jiulongdian	1958	63	5.74	9.1
Dahaibo	1958	33	7.8	23.6
Qingfeng	1969	11.2	8.93	79.7
Menglian	1983	10.4	7.5	72.1
Bingjian	1982	17.8	2	11.2
Zhongtun	1991	11	1.1	10.0
Laochanghe	1992	15.7	3.83	24.4
Xijinhe	1999	11.2	5	44.6

Table 4 Soil erosion rate estimated in runoff plot (after Yunnan Hydraulic Bureau, 1987).

Site	Plot size (m^2)	Slope (degree)	Land use	Erosion rates (t km^{-2} year^{-1})
Lufeng	5 × 5	39	Wheat & bean	6 910
Dayao	10 × 10	30	Wheat & bean	5 400
Yuanmou	10 × 10	20	Barren land	34 140
Yuanmou	5 × 5	19	Barren land	28 000
Yuanmou	10 × 10	23	Barren land	18 900
Yuanmou	10 × 10	16	Barren land	21 600
Yuanmou	10 × 10	41	Barren land	22 950

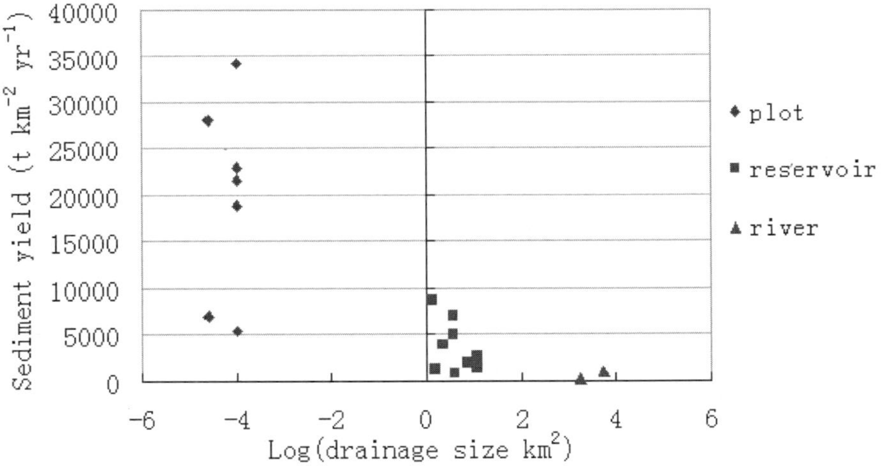

Fig. 5 Sediment yield measured at plots, reservoirs and river.

CONCLUSIONS

This study indicates that Longchuanjiang Basin, a typical hot-dry valley in southwest China, has experienced a significant increase in sediment discharge, notably in the lower part of the basin, as a result of human activities. Anthropogenic activities related to land surface disturbance such as deforestation and afforestation, agriculture land expansion, and road construction play an important role in this context. Although a significant amount of sediment was trapped by the reservoirs constructed since the 1950s, the increase in sediment discharge is still obvious. Though there has been a tremendous effort on reforestation within the river basin, this increase has become more pronounced since the 1990s. At Xiaohekou, the average annual sediment yield increased from 240.3 t km^{-2} $year^{-1}$ before 1990 to 309.1 t km^{-2} $year^{-1}$ after 1990. At Xiaohuangguayuan, it increased from 725.9 t km^{-2} $year^{-1}$ to 1453 t km^{-2} $year^{-1}$. The profound changes in hydrological regimes resulting from such human activities have significant implications for effective management of soil erosion control.

This study provides an example of time series changes in sediment discharge within a dry-hot valley in the Lower Jinsha tributary of the Upper Yangtze River over the past decades. Attempts have been made to evaluate the impacts of climate change and human activity on seasonal changes. While some of the changes can be explained, some other issues such as climate variation and detailed land use changes and their impact on the water discharge and sediment flux require further attention. A further quantitative analysis of contributions to sediment discharge from different factors is in process.

Acknowledgements This research is funded by National University of Singapore and Chinese National Basic Research Program (also called 973 programme) (project no: 2003CB415105-6). Acknowledgement goes to Mrs Huang Ying and Mr Guo Youan, Yunnan Hydrological and Water Resources Bureau for their support during field work. Moshe Inbar and Adolfo Calvo are thanked for a thorough review of the original manuscript.

REFERENCES

Burn, D. H. & Hag-Elnur, M. A. (2002) Detection of hydrologic trends and variability. *J. Hydrol.* **255**, 107–122.

Calder, I. R.(1993) Hydrologic effects of land-use change. In: *Handbook of hydrology* (ed. by D. R. Maidment). McGraw-Hill, New York, USA.

Chen, G. W. (1999) Forest hydrological effects in the upper Yangtze River valley. *Ambio* **28**, 456–459.

Chuxiong Hydraulics Bureau (2003) *Report on the soil and water conservation project in Longchuanjiang River* (in Chinese).

Dai, D. Z.& Tan, Y. (1996) Soil erosion and sediment yield in the Upper Yangtze River Basin. In: *Erosion and Sediment Yield: Global and Regional Perspectives* (ed. by D. E. Walling & B. W. Webb) (Proc. Exeter symposium, July 1996), 191–203. IAHS Publ. 236. IAHS Press, Wallingford, UK.

Hirsch, R. M., Helsel, D. R., Cohn, T. A. & Gilroy, E. J. (1993) Statistical treatment of hydrologic data. In: *Handbook of Hydrology* (ed. by D. R. Maidment). McGraw-Hill Inc., New York, USA.

Helsel, D. R. & Hirsch, R. M. (1992) *Statistical Methods in Water Resources.* Elsevier Publishers, Amsterdam, The Netherlands.

Kendall, M. G. (1938) A new measure of rank correlation. *Biometrika* **30**, 81–93.

Lu, X. X. & Higgitt, D. L. (1998) Recent changes of sediment yield in the Upper Yangtze, China. *Environmental Management* **22**, 697–709.

Lu, X. X. & Higgitt, D. L. (2001) Sediment delivery to the Three Gorges. 2: Local response. *Geomorphology* **41**, 157–169.

Mann, H. B. (1945) Nonparametric test against trend. *Econometrica* **13**, 245–259.

Xiong, L. H. & Guo, S. L. (2004) Trend test and change-point detetion for the annual discharge series of the Yangtze River at the Yichang hydrological station. *Hydrol. Sci. J.* **49**, 99–112.

Yunnan Hydraulic Bureau (1987) *Soil erosion investigation report of Chuxiong Region* (in Chinese).

Yue, S. & Pilon, P. (2004) A comparison of the power of the t test, Mann-Kendall and bootstrap tests for trend detection. *Hydro. Sci. J.* **49**, 21–37.

Zhang, J., Yang, Z., Wang, D. & Zhang X. (2001) Climate change and causes in the Yuanmou dry-hot valley of Yunnan, China. *J. Arid Environ.* **51**, 153–162.

The change in flood regime along the Lower Jordan River: its influence on flood plain land use

MICHA KLEIN

Department of Geography and Environmental Studies, Haifa University, Mount Carmel, Haifa 31905, Israel

m.klein@geo.haifa.ac.il

Abstract Cooperation among riparian parties along the Lower Jordan River (LJR) may promote development of the region's natural resources. Cooperation requires understanding of the river's present flow regime, water quality and current land use in the flood plain. The average annual flow in the LJR has decreased from 1166 Mm^3 $year^{-1}$ (1926–1945) to 167 Mm^3 $year^{-1}$ (1978–2000), of which about 70% (120 Mm^3), are of good water quality. Due to the extremely dry years, the allocation of water to the state of Jordan of 50 Mm^3 $year^{-1}$ after the 1994 Peace Agreement and the barrage built on the Yarmouk (YR) at Adasia, the average yearly flow of the last decade was less than 120 Mm^3. At present the river's cross section area is in a re-adjustment phase and is becoming much smaller. The LJR flood plain is not inhabited and the population lives outside the area, yet the flood plain is used for agricultural purposes.

Key words discharge; land use; Lower Jordan River; riparian cooperation

THE JORDAN BASIN

The Jordan River is situated in the Jordan Rift Valley and drains an area of 17 665 km^2 (Schattner, 1962; Fig. 1). The Rift Valley trends north–south and constitutes a 15 km wide depression in which the Dead Sea acts as a regional base level. The depression is bound on both sides by normal faults (Freund *et al.*, 1970; Begin, 1974). The average annual precipitation decreases from 1600 mm in the north (Mount Hermon) to 250 mm south of the Sea of Galilee (SOG), and to about 100 mm near the Dead Sea. Most of the valley bottom south of the SOG is covered with Lisan marl, deposited in Lake Lisan, which occupied the region during the last glaciations, 60 000 to 18 000 years B.P. (Begin, 1974; Begin *et al.*, 1985). The drainage basin is divided into two parts. The upper Jordan River drains into the SOG while the Lower Jordan River (LJR) is situated between the SOG and the Dead Sea. Hence, floods occurring in the upper Jordan River have no direct impact on the LJR (Marwan & Klein, 2002). The meandering LJR is incised in fine sediment of the Lisan formation.

The Lower Jordan River flow regime

The SOG and YR are the two main sources of water to the LJR. The majority of the tributaries are *wadis* or seasonal streams that carry water only during a limited number of days each winter. The flow regime of the river has been altered several times. In 1932, a barrage was built at the exit of the river from the SOG to regulate the outflow

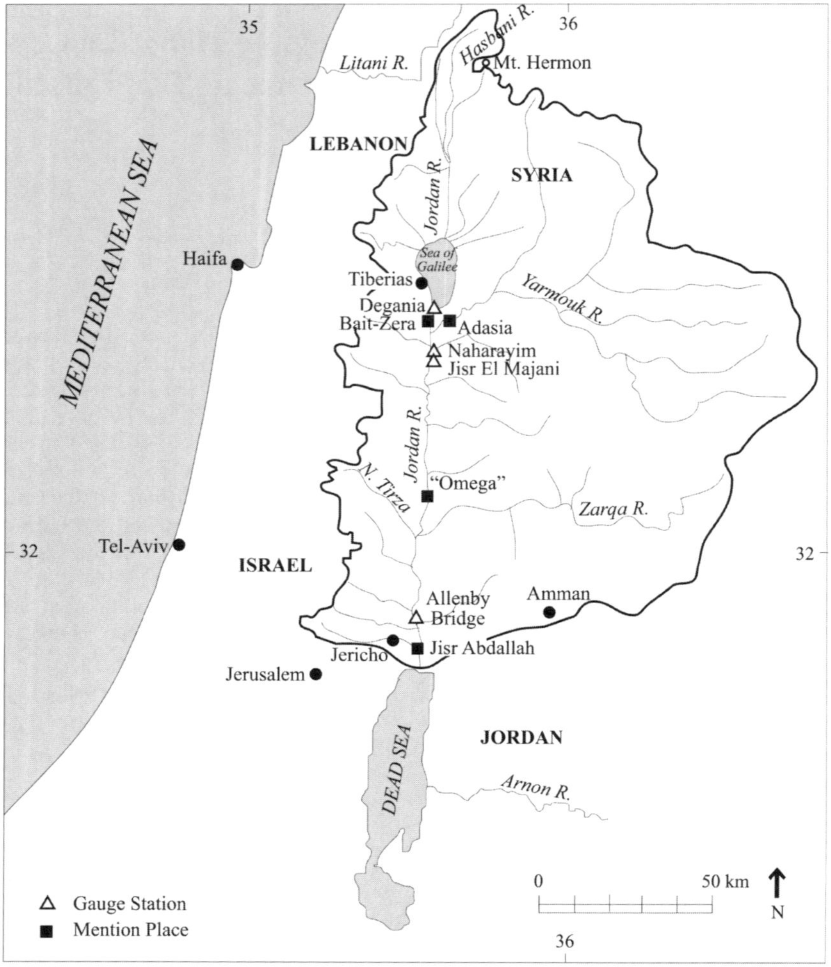

Fig. 1 Map of the study area.

for hydropower purposes. The main goals of the dam were to reduce the winter flows and to store water for summer release. After 1948, the power station was abandoned and the barrage became part of the irrigation system of the LJR (Schattner, 1962). In 1964, the Israeli National Water Carrier was completed which diverted water from the SOG. In the late 1960s and early 1970s, Jordan diverted a large proportion of the base flow from the YR. In the late 1980s, Syria built several dams on the tributaries of the YR, retaining a large proportion the YR water. In the 1990s further reductions occurred due to 50 Mm^3 $year^{-1}$ supply of water to the Jordanian state and the barrage built by Jordan on the YR at Adasia.

Flow data have been collected for various periods along the LJR. The Daganya station, located near the outlet of the SOG, has operated since 1926. The other stations are Naharayim (1977/8–1999/00), Jisr El Majama (1934–1944), and Allenby Bridge (1926/7–1945/6). The Naharayim and Jisr El Majama stations can be considered jointly because of their proximity (6 km) to each other and because there is no tributary between them. Although the stations cover two different periods of record, they can be used to assess changes in the flow regime of the LJR due to water utilization.

Table 1 The Lower Jordan River flow data. (The 1926/6–1945/6 data measured at Allenby Bridge (Schattner, 1962); 1978–2000 measured at Naharayim.)

Year	Q annual (Mm^3)	Rank	Q peak $m^3 s^{-1}$	Rank
1977/78			357	8
1978/79	67	13	40	14
1979/80	476	2	469	3
1980/81	471	3	289	9
1981/82	117	10	450	4
1982/83	383	5	646	2
1983/84	86	11	384	6
1984/85	125	9	358	7
1985/86	41	16	108	13
1986/87	133	8	268	10
1987/88	236	6	443	5
1988/89	32	18	2	22
1989/90	30	20	5	20
1990/91	25	22	3	21
1991/92	647	1	723	1
1992/93	412	4	192	11
1993/94	51	14	6	18
1994/95	145	7	154	12
1995/96	42	15	6	19
1996/97	41	17	36	17
1997/98	76	12	37	15
1998/99	31	19	1	23
1999/00	27	21	37	16
Average	167			
1926/7	1419	6		
1927/8	890	16		
1928/9	1680	1		
1929/30	1090	12		
1930/1	1550	3		
1931/2	770	19		
1932/3	800	18		
1933/4	648	20		
1934/5	1494	4		
1935/6	823	17		
1936/7	1289	7		
1937/8	1198	9		
1938/9	1182	10		
1939/40	1082	13		
1940/1	1105	11		
1941/2	1279	8		
1942/3	1650	2		
1943/4	1001	14		
1944/5	1444	5		
1945/6	919	15		
Average	1166			

The LJR flow data are given in Table 1. The mean annual flow has been altered significantly. Prior to the water diversions, the annual discharge of the LJR ranged between 648 and 1680 Mm^3 $year^{-1}$ with a mean discharge of 1166 Mm^3 $year^{-1}$, for the years 1926–1946 (Schattner, 1962). Between 1964 and 1970, Israel diverted about 200 Mm^3 $year^{-1}$ from the basin from the SOG. Since 1971, this amount has increased to about 350 Mm^3 $year^{-1}$, in addition to some 60 Mm^3 $year^{-1}$ pumped for local use (Klein, 1998). As a result, the annual discharge at Daganya has decreased from 480 to about 70 Mm^3 $year^{-1}$ (Klein, 1998). The mean annual discharge of the YR is estimated at 445 Mm^3 $year^{-1}$, of which Jordan initially diverted during the mid 1960s about 70 Mm^3 $year^{-1}$. These quantities were subsequently increased to about 200 Mm^3 $year^{-1}$ during the 1970s. In addition, about 150 Mm^3 $year^{-1}$ have been diverted by Syria (Klein, 1998, 1999). The overall flow dropped from 1166 Mm^3 $year^{-1}$ in the early 1960s to less than 120 Mm^3 $year^{-1}$ in the 2000s. From 1981 to 1993 there were eight occasions when the gates of the Daganya dams were opened for water release to the Dead Sea as a result of limited storage in the SOG. No water was released during 1994–2004. The period of 1996–2002 was a succession of dry years during which the level of the SOG was lowered considerably.

Figure 2 presents the recurrence interval of annual flow in the LJR before and after water diversion. As the data before and after are not from the same station we have to reduce some 150 Mm^3 which is the estimated natural flow into the river between Naharayim and Allenby Bridge. The recurrence interval line for the "new" flow regime is lower but parallel to the "old" line. However in low flows the "new" line has a much lower slope than the "old" line because the low flow of the LJR is controlled by water from the saline aqueduct which drains saline water directly to LJR, and by sewage water which is not related to rain.

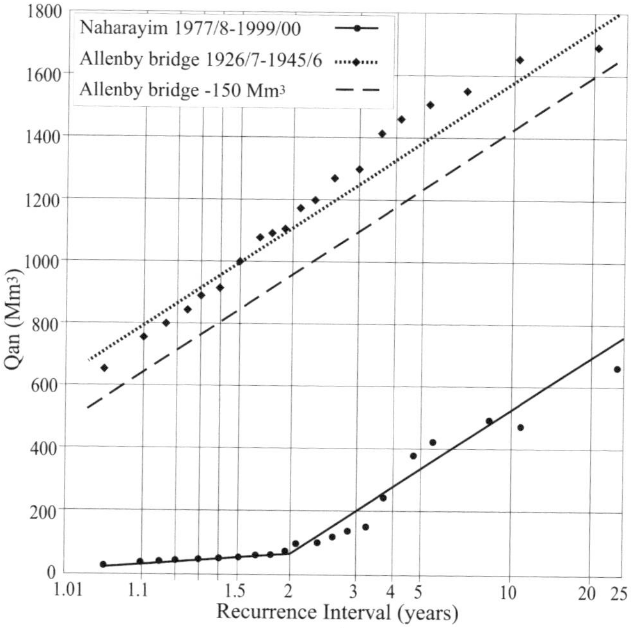

Fig. 2 The recurrence interval of annual flow of the LJR before and after water diversion.

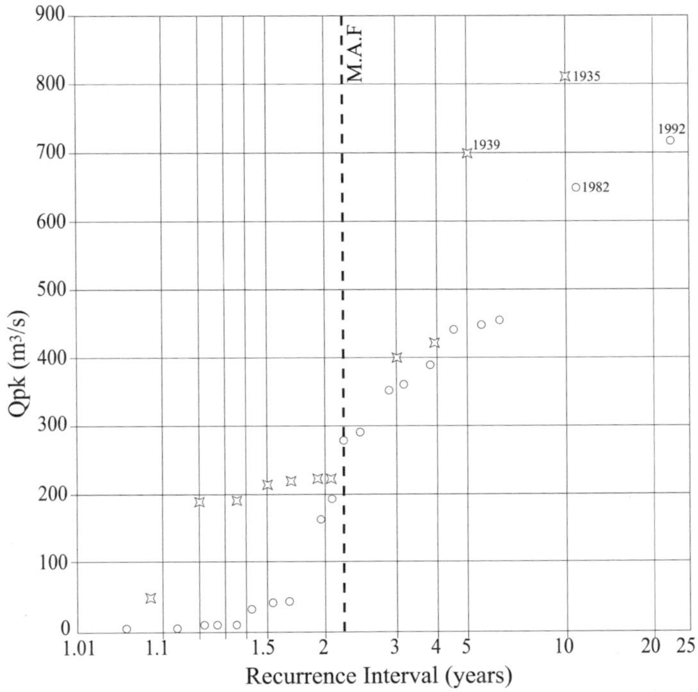

Fig. 3 The recurrence interval of annual flood of the LJR before and after water diversion (note that the slope of the "regression line" above the M.A.F. is steeper for the "old" data).

Figure 3 presents the frequency of annual peak flows of the LJR at Naharayim (1978–2000), and at Jisr El Majama (1934–1945). Most contemporary floods in the LJR are due to contributions from the YR only. During years of heavy rainfall the flow from the SOG may contribute up to about 150 $m^3 s^{-1}$ (the maximum discharge through the Daganya dam).

The Lower Jordan River peak flows

The highest measured peak in the LJR at Naharayim was 829 $m^3 s^{-1}$ in 1935 (Ministry of Agriculture, 1993 data correction). Since 1978 the highest peak was 723 $m^3 s^{-1}$ in 1992 (Schattner, 1962 refers to 1700 $m^3 s^{-1}$ as the highest estimate peak). The two largest floods in the LJR were recorded during the 1934–1944 period. However, the flood data with a recurrence interval of 2–5 years for the two periods overlap (Fig. 3). Both periods yielded about the same mean annual flood, recurrence intervals of 2.33 years, of 250–300 $m^3 s^{-1}$. Comparison of equal spans of time from 1934 to 1945 to the recent span from 1989 to 2000 (11 years is the length of record in at Jisr El Majama), shows a very sharp decrease in the 2.33 recurrence interval flood, from 250 to 300 $m^3 s^{-1}$ to about 40 $m^3 s^{-1}$. The effect of water utilization becomes evident for flows with recurrence interval less than 1.8 years, where data from Jisr El Majami station exceed those from Naharayim by a factor of seven.

The Lower Jordan River water quality

The relationship between discharge and Cl concentrations in the LJR at Naharayim is shown in Fig. 4. Up to a discharge of 2 m^3 s^{-1} the Cl concentrations are around 2000 mg l^{-1}, meaning that the Saline Water Aqueduct (diversion of saline spring from the SOG to the LJR) is the only contributor to the river flow. In discharges higher than 10 m^3 s^{-1} the Cl concentrations are less than 400 mg l^{-1}, and in most cases lower than 200 mg l^{-1}. Hence the flood water is useable for agriculture. A detailed study of the LJR water chemistry was published recently (Farber *et al.,* 2004).

Flood plain land-use

Land-use of the LJR during the 19th century has been described by Lynch (1849). The area was covered by natural forest mainly form by *Tamarix* trees (Fig. 5). A survey of flood plain land use south of Beit-She'an valley was done using air photo from 1986. The river divides the flood plain unevenly; two thirds of the area extends to the east of the river. The main land uses of a total area of 70 km^2 are 26 km^2 (37%) cultivated east of the river, 14 km^2 (20%), cultivated west of the river; the channel accounts for 7 km^2 (10%), and 22 km^2 (32%) are uncultivated areas. The "Omega" case aptly demonstrates the role of flooding and politics on land use. Figures 6(a) to 6(d), illustrate the change in river morphology, and changing land uses in the LJR flood plain. Figure 6(a) was taken in December 1968 during a very high flood. The time is a year and a half after the 1967 war, and the land use still represents the period when both sides of the river were under Jordanian rule. Both sides of the river are cultivated. Figure 6(b) is dated March 1974,

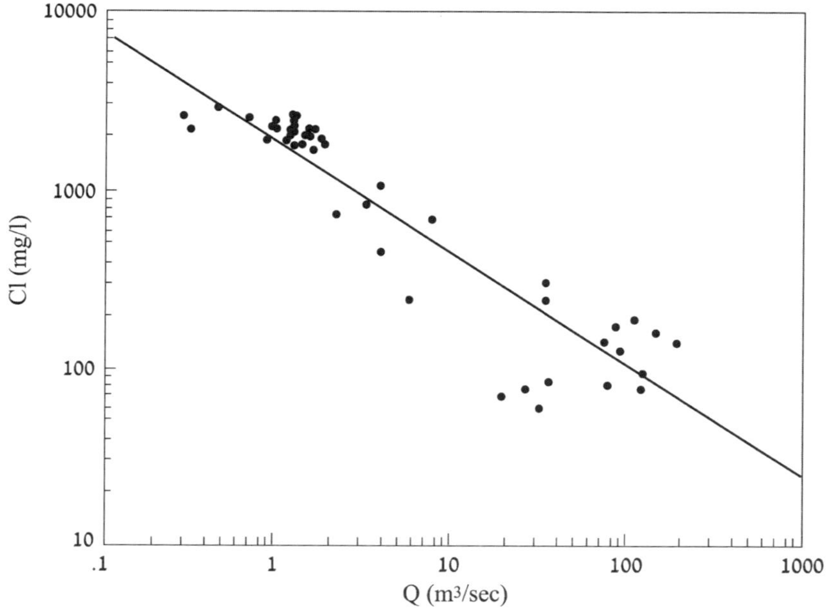

Fig. 4 The relationship between discharge and Cl concentration in the LJR at Naharayim.

seven years after the 1967 war. The river changed its course and developed a classic Omega-shape meander. The west–east segment of the water line (the north side of the Omega) shifted to the south by 250 m. Since 1967 the river is defined as the cease-fire line between Israel and Jordan. Nevertheless Israel does not cultivate the flood plain areas; hence the area is covered mainly by natural vegetation (*Tamarix*). The land on the east side is cultivated. Figure 6(c) is from 1980. Vegetation cover on the west side is very dense, while farmers on the east side cultivate the new areas exposed by the shifting of the river. The photo was taken during flood and it can be seen that because of the different land use the flooding covers a larger area on the east side. In the early 1980s the Israeli government decided to utilize the flood plain for agriculture. The vegetation cover on the west side was removed and a patrol road was built along the river (Fig. 6(d), 1982). As a result of the March 1983 flood, with a peak discharge of 646 m^3 s^{-1}, which caused flooding on both sides of the river, a survey of the damage along the LJR was prepared (Fig. 7). In this segment of the flood plain 2648 dunam (1 dunam = 1000 m^2) were prepared for cultivation in the years 1981–1982. The flood water covered about 1300 dunam, of which 110 dunam were severely eroded. The 2285 m of the security fence along the river, which originally had a length of 6460 m in this study segment, were destroyed at 14 points, mainly by collapse of the fence. The fence acted as a trap for the eroded vegetation from the flood plain during the falling stage of the flood, as a result of the pressure on the fence toward the river. Fence damage was also caused by undermining. The patrol road was eroded at four points along a total length

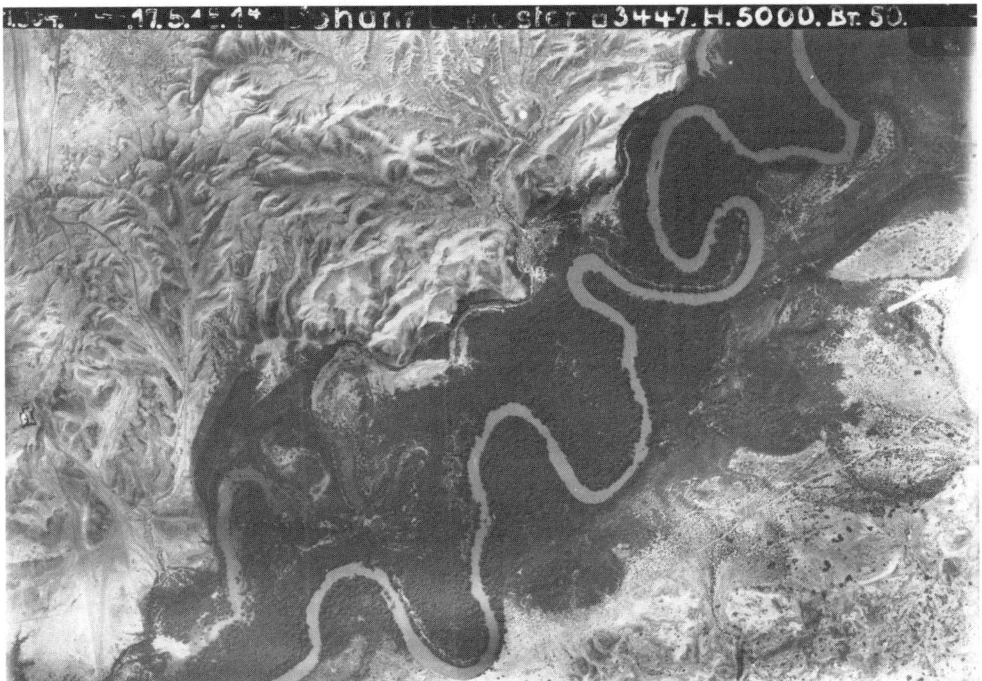

Fig. 5 Natural forest cover along the LJR. Air Photo 22 March 1918. The Air Photo was taken from the collection of the German Air Force Air Photo. Munich: Bayer. Hauptataatarchiv: 1918 (in possession of the library of Haifa University).

of 465 m. Work on the banks of the LJR remains problematic since the right bank is under Israeli, and the left bank under Jordanian control. Bank protection is usually done by building a groin or a wall, often without taking into account the effect on the opposite side. In the study section eight walls were built on the right bank. Owing to repeated damage to the fence and the patrol road which is not transit-able for long periods, the new patrol road was straightened and paved away from the river (Fig. 8).

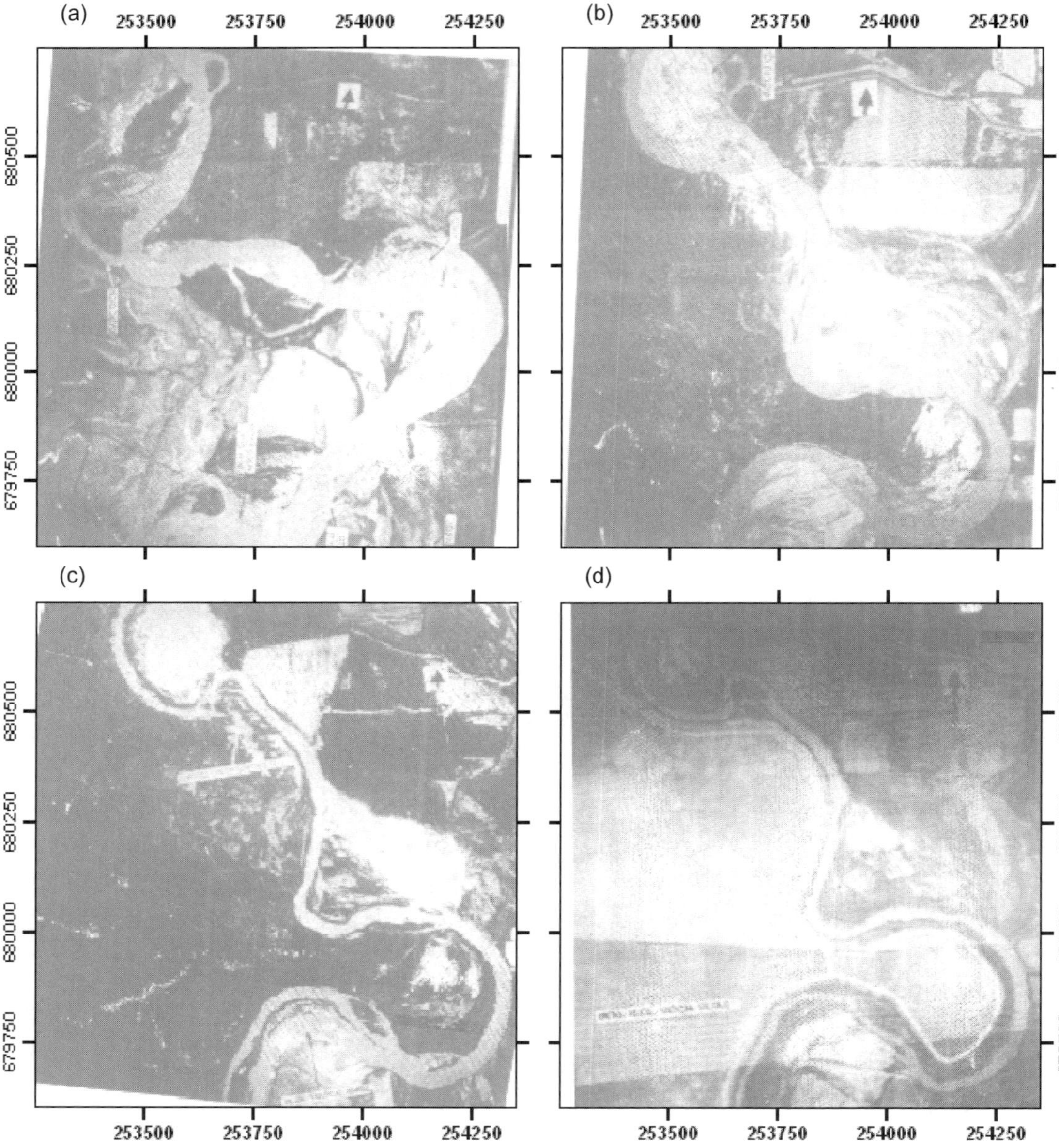

Fig. 6 Land use changes at the "Omega". (a) 1968; (b) 1974; (c) 1980; (d) 1982.

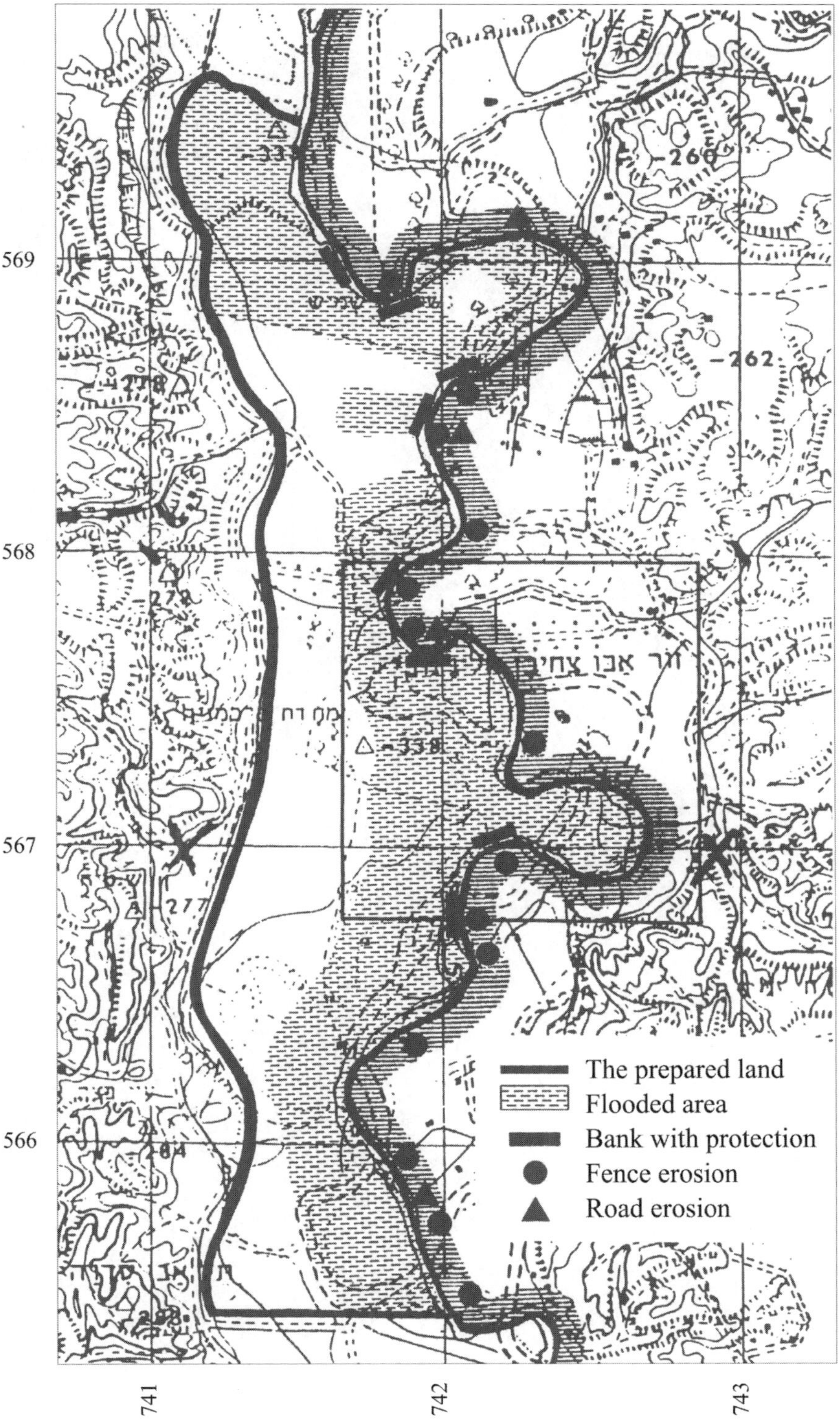

Fig. 7 Map of damages in the study area following the 1983 flood.

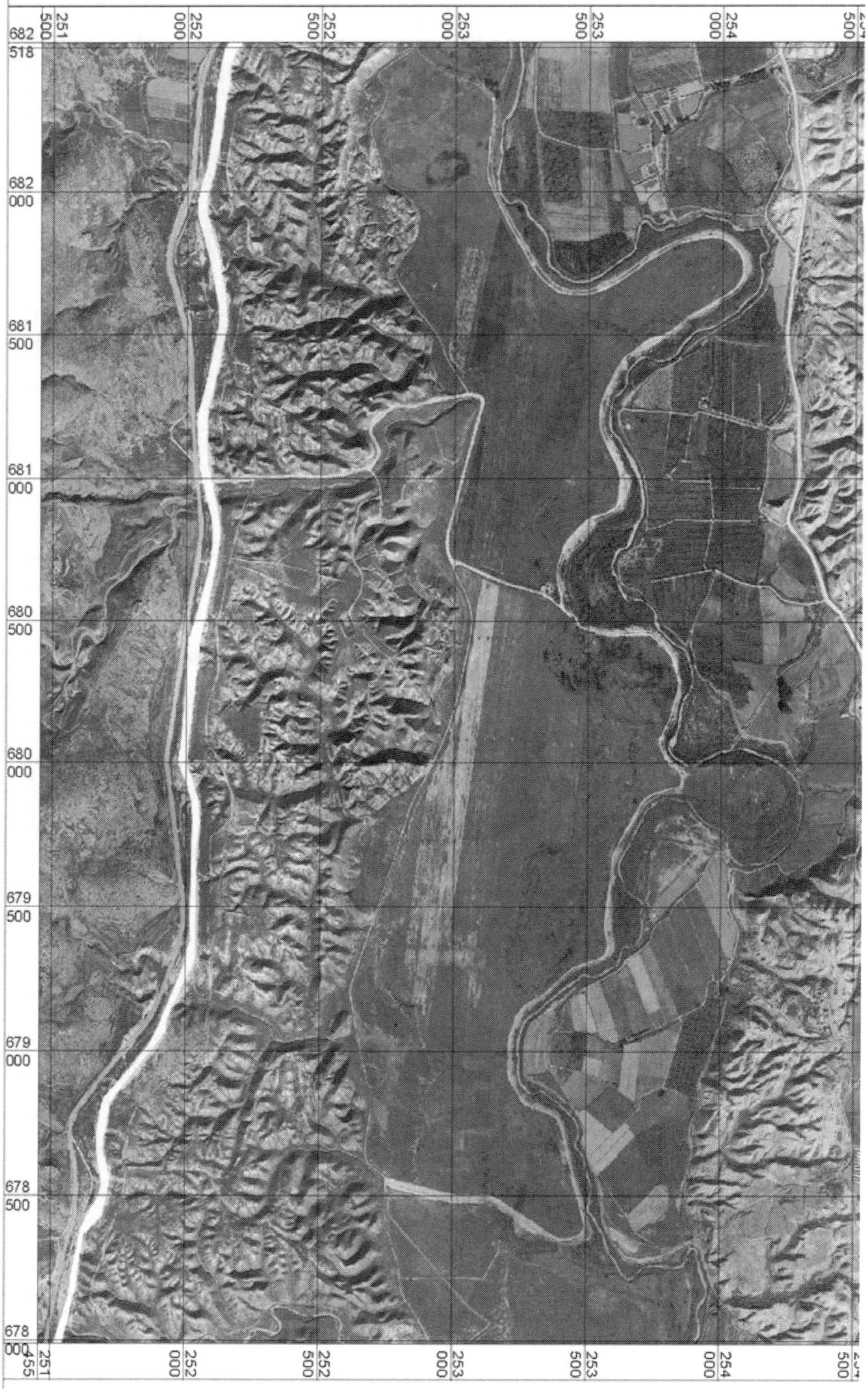

Fig. 8 The "Omega", 2003. Notice the new road away from the river.

PLANNING RECOMMENDATIONS

Cooperation between Israel and Jordan, including the diversion of 50 Mm^3 $year^{-1}$ from the SOG and construction of the barrage at Adasia, both after the 1994 Peace Agreement, brought about a further decrease in the annual flow and a reduction in the annual peak flows and flood plain flooding. Further cooperation for flood plain development would entail adopting a discharge rate which would be accepted as the "new" desired bank-full discharge. Flooding could then be minimized by controlling the river cross-section and controlling the river's slope by a man-made artificial "cut-off" and increased flow velocity. An artificial meander "cut-off" may support agricultural development by enabling land transfer between the two parties, and facilitating large scale cultivation. Near Beit-Zera in the northern section, where both banks are in Israel, five man-made artificial "cut-offs" were excavated along a 4-km river section. Development of recreation activities and tourism along the LJR (e.g. baptizing in the Jordan River) implies the improvement of water quality. For the time being the saline summer flow and sewage water does not encourage such activities. Recreation along the river requires the release of good-quality water not merely for agriculture. If areas along the river are declared as natural reserves, the river banks will not be artificially controlled, because the dynamics of a river that changes its course characterizes a natural process, especially along a meandering river.

The area from Beit-She'an valley to the Dead Sea (about 70 km^2) should be divided according to the demands of both parties for the following purposes:

(a) Agricultural land use—in large parts the farmers possess cultivation rights.
(b) Recreation areas—such as camping, river boating, and baptismal site along the LJR.
(c) Natural reserves—walking areas in which no other activity is permitted.

Owing to the characteristics of a pluvial landscape system, migrating laterally across the entire flood plain width, land uses on both sides of the river must be compatible. Land use planning, based on cooperation between Israel and Jordan (like the cooperation that took place in water allocation) should be done across the entire width of the flood plain and not along the channel. This is required in order to prevent conflicting uses, e.g. cultivation with the use of chemical fertilizers on one side *vs* a natural reserve which attempts to avoid chemicals on the other side.

Acknowledgements Author would like to thank Olav Slaymaker for the helpful comments while reviewing the paper.

REFERENCES

Begin, B. Z. (1974) The geology in the vicinity of Jericho. *Geol. Surv. Isr. Bull.* **67**.

Begin, B. Z., Broecker, W., Buchbinder, B., Druckman, Y., Kaufman, A., Magaritz, M. & Neev, D. (1985) Dead Sea and Lake Lisan levels in the last 30 000 years: a preliminary report. *Report Geol. Survey of Israel No. 29/85,* 18 pp.

Farber, A., Vengosh, A., Gavrieli, I., Marie, A., Bullen, T. D., Mayer, B., Holtzman, R., Segal, M. & Shavit, U. (2004) The origin and mechanisms of salinization of the Lower Jordan River. *Geochimica et Cosmochimica Acta* **68**(9), 1989–2006.

Freund, R., Garfunkel, Z., Zak, I., Goldberg, M., Weissbord, T. & Derin, B. (1970) The shear along the Dead Sea Rift. *Philos. Trans. R. Soc. London ,Ser. A*, 107–130.

Klein, M. (1998) Water balance of the Upper Jordan River basin. *Water Int.* **23**, 244–248.

Klein, M. (1999) Water balance and water resources developments in the upper Jordan basin. *Water and Irrigation* **394**, 22–27 (in Hebrew).

Lynch, W. F. (1849) *Narrative of the United States' expedition to the River Jordan and the Dead Sea* (ed. by R. Zeevy).. MOD Publishing house., Israel (in Hebrew).

Marwan, A. H. & Klein, M. (2002) Fluvial adjustment of the Lower Jordan River to a drop in the Dead Sea level. *Geomorphology* **45**, 21–33.

Ministry of Agriculture (1993) *Hydrological Year Book of Israel.* Water Commission, Hydrological Service, Israel.

Schattner, I. (1962) The Lower Jordan Valley. *Scr. Hierosolymitana* **XI**.

Downstream effects of river impoundment on hydrological and geomorphological aspects of bedrock rivers (Golan Heights, Israel)

NURIT SHTOBER-ZISU[1] & MOSHE INBAR[2]

1 *Department of Israel Studies, University of Haifa, Haifa, 31905, Israel*
nshtober@research.haifa.ac.il

2 *Department of Geography, University of Haifa, Haifa, 31905, Israel*

Abstract Construction of dams along rivers affects the hydrological regime and the geomorphology. Two similar catchments were analysed: the impounded Daliyyot River (with impounded runoff index IR = 37%) and the non-impounded Meshushim River (IR = 1%). In the impounded basin, the specific average peak discharge and the annual average water volume discharge decreased by more than 50%; the recurrence interval time period increased for highest peak flows and the bankfull flow area decreased as peak discharges fell. Minor geomorphic changes occurred in the morphological cross sections only after major floods, with a recurrence interval of 1:5 years for the Meshushim River and 1:20 years. Small ponds downstream of the dams were partially filled with fine sediments.

Key words dams; downstream effects; Golan Heights; hydrological regime

INTRODUCTION

Dams and reservoirs cause drastic changes to river systems, including hydrological, morphological, biological and aesthetic aspects (Petts, 1984). Most studies deal with qualitative and quantitative aspects of the river's response to the building of dams. Among others, Williams & Wolman (1984) have described a general decreasing of flood peaks, sediment concentration and suspended loads after construction of 21 dams on alluvial rivers, as well as bed degradation. Inbar (1990) documented the channel geometry changes of the Golan Heights rivers. The increasing return period and the reduced dam effects for downstream reaches where the percentage of the impounded area decreases were studied in the Alperhein River (Pfaundler, 1998). Major changes in the channel morphology downstream of reservoir were also found in the Piave River in Italy (Surian, 1999). There the channel underwent a general narrowing with a decrease in average width to 35% of its initial value, while the braiding index decreased from about 3 to 1.5. The meandering Green River in the USA went from a deep, to a shallow, braided channel over the 37 years since construction of Flaming Gorge Dam (Merritt & Cooper, 2000). The aims of the present study were: (1) to assess the hydrological impact of dams on a small impounded bedrock river, in comparison with a non-impounded catchment; (2) to analyse quantitatively changes in several hydrological and geomorphological factors affected by the dams.

GEOGRAPHICAL SETTING

The Golan Heights is a relatively humid area (500–1000 mm year^{-1}). Nevertheless, the structure of the basaltic lava flows restrains the development of large aquifers and springs. Between 1968 and 1985, 11 water reservoirs were built for irrigation purposes. Channels are mainly ephemeral and about three quarters of the total runoff is rainstorm-induced floods during winter storms.

The study covered two river basins (Fig. 1): (1) the Meshushim River, a non-impounded catchment covering 160 km^2 and (2) the Daliyyot River, an impounded catchment with high impounded runoff index (IR = 37%) [IR is the ratio of reservoir capacity divided by mean annual runoff (Batalla *et al.,* 2004)]. The catchment area of the Daliyyot River is 114 km^2, of which 50% is regulated by two reservoirs: Dvash (3.6×10^6 m^3) and Daliyyot (1.5×10^6 m^3). The two basins are very similar in lithology and morphology (Table 1).

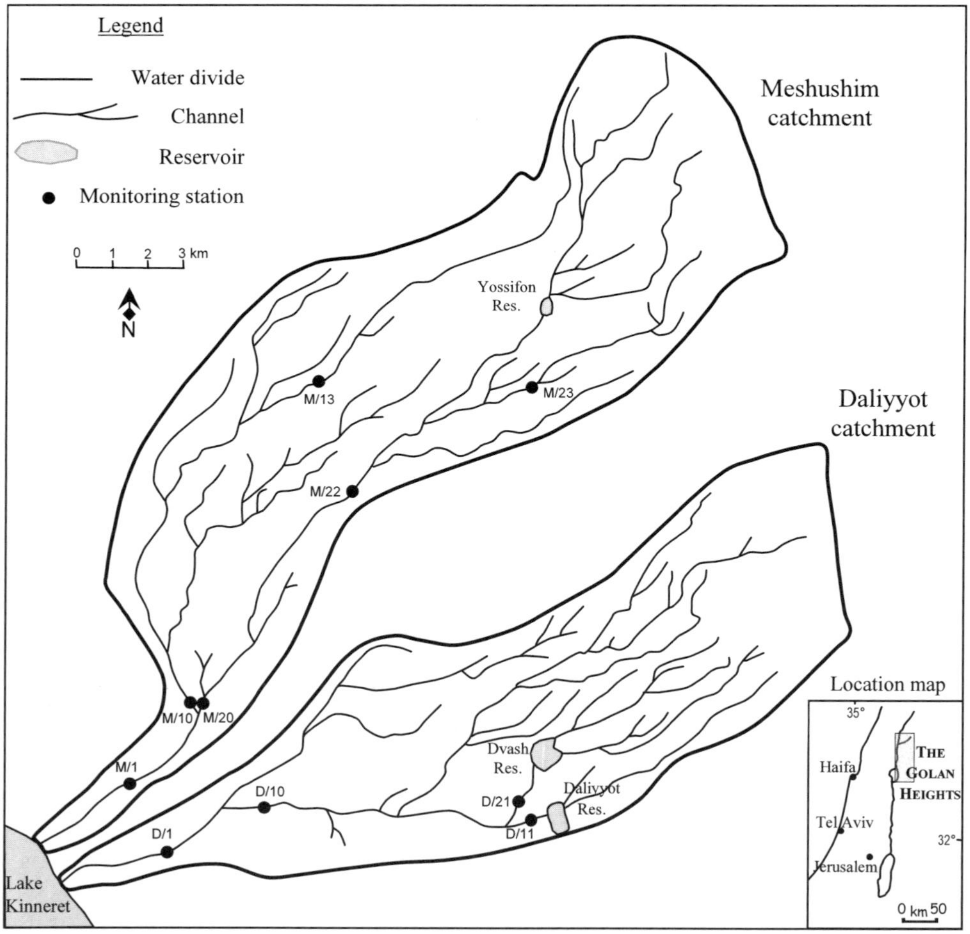

Fig. 1 The studied catchments, river network, monitoring stations, and reservoirs.

Table 1 Physical and hydrological parameters of the catchments (1984/1985–2003/2004).

Parameter	Meshushim catchment	Daliyyot catchment
Area	160 km^2	113 km^2
Relief	1413 m	1367 m
Slope	5.13%	5.7%
Elevation of canyons' head	400 m to 510 m a.m.s.l.	410 m, 460 m and 400 m a.m.s.l.
Percentage of basaltic lithology	95.5%	95.15%
Percentage of pyroclastic lithology	4.5%	1.65%
Percentage of carbonate lithology	0%	3.2%
Drainage density	0.84 km km^{-2}	0.96 km km^{-2}
Average annual precipitation	648 mm $year^{-1}$	607 mm $year^{-1}$
Average annual peak discharge	89 $m^3 s^{-1}$	27 $m^3 s^{-1}$
Maximum recorded peak discharge	211 $m^3 s^{-1}$ (2/97)	47.6 $m^3 s^{-1}$ (1/92)
Specific average annual peak discharge	0.57 $m^3 s^{-1} km^{-2}$	0.24 $m^3 s^{-1} km^{-2}$
Average annual flow volume	26.5 10^6 m^3 $year^{-1}$	13.7 10^6 m^3 $year^{-1}$
Maximum recorded flow volume	90 10^6 m^3 $year^{-1}$ (1991/92)	65.5 10^6 m^3 $year^{-1}$ (1991/92)
Specific average annual flow volume	0.17 $10^6 m^3 km^{-2}$	0.12 $10^6 m^3 km^{-2}$
Percentage of impounded catchment	22%	50%
Reservoir water volume	0.3 10^6 $m^3 year^{-1}$	5.1 10^6 $m^3 year^{-1}$
IR*	1%	37%
Runoff/rainfall coefficient	25.6%	20%

*IR, Impounded Runoff Index: ratio of reservoir capacity divided by mean annual runoff (after Batalla *et al.*, 2004).

METHODS

This study was based on a project started in 1985, when 11 monitoring stations were established downstream of the reservoirs (Fig. 1): six stations in the Meshushim River, and five stations in the Daliyyot River (Inbar *et al.,* 1986). Permanent bench marks consisting of painted stakes were used for repeated surveys of cross section measurements. The main features around each station were marked and photographed in order to check changes. Measurements were carried out twice a year: at the end of the winter and at the end of the summer. Data collection includes:

Geomorphology Longitudinal profiles were performed in order to measure stream gradient. The bankfull level was established each year after the flood season from morphological observations. The highest levels of the seasonal peak flood were determined from flood marks in the lower channel reaches. Several pools were surveyed in order to monitor probable rates of sedimentation.

Hydrology Two of the monitoring sites are located near hydrometric gauging stations of the Hydrological Service and therefore there is a continuous hydrological record available. For the other monitoring stations, the annual peak flood was calculated using the Manning's equation.

RESULTS

Notwithstanding the fact that the two basins have very similar geological and geomorphic characteristics and quite similar size, their hydrological behaviour has been very different in the last years. Since the construction of the dams the specific annual average water volume discharge in the Meshushim catchment has been double that in the Daliyyot catchment (Table 1). The reservoirs are the main dominant factor affecting the hydrological regime of the Daliyyot River , with an impounded runoff index (IR) of 37%. The recurrence interval of specific peak flows in the Meshushim River is shorter than in the Daliyyot River and increases with higher peaks. For example, a specific peak flow of 0.4 m^3 s^{-1} km^{-2} has a return period of 1:2 years in the Meshushim River, while the same magnitude in the Daliyyot has a return period of 1:4.3 years. For a specific peak flow of 1.0 m^3 s^{-1} km^2 the return period in the Meshushim River is 1:4 years, while the same magnitude in the Daliyyot River has a return period of 1:50 years (Fig. 2). As the high floods are the shaping factor of the fluvial system, an increase of one order of magnitude for the return period of peak flows considerably affects the geomorphological processes of the impounded channels, e.g. the transport of big boulders, shifting of clusters and bars, and bank erosion. No major geomorphological changes were observed during drought and average water flow years. Only after major floods, with a recurrence interval of 1:5 years for the Meshushim River and 1:20 years for the Daliyyot River, did minor geomorphic changes occur in the morphological cross sections.

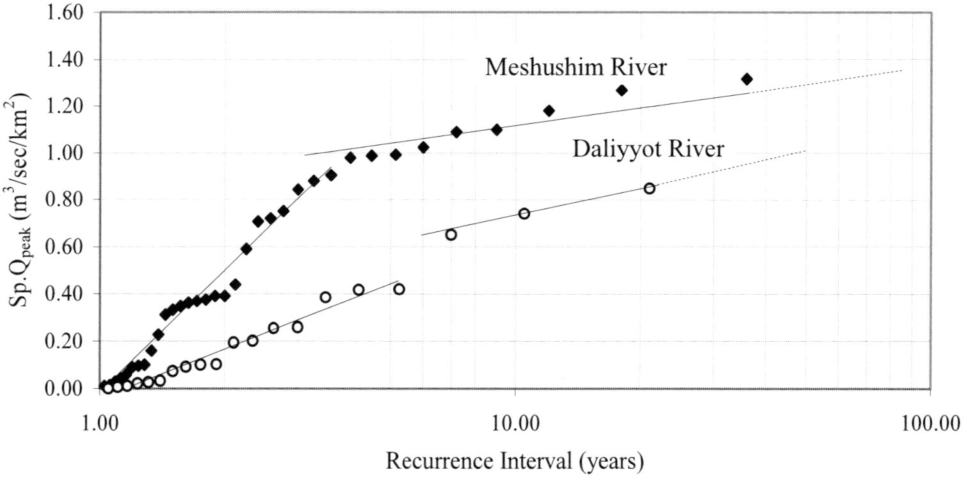

Fig. 2 The recurrence interval for specific peak discharges. Meshushim River: 1967/70–2003/4; Daliyyot River: 1984/5–2003/4.

The correlation between the bankfull flow area and the drainage basin area emphasizes the decreasing peak floods in the impounded channels, showing a lower bankfull flow area for the Daliyyot River (Fig. 3). The flood plain is invaded by perennial shrubs consolidating the loose boulders and gravels; their transport is therefore restricted to low frequency high peaks in the impounded catchment. Small ponds downstream of the dams in the Daliyyot River were partially filled with fine sediments, showing a gradual alteration from a running fresh water habitat into a swamp habitat (Fig. 4).

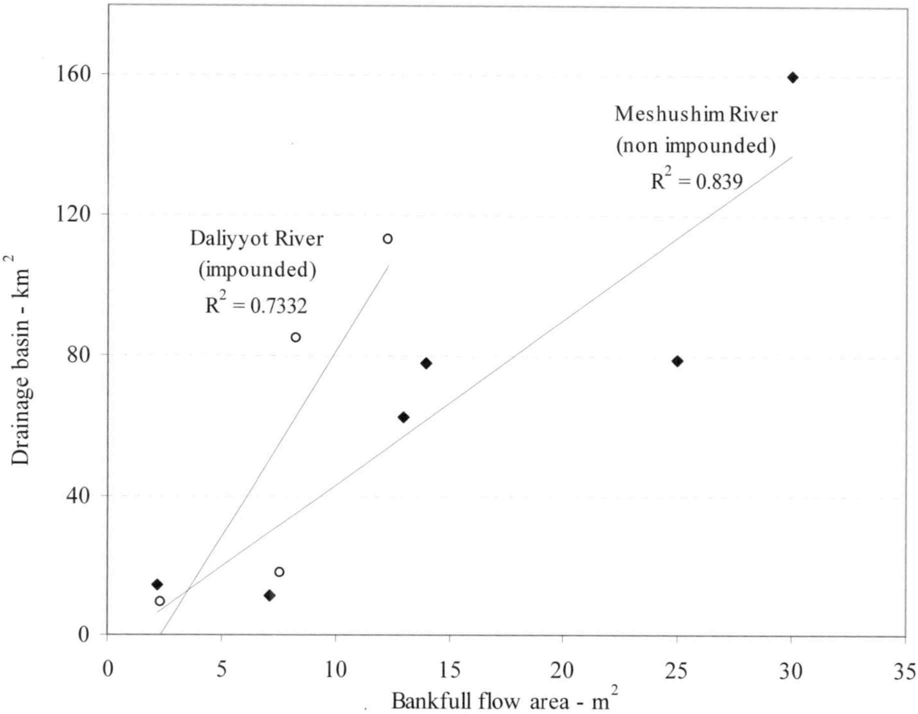

Fig. 3 Correlation between bankfull flow area and drainage basin area. In the impounded basin the channel bankfull area decreases allowing riparian vegetation to invade the whole channel area.

Fig. 4 Monitoring station downstream of Dvash Reservoir in the impounded Daliyyot River basin. In the absence of floods, riparian vegetation develops into the channel.

Acknowledgements The authors thank D. Kaplan and H. Glassman from the Israel Nature & National Parks Protection Authority. We also thank Nicola Surian for useful comments on the manuscript.

REFERENCES

Batalla, R. J., Gomez, C. M & Kondolf, G. M. (2004) Reservoir-induced hydrological changes in the Ebro River basin (Northeastern Spain). *J. Hydrol.* **290**(1–2), 117–136.

Inbar, M. (1990) Effect of dams on mountainous bedrock rivers. *Physical Geogr.* **11**(4), 305–319.

Inbar, M., Kaplan, Y., Glassman H. & Ortal, R. (1986) Water reservoirs impact on the downstream hydrological environment of the Golan Heights Rivers. In: *Environmental Quality and Ecosystem Stability* (ed. by Z. Dubinsky & Y. Steinberg), 287–294. 3. A/B. Bar Ilan University Press, Bar Ilan, Israel.

Merritt, D. M. & Cooper, D. (2000) Riparian vegetation and channel change in response to river regulation: A comparative study of regulated and unregulated streams in the Green River Basin, USA. *Regulated Rivers: Research and Management* **16**(6), 543–564.

Petts, G. E. (1984) *Impounded Rivers: Perspectives For Ecological Management*. Wiley, Chichester, UK.

Pfaundler, M. (1998) The influence of reservoirs on the flood-regime: a case study of the River Alpernhein. *Z. fur Kulturtechnik und Landentwicklung* **39**(1), 27–31.

Surian., N. (1999) Channel changes due to river regulation: the case of the Piave River, Italy. *Earth Surf. Processes Landf.* **24**(12), 1135–1151.

Williams, G. P. & Wolman, M. G. (1984) Downstream effects of dams on alluvial rivers. *US Geol. Survey Professional Paper, 1286.*

A proposed index of channel sensitivity to riparian disturbance

CHAD E. BAILEY[1], ROBERT G. MILLAR[1] & MICHAEL MILES[2]

1 *Department of Civil Engineering, University of British Columbia, Vancouver, British Columbia, Canada*
millar@civil.ubc.ca

2 *M. Miles and Associates Ltd., Victoria, British Columbia, Canada*

Abstract Anecdotal and empirical evidence has long suggested that riparian vegetation can exert an important control on channel width and stability in alluvial rivers. Despite this evidence, the importance of riparian vegetation has yet to gain widespread acceptance. This appears due to several reasons, including the difficulty in quantifying vegetation effects and variation in the sensitivity of alluvial channels to riparian disturbance. In this study we develop a theoretical riparian disturbance sensitivity index, which is tested using 15 before-and-after case studies of gravel-bed rivers in British Columbia, Canada. The majority of the study sites have been subjected to direct riparian disturbance. Observed changes in width, expressed as a percentage of the initial, undisturbed width, correlate strongly with the proposed riparian disturbance index. The three undisturbed reaches show no systematic change in width. We conclude that the riparian disturbance sensitivity index provides a good measure of expected channel sensitivity and change in width following riparian disturbance for gravel-bed rivers in southern British Columbia.

Key words erosion; riparian vegetation; rivers; stability

INTRODUCTION

Anecdotal and empirical evidence has long suggested that riparian vegetation can exert an important control on channel width and stability in alluvial rivers. For instance, Mackin (1956) described a dramatic change or metamorphosis in channel form from meandering to braiding in instances where a reduction in riparian vegetation appears to be the only obvious change in channel control. Hey & Thorne (1986) demonstrated that gravel bed rivers with weakly or thinly vegetated banks can be on the order of 1.85 times wider than densely vegetated channels. Millar & Quick (1993) analysed the same data from Hey & Thorne, and found that for unvegetated channels, average widths, depths and slopes are of the order of 1.6, 0.7 and 1.1 times their respective heavily vegetated counterparts. Millar (2000) extended the analysis to develop a meandering-braiding index that includes a parameterization of bank vegetation, and demonstrated that a reduction in bank vegetation strength alone can result in substantial widening, and even a metamorphosis in channel form from a single-thread to a braided planform. Gran & Paola (2001) cultivated alfalfa seedlings on a pre-existing experimental braided channel, and documented a narrowing and transformation to a wandering planform, which is considered intermediate between braided and meandering. Thus, there exists strong field, analytical and experimental evidence to support the importance of riparian vegetation as a control on channel width and morphology.

Despite this evidence, the importance of riparian vegetation has yet to gain widespread acceptance (Bridge, 1993). This appears to be due to several reasons including the difficulty in quantifying vegetation effects (Hickin, 1984) and variation in the sensitivity of alluvial channels to riparian disturbance (Simon & Downs, 1995). Casual observation tells us that there are many rivers that exhibit little apparent widening or instability following riparian clearing. The role of vegetation is further complicated by scaling effects (Zimmerman *et al.*, 1967; Eaton & Millar, 2004). Trimble (1997) describes instances where small rivers and streams with forested banks can be wider than those running through pasture or with grassy banks. Eaton & Millar (2004) demonstrate that bank vegetation effects become insignificant in larger gravel rivers with bankfull discharges greater than about 400 $m^3 s^{-1}$. Evidently, the degree to which riparian vegetation controls channel width and morphology is highly variable between different rivers and streams. It would be desirable therefore, to develop a quantitative measure or index that characterizes the sensitivity of a particular channel to riparian disturbance or change (Simon & Downs, 1995). The primary objective of this study is to develop a theoretical riparian disturbance sensitivity index based on an extension of theory developed by Millar & Quick (1993) and Millar (2000). We test the proposed index using 15 before-and-after case studies from British Columbia, Canada, the majority of which (12 out of the 15) have been subjected to direct riparian disturbance.

Proposed Riparian Disturbance Sensitivity Index

Millar (2000) developed a theoretical Leopold & Wolman (1957)-type meandering/–braiding threshold slope relation:

$$S^* = 0.0002 D_{50}^{\;0.61} \phi'^{1.75} Q^{-0.25} \tag{1}$$

where S^* is the threshold (or transitional) slope between meandering and braiding, D_{50} is the median sediment diameter for the banks and bed surface (m), ϕ' is the bank sediment friction angle (°), which is directly related to the critical dimensionless shear stress, and Q is the bankfull discharge ($m^3 s^{-1}$). Values of the channel slope $S \ll S^*$ predict single thread or meandering channels (in a general sense), while values of $S \gg S^*$ predict braided channels. The parameter ϕ' is used to account for the stabilizing influence of bank vegetation and/or interstitial cohesion. Equation (1) is dimensional. In a recent paper, Millar (2005) develops an equivalent dimensionless relation for S^*:

$$S^* = 0.0957 Q^{*-0.25} \mu' \tag{2}$$

where dimensionless discharge, $Q^* = Q \Big/ \left(D_{50}^2 \sqrt{g D_{50} (s-1)}\right)$, and relative bank strength $\mu' = \tau_{bankc}/\tau_{bedc}$, where τ_{bankc} and τ_{bedc} refer to the dimensional critical shear stresses (Pa) for the bank and bed sediments, respectively. When the banks and bed are both composed of loose, noncohesive gravel they have the same values for critical shear stress, and therefore $\mu' = 1.0$. Values for $\mu' > 1$ indicate that the banks are more resistant to erosion than the bed sediment. For instance, a value of $\mu' = 1.5$ indicates that gravel in the banks is 1.5 times more resistant to fluvial erosion than in the bed (Millar, 2005). This allows the influence of different bank strengths to be incorporated into the meandering-braiding transition slope, S^*.

We propose that the ratio of the channel slope S to the meandering braiding threshold slope, S/S^*, provides a quantitative index of the channel planform. Values of $S/S^* \ll 1$ indicate a laterally-stable, single-thread or meandering planform, while $S/S^* \gg 1$ indicates a laterally-unstable braided planform. Our proposed riparian disturbance sensitivity index (ζ) is based on the ratio S/S^* calculated with $\mu' = 1.0$:

$$\zeta = \frac{S}{S^*} = 10.5 Q^{*0.25} S \qquad (3)$$

where S^* is obtained from equation (2). This reflects the value of the ratio S/S^* following complete removal of any riparian vegetation effects, leaving the banks composed only of loose, noncohesive gravel. It does not apply to rivers where the bank sediment is appreciably stabilized by cohesive silt and clay. Values of ζ close to zero should indicate insensitivity to riparian disturbance, with little increase in width following riparian clearing. Conversely, values of $\zeta > 1.0$ would indicate extreme sensitivity, and potential transformation from a meandering to a braided channel. Intermediate values would span a range of sensitivities.

METHODOLOGY

Study methodology includes field surveys, measurement of historic channel widths from aerial photographs, and hydrologic analysis. These are discussed briefly here, with details provided in Bailey (2003). Wolman pebble counts with a sample size of 100 stones were performed at riffle sections and along banks to determine the surface median particle diameters (D_{50}). Longitudinal and cross sectional surveys were performed at each study reach, which is between 10 and 20 channel widths in length and includes at least three riffle-pool sequences to provide estimates of current reach-average channel slope and channel width. Measurements of unvegetated channel widths from historic and recent aerial photographs were made perpendicular to flow direction at approximately 50 m intervals, from which the average was then determined. Average widths obtained from recent aerial photographs generally agree well with field measurements, usually within one standard error. The channel forming morphological bankfull discharge is assumed here to be equivalent to the mean annual flood (Millar, 2000), which by definition has an average recurrence interval of close to two years. Our study sites are not located at hydrometric gauges, and therefore an estimate of the mean annual flood was obtained using a scaling and regionalization technique developed for British Columbia (Eaton *et al.*, 2002). Basic hydrological and field data are summarized in Table 1.

RESULTS

For each study site, the change in average width was determined and is expressed as a percentage of the earliest (and generally undisturbed) width, $\Delta W\%$. Data are stratified

Table 1 Summary of Hydraulic and Geomorphic Data.

Reach #	D_{50} (m)	S	Q ($m^3 s^{-1}$)	Historic W (m) (Year)	Recent W (m) (Year)	$\Delta W\%$	ζ	Condition
Big Horn	0.035	0.0064	19	18 (1962)	39 (2000)	114	0.80	Disturbed
Bonaparte	0.021	0.0029	32	13 (1959)	29 (2002)	127	0.57	Disturbed
Coldwater 1	0.065	0.0049	77	26 (1953)	62 (2002)	135	0.59	Disturbed
Coldwater 2	0.037	0.0033	31	28 (1953)	34 (2002)	23	0.45	Intact
Coldwater 3	0.03	0.0047	28	15 (1953)	16 (2002)	4	0.71	Intact
Deadman 1	0.05	0.0027	22	17 (1959)	30 (2002)	76	0.28	Disturbed
Deadman 2	0.04	0.0047	22	14 (1959)	37 (2002)	164	0.56	Disturbed
Elk 1	0.05	0.0063	124	23 (1930)	88 (2002)	274	1.01	Disturbed
Elk 2	0.062	0.0104	61	18 (1930)	27 (2002)	49	1.22	Intact
Eve	0.034	0.0107	152	45 (1953)	69 (2002)	68	2.29	Disturbed
Narrowlake	0.063	0.0070	30	29 (1946)	58 (1997)	100	1.67	Disturbed
Salmon	0.042	0.0018	603	138 (1946)	179 (2002)	30	0.48	Disturbed
Slesse	0.133	0.0210	88	28 (1936)	145 (1997)	418	0.68	Disturbed
Tahsis	0.041	0.0039	196	29 (1954)	60 (2002)	106	0.79	Disturbed
West Kettle	0.078	0.0028	127	38 (1951)	44 (1998)	15	0.34	Disturbed

into disturbed and intact reaches depending upon the current condition of the riparian zone (Table 1). Disturbed indicates that the riparian zone has been directly impacted by clearing as a consequence of logging or agricultural activities, whereas intact indicates little or no direct riparian disturbance. For the disturbed channels (excluding Eve River), there is strong linear correlation between $\Delta W\%$ and ζ over the range of observations:

$$\Delta W\% = 273\zeta - 52.4 \; (r^2 = 0.84, p < 0.0001, n = 11) \tag{4}$$

The undisturbed reaches ($n = 3$) show variable $\Delta W\%$ between 4% and 50%, and do not appear to correlate with ζ, although the sample size is small. Eve River is regarded as an outlier, which may reflect a disturbed bankfull channel capacity that is currently much less than the mean annual flood. We examine the possibility that the single data point representing Slesse Creek could potentially be an outlier and over represented in the regression statistics (Fig. 1, equation (4)). When Slesse Creek is excluded from the regression, $\Delta W\%$ for the disturbed channels remains strongly correlated with ζ ($r^2 = 0.56$, $p = 0.013$, $n = 10$). There is no apparent physical justification for excluding Slesse Creek. The non-zero intercept in equation (4) indicates that the expected change $\Delta W\%$ approaches zero for values of ζ less than 0.19. We expect that rivers in southern British Columbia with values of ζ less than about 0.2 would be essentially insensitive to riparian disturbance. However, no reaches in our study fall into this range.

DISCUSSION

This analysis indicates that the proposed riparian disturbance sensitivity index explains most of the post-disturbed change in channel width for the disturbed reaches in these case studies. With the exception of Eve River, correlation between $\Delta W\%$ and ζ is strong, with a high level of statistical significance. This relation is further strengthened by examining the individual case studies (Bailey, 2003). Because of length restrictions, we consider here only the Coldwater River study Reaches 1 and 3. Basic hydraulic and index values are summarized in Table 1.

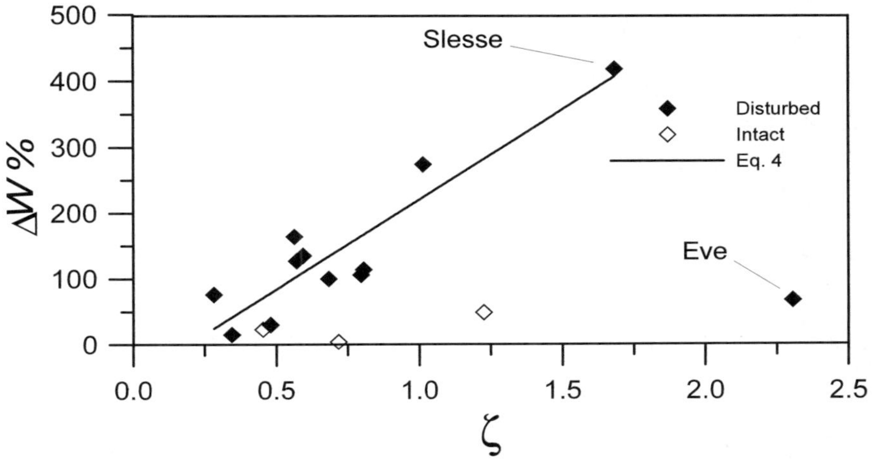

Fig. 1 Percentage change in width *vs* riparian disturbance sensitivity index (□). Points corresponding to Slesse Creek and Eve River are indicated.

Coldwater River case study

Coldwater River Reach 1 is located 15 km upstream from the confluence with the Nicola River and has a catchment area of 735 km^2. Since the 1960s, the valley flat and riparian areas have been extensively cleared for logging and agriculture, and there is currently extensive cattle grazing. The predisturbance channel in the 1953 aerial photograph is narrow and meandering, with an average unvegetated channel width of 26 m (Fig. 2(a)). In 2001, width had increased by 135% to 62 m, and the current channel is wide and sinuous, with meander cutoffs and other instabilities (Fig. 2(b)). Reach 3 of the Coldwater River is located within an essentially undisturbed floodplain approximately 35 km upstream of Reach 1, with a drainage area of 82 km^2. The riparian zone is intact, although it was subject to a wildfire in 1938 (Fig. 3(a)). Since 1953 there has been essentially no significant change in channel width or morphology, and the channel remains narrow and meandering (Fig. 3(b)). Estimated width measured from the 1953 aerial photograph is 15 m, which agrees well with an average value of 16 m measured in the field in 2001.

Reaches 1 and 3 both have intermediate values of ζ (0.59 and 0.71, respectively), which indicate moderate to high sensitivity to riparian disturbance, with roughly a doubling (100% increase) in channel width expected after riparian disturbance based on equation (4). Although there would be some scaling effects due to the differences in catchment areas, both reaches have been subject to essentially the same flood history. Reach 3 has an intact riparian zone, in contrast to Reach 1, which has a heavily disturbed riparian zone as a consequence of land clearing and ongoing grazing

Fig. 2 Coldwater River Reach 1. (ζ = 0.59). (a) 1953. (b) 2000. (British Columbia Government Aerial Photography: BC1745:92 BCC00036:170). Flow is from left to right.

pressures. The width and morphology of Reach 3 is relatively unchanged between 1953 and present (Fig. 3), whereas within Reach 1 width has increased by some 135% (Fig. 2). We conclude that the current difference in morphology and stability between the two reaches is a direct consequence of the riparian disturbance in Reach 1, and the magnitude of the width increase is consistent with our proposed riparian disturbance sensitivity index, ζ.

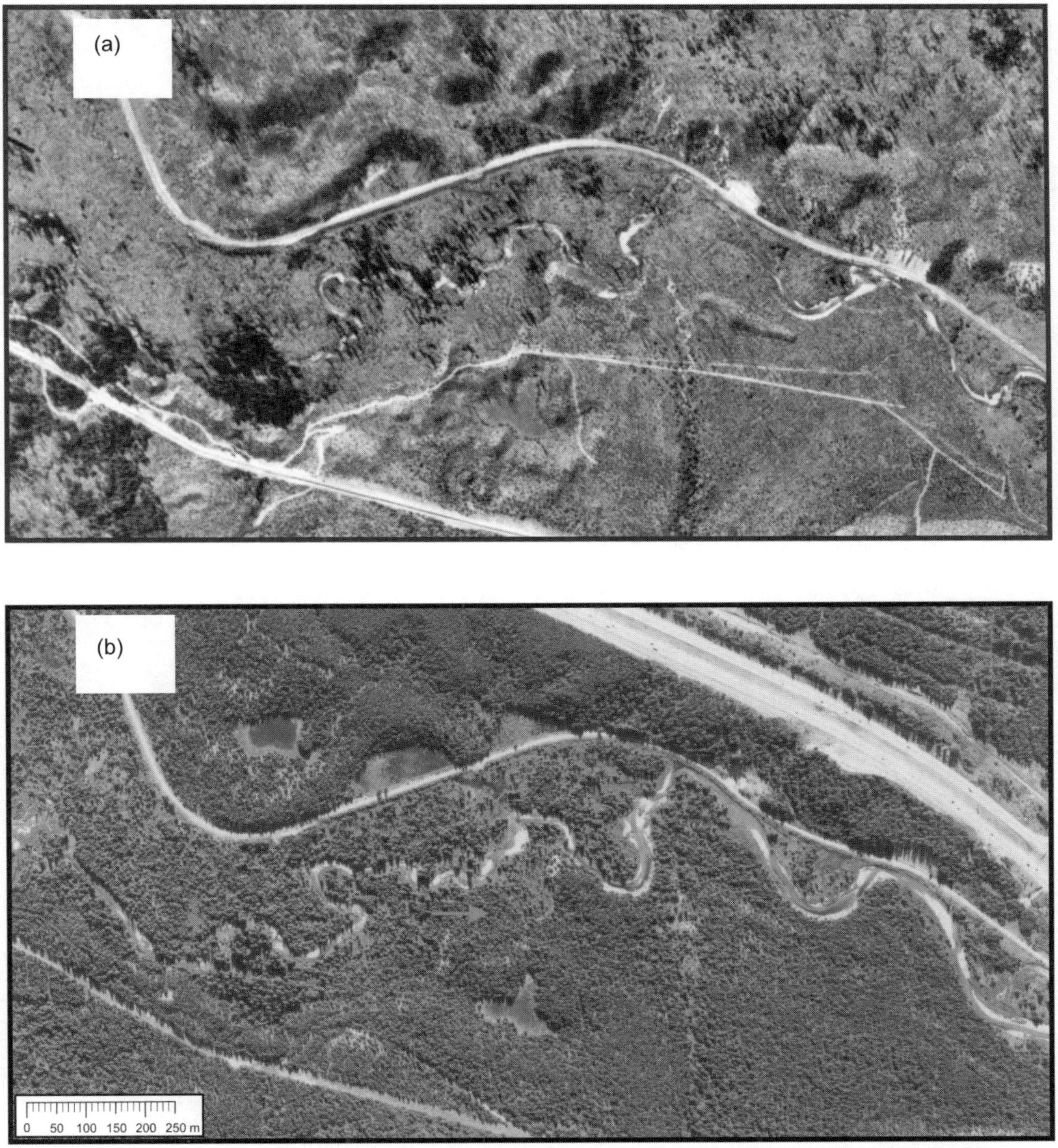

Fig. 3 Coldwater River Reach 3. (ζ = 0.71) (a) 1953. (b) 2000. (British Columbia Government Aerial Photography BC1746:8, BCC00033:19). Flow is from left to right.

CONCLUSIONS

A proposed theoretical riparian disturbance sensitivity index was developed and tested using 15 gravel-bed river study reaches from southern British Columbia. Observed changes in width, expressed as a percentage of the initial, undisturbed width, correlate strongly with the proposed riparian disturbance sensitivity index. The three undisturbed reaches show no systematic change in width. On the basis of this study we conclude that the proposed riparian disturbance sensitivity index (ζ) provides a good predictor of expected channel sensitivity and change in width following riparian disturbance for gravel rivers in southern British Columbia. The relationship between $\Delta W\%$ and ζ, equation (4), is not universal, as the slope and probably the intercept will vary regionally as a function of riparian vegetation characteristics.

Acknowledgements The authors gratefully acknowledge funding provided by Forest Renewal BC under the British Columbia Watershed Restoration Program. For assistance in the field, thanks are due to Ted Tedford and Andrew Wu. Don Ignace of the Skeetchestn Indian Band allowed access to private lands bordering the Deadman River. This paper benefited from the review and comments received from Steve Darby.

REFERENCES

Bailey, C. E. (2003) Riparian disturbance sensitivity index for gravel-bed river morphology. MASc. Thesis, University of British Columbia, Vancouver, Canada.

Bridge, J. S. (1993) The interaction between channel geometry, water flow, sediment transport and deposition in braided rivers. In: *Braided Rivers* (ed. by J. L. Best & C. S. Bristow), 13–71. Geol. Soc. Special Pub. 75.

Eaton, B. C. & Millar, R. G. (2004) Optimal alluvial channel width under a bank stability constraint. *Geomorphology* **62**, 35–45.

Eaton, B., Church, M. & Ham, D. (2002) Scaling and regionalization of flood flows in British Columbia, Canada. *Hydrol. Processes* **16**(16), 3245–3263.

Gran, K. & Paola, C. (2001) Riparian vegetation controls on braided stream dynamics. *Water Resour. Res.* **37**(12), 3275–3283.

Hey, R. D. & Thorne, C. R. (1986) Stable channels with mobile gravel beds. *J. Hydraul. Eng.* **112**(8), 671–689.

Hickin, E. J. (1984) Vegetation and river channel dynamics. *Can. Geogr.* **28**(2), 111–126.

Leopold, L. B. & Wolman, M. G. (1957) River channel patterns: braiding meandering and straight. *US Geol. Survey Professional Paper* **262**-B, 39–85.

Mackin, J. H. (1956) Cause of braiding by a graded river. *Geol. Soc. Am. Bull.* **67**, 1717–1718.

Millar, R. G. (2000) Influence of bank vegetation on alluvial channel patterns. *Water Resour. Res.* **36**(4), 1109–1118.

Millar, R. G. (2005) Theoretical regime equations for mobile gravel-bed rivers with stable banks. *Geomorphology* **64**(3–4), 207–220.

Millar, R. G. & Quick, M. C. (1993) Effect of bank stability on geometry of gravel rivers. *J. Hydraul. Eng.* **119**(12), 1343–1363.

Simon, A., & Downs, P. W. (1995) An interdisciplinary approach to evaluation of potential instability in alluvial channels. *Geomorphology* **12**, 215–232.

Trimble, S. W. (1997) Stream channel erosion and change resulting from riparian forests. *Geology* **25**, 467–469.

Zimmerman, R. C., Goodlett, G. C. & Comer, G. H. (1967) Influence of vegetation on channel form of small streams. In: *Symposium on River Morphology*, 255–275. IAHS Publ. 75. IAHS Press, Wallingford, UK.

Channel morphology and sedimentology in a mountainous channel: a comparison between disturbed and undisturbed reaches

LEA WITTENBERG[1] & NOAM GREENBAUM[1,2]

1 *Department of Geography and Environmental Studies, University of Haifa, Haifa 31905, Israel*
leaw@geo.haifa.ac.il

2 *Department of Natural Resources and Environmental Management, University of Haifa, Haifa 31905, Israel*

Abstract River adjustment to an artificial disturbance is examined in the ephemeral mountainous stream—Nahal Oren, Mt. Carmel, Israel. We compare sedimentological and morphological characteristics of two sections, 50 and 500 m downstream of an artificial pond. Our results indicate that the deprivation of bed load floodwater in the upper section generates high calculated entrainment potentials as indicated by the large differences between measured and calculated D_{50} which exceed 80%. Each flow produces an effective flow ($Q > 0.125$ m^3 s^{-1}) of between 17 and 58% of the flood duration to entrain pebbles of the D_{50} size (50 mm). In a large flow of 12 m^3 s^{-1} (recurrence interval of 10 years), resulting bank erosion, in the upper section exceeded 4 tons per 20 m long section, whereas during the period 1999–2004 only minor degradation of <40 cm occurred at the lower section. The eroded material from the upper reach is loosely deposited, transported downstream and selectively deposited in the step-pool section. The impact of the pond on the channel at the upper section disappears 500 m downstream, where the site shows relative stability according to the lower calculated entrainment potential, the step-pool morphology of the reach and the higher ratio of cluster arrangements.

Key words bedforms; entrainment potential; Mediterranean streams; river adjustment

INTRODUCTION

The mechanism of river adjustment to disturbances such as dam construction has been widely addressed with regards to the impact on hydrology, channel morphology and sedimentology (Church, 1995). Numerous studies of rivers downstream of reservoirs documented changes in bed material size indicating that the effects of the disturbance differ greatly (Petts, 1980). Mountainous gravel bed streams are often associated with high rates of erosion, transport and deposition of coarse bed load in response to large floods or changes in sediment supply. The surface of most gravel-bed rivers is characterized by morphological features of different scales, depending on the sedimentological characteristics, flow regime, and general morphology of the channel (Hassan & Church, 2000). Three scales of roughness are commonly addressed: (a) plane bed of gravel size material, (b) small-scale bedforms including pebble cluster, and (c) unit forms: large-scale bed undulations such as step pools (Lawless & Robert, 2001a). These scales reflect, to different extents, the stability of the channel and may indicate changes in the sedimentological or the hydrological regime. A common measure of channel bed stability is bedforms (Wittenberg, 2002). Cluster bedforms are probably the most widespread type of structures in humid perennial gravel-bed rivers occupying 10–50% of the bed material (Reid *et al.*, 1992; Wittenberg, 2002). Clusters involve accumulations of sediment on either or both the lee and the stoss of an obstacle clast (Laronne & Carson, 1976; Reid *et al.*, 1992). The formation of bedforms reduces the mobility of the bed sediments (Hassan & Church, 2000) and reflects on sediment continuity. Feedback processes between sediment deposition and sediment entrainment control the micro-morphology of the bed and thus the stability of the channel. The shape of the hydrograph, as indicative of flow regimes, plays an important role in clusters formation. Field and laboratory experiments demonstrate that surface texture in gravel-bed rivers respond to changes in sediment supply (Buffington & Montgomery, 1999). Reservoirs which trap all bed load, release sediment-starved or "hungry" water which may transport sediment of various sizes downstream without replacement from upstream, resulting in coarsening of the surface layer, incision, and bank erosion (Kondolf, 1997). In this study we examine the effect of an impoundment within an upland Mediterranean stream by comparing two sections of the mountainous ephemeral gravel bed channel of Nahal Oren, located 50 and 500 m downstream of a small seasonal reservoir (from hereinafter referred to as the Pond) (Fig. 1). This

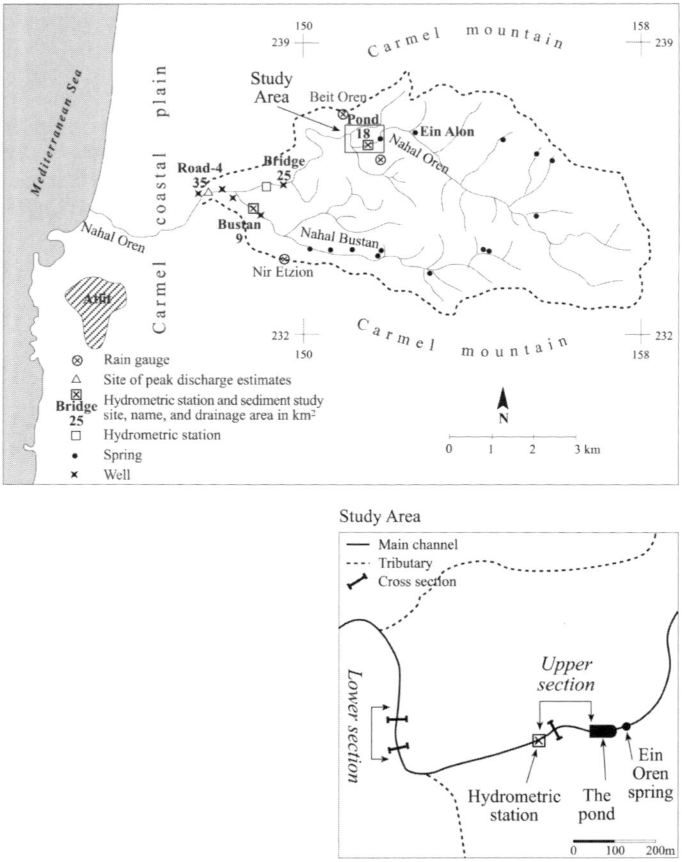

Fig. 1 Nahal Oren drainage basin, hydrometric stations, raingauge, and the study sites, Mt Carmel, Israel

reservoir is a large artificial pool which traps all bedload and part of the suspended load, but has only a little effect on stream flow. We analyse the effect of the pond on channel adjustment and stability downstream as indicated by sedimentological and morphological characteristics.

Study site

Mt. Carmel is a distinctive mountain ridge along the NW coast of Israel with its highest peak at 546 m a.m.s.l. (Fig. 1). The upper Cretaceous carbonate rocks of Mt. Carmel are covered with shallow soils. The vegetation is a Mediterranean forest, composed mainly of a complex of pine (*Pinus halepensis*), oak (*Quercus calliprinos*), *Pistacia Lentiscus* and associations. The climate is typical Mediterranean: rainy winters with hot and dry summers. Most rainstorms occur between November and March and are often convective during the autumn while winter precipitation (December–February) results from Eastern Mediterranean frontal systems. Mean annual precipitation increases from 550 mm at the coastal plain to 750 mm at the top. Flow events at the ephemeral stream systems of Mt. Carmel are sporadic and occur following either high intensity rainstorms or during the winter after the soil is well saturated. Large flow events are initiated following rainstorms of over 100 mm. Flow durations range from several hours to a few days and are characterized by a steep rising limb and a long recession. Only in a few short channel reaches local springs discharge sustain base flows, which may last for a few days or up to several weeks following the storms.

Nahal Oren is a mountainous ephemeral stream with a drainage area of 35 km^2 in the western part of Mt. Carmel (Fig. 1) and with an average gradient of 3%. It has a bedrock channel with several alluvial sections in between and step pool morphology in general. The partial measured hydrological record from 1957 to 1964 and from 2001 to 2004 is derived from three hydrometric stations (Fig. 1). Additional data from 1991 to 2004 is available from the Bridge station. Between these periods only extreme events were indirectly measured. During the last 3 years four events with $Qp > 5$ m^3 s^{-1} (calculated recurrence interval 5 years), occurred in the Nahal Oren basin. The largest recorded flood had a peak discharge of 90 m^3 s^{-1} (January, 1969). The channel at the study area is alluvial and is entrenched into the gravelly-sandy-clay sediments of the alluvial flood plain. The artificial pond was established in 1973. The upper gauging station, the Pond, (18 km^2) provides hydrological data for the present study. Sedimentological measurements were carried out following the two largest floods: December 2001 ($Qp = 12$ m^3 s^{-1}; recurrence interval 10 years) and January 2002 ($Qp = 5$ m^3 s^{-1}). Maximal water depth was 1.1 m and 0.8 m, respectively. The 2001 flood had a 3-peak hydrograph (Fig. 2(a)) with maximal water depth of 1.1 m, flow duration of 114 h and was followed by 3 weeks of base flow. The 2002 flood had a 2-peak hydrograph; maximal water depth was 0.8 m, flow duration of ~100 h and was followed by 2 months of base flow (Fig. 2(b)). The upper section is a plane bed channel that lacks rhythmic unit-scale bedforms. Its gradient, 0.022, falls within the frequent range of low gradient channels plane bed (Chartrand & Whiting, 2000), which are characterized by long stretches of relatively featureless bed. Plane bed channels indicate a transport capacity greater than sediment supply (Montgomery & Buffington, 1997). The lower site has step-pool morphology. We analysed a sequence of four steps and three adjacent pools. The average channel gradient is 0.038; pools average gradient is 0.008 whilst the steps average gradient is 0.046. Channel average width varies from 4 m (steps) to 6 m (pools). Sites characteristics of the study sections are summarized in Table 1.

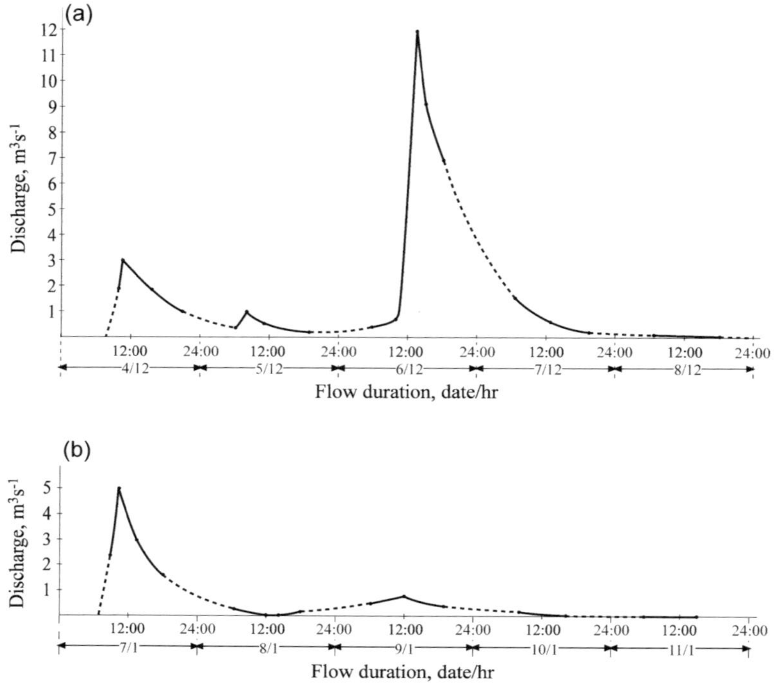

Fig. 2 Hydrographs of the (a) December 2001 and (b) the January 2002 floods at the Pond hydrometric station. The broken line represents reconstructed segments of the hydrograph.

Table 1 Channel characteristics of the upper and the lower sections.

	Upper section	Lower section
Section length (m)	100	120
Dist. from pond (m)	50	500
Channel morphology	Plane bed	Step-pool
Gradient (mm^{-1})	0.022	0.008–0.046 (average 0.038)
Avg. width /depth ratio	8.4	4.6
Hydraulic roughness	0.04	0.035–0.05

Methodology

In both reaches sediments were sampled following the large flow events. In the step-pool segment successive sedimentological data was collected also during 1994–2001. At each site (50 m long), 8–10 cross-sections were set at 5–10 m intervals. Following each flood, bed material was sampled along each cross-section. Additionally, a large number of Wolman grain size distributions were assessed. All grain size data were combined and analysed. Bed material was divided into three major scales: clast/grain scale which includes surface grain size distribution, bedform scale distribution and site scale—general information of channel gradient, width–depth ratio, and major morphological controls (Table 1). To estimate the entrainment potential (Milan *et al.*, 2001) local boundary shear stress was calculated for each section based on the depth-slope products (DuBoys equation):

$$\tau_0 = \rho g h s \tag{1}$$

ρ = fluid density; g = gravity acceleration, s = reach average channel slope, h = reach average channel water depth (Buffington & Montgomery, 1999). DuBoys equation is not very appropriate for the highly non-uniform flows associated with step-pool systems and is unlikely to give reliable estimates of bed shear stress available for sediment transport since the equation is based on the assumption of uniformly distributed energy losses. However, the equation gives an upper bound estimate of shear stress (Zimmerman & Church, 2001). Consequently, albeit its limitation, the equation is widely used for estimating shear stress in step pool section (e.g. Chartrand & Whiting, 2000; Marion & Weirich, 2003); it is used in this study to compare between the two study sections. The competent, median bed-surface grain size (D_{50}) can be predicted from the Shields equation. Shields related the critical tangential fluid stress at the boundary responsible for grain movement (τ_0) to the weight per unit area of the particles forming at the bed surface by defining the dimensionless entrainment function τ^*_c (Powell, 1998):

$$\tau^*_c = \tau_o/[(\rho_s - \rho) g D_{50}] \tag{2}$$

ρ_s is sediment density and D_{50} the median particle diameter. Since critical shear stress is proportional to particle size, it is possible to estimate the size of the competent particles that would be entrained at certain discharges (after the relation between discharge and shear stress is assessed):

$$D_{50} = \tau_o/[\tau^*_c(\rho_s - \rho) g] \tag{3}$$

Buffington and Montgomery (1999) set the dimensionless critical shear stress for the competent D_{50} equal to 0.03 for visually based studies. This value is commonly used for individual unconstrained grains on the bed surface. A higher value of 0.08 was used by Zimmerman & Church (2001) for constrained materials. Consequently, we used $\tau^*_c = 0.03$ for the pools and for the upper section and $\tau^*_c = 0.08$ for the steps at the lower section.

RESULTS

Grain size distribution

Upper section grain size distribution (GSD) underwent minor modification following various flow events (Fig. 3). In comparison to the upper site, the pool sediments at the

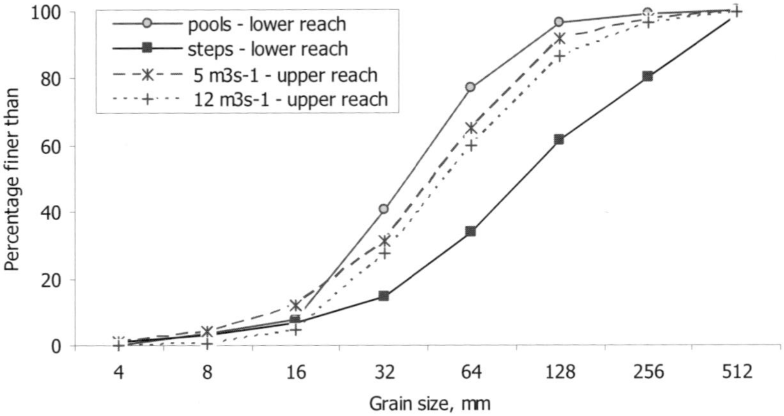

Fig. 3 Grain size distribution following 5 and 12 $m^3 s^{-1}$ floods at the upper section and in the pools and steps at the lower section following the 5 $m^3 s^{-1}$ flood.

Table 2 Grains size distribution—the upper and the lower sections following floods with Qp of 5 $m^3 s^{-1}$ and 12 $m^3 s^{-1}$.

	Lower section Pools	Upper section Steps		Lower section Pools
D_{16} (mm)	21	35	D_{16} (mm)	21
D_{50} (mm)	39	90	D_{50} (mm)	39
D_{84} (mm)	87	312	D_{84} (mm)	87
D_{90} (mm)	112	338	D_{90} (mm)	112
SI (D_{84}/D_{16})	4.1	8.9	SI (D_{84}/D_{16})	4.1

lower site are relatively finer while step sediments are coarser and have a wider textural range, as also reflected in the considerably higher sorting index (Table 2).

Clusters distribution

Comparing the distribution of clusters *vs* open-bed particles (Fig. 4) reveals that 40–50% of the bed surface at the upper site is clustered. This ratio resembles the distribution of clustered particles in perennial streams elsewhere (Wittenberg, 2002) and indicates a relative stability of the bed surface to entrainment. In the pools, more than 80% of the particles are deposited as open-bed. Over 60% of the step particles are arranged in clusters, which reduces the entrainment and mobility of the bed material and thus increase the stability of the segment.

Cross section modifications

The upper segment is affected by substantial bank erosion. Following the largest flood of December 2001 ~ 4 t of sediment were eroded from a channel reach of 20 m long (left bank, Fig. 5). At the same time, no significant changes in channel geometry were identified at the lower section (Fig. 6). The cumulative degradation, following more than 20 floods between 1999 and 2004, was less than 0.4 m at the pool whilst the step endured minor changes of merely a few centimeters (Fig. 6(a) and (b)). In contrast to the upper section, no bank erosion was observed either in the pools or in the steps.

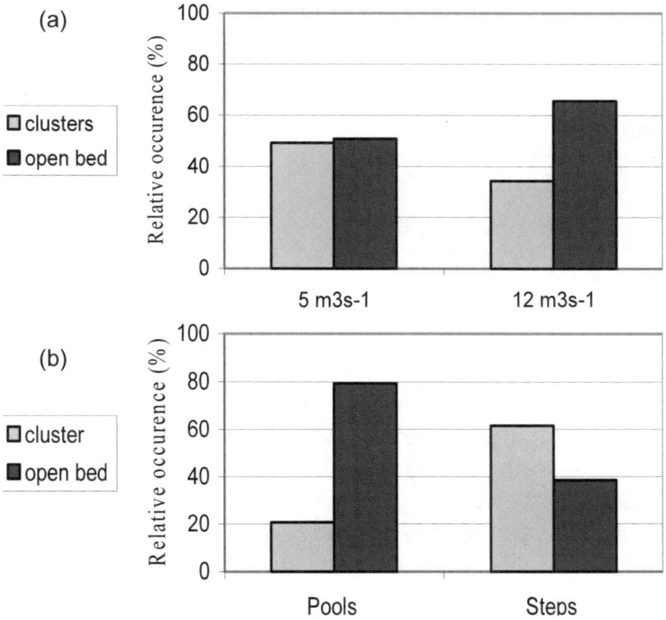

Fig. 4 (a) Relative occurrence of clusters vs. open bed structures at the upper section, following peak discharges of 5 $m^3 s^{-1}$ and 12 $m^3 s^{-1}$ (b) Relative occurrence of clusters *vs* open bed structures at the lower site (pools and steps) following a peak discharge of 5 $m^3 s^{-1}$.

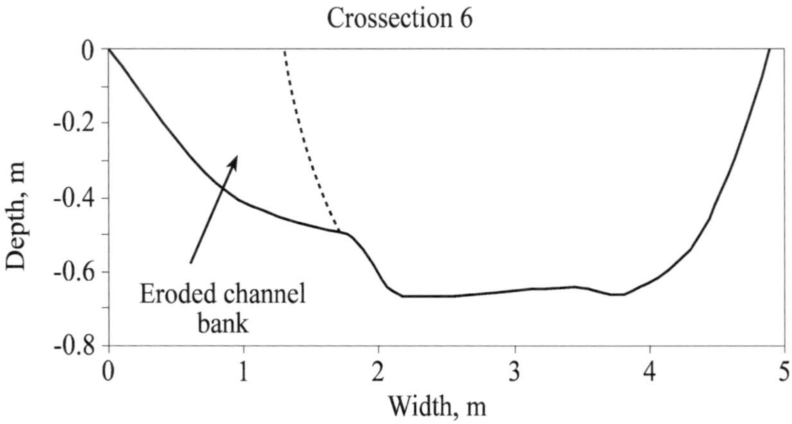

Fig. 5 cross section modification at the upper site due to bank erosion during December 2001 flood (Qp = 12 $m^3 s^{-1}$).

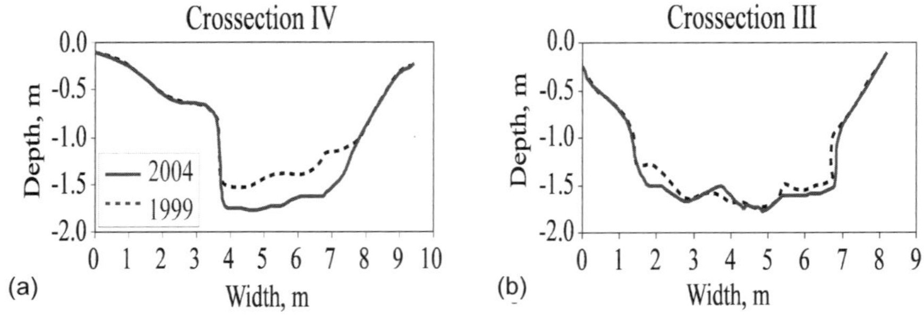

Fig. 6 Cross section modification at the lower site during the period 1999–2004 (a) pool, (b) step.

Entrainment potential

Entrainment potential (Milan *et al.*, 2001) was estimated by subtracting the critical boundary shear stress from local bed shear stress values that were calculated using the hydraulic data. The amount by which τ_0 exceeds τ_c indicates the sediment transport potential. Tables 1–3 summarize the morphologic and hydraulic data for the upper and lower sections. Owing to the differences in grain size distribution, cluster frequencies and channel geometry, the entrainment potential at the upper section, 126 N m^{-2} is considerably higher than that of the steps, 60 N m^{-2} (τ^*_c =0.08) and the pools, 20 N m^{-2} at the lower section. Using $\tau^*_c = 0.03$ for the steps, yields an entrainment potential of 128 N m^{-2}, similar to that of the upper reach. The observed/predict D_{50} ratio is 17% at the upper site, 67% in the steps and 25% in the pools, at the lower site. However, since the pool gradient is the lowest gradient along a stream reach through which mobilized sediment must pass (Zimmerman & Church, 2001) this measure is more appropriate for demonstrating the transport capacity of the step-pool section. Using the observed D_{50} and the corresponding shear stress at the upper site, we calculated the effective flow discharge 0.125 $m^3 s^{-1}$ (the minimal discharge for entrainment of the D_{50} sediment size). This discharge comprises between 17–58% of all flow durations, indicating that each flow event at the upper site has a sufficient effective discharge for transporting up to the median size (~50 mm) of the bed material.

DISCUSSION

Two sections of the mountainous gravel bed Nahal Oren stream were compared to estimate the effect of an artificial reservoir which traps all bedload on the channel bed and to assess the morphological and sedimentological indications for river bed adjustment. The upper section, in addition to the floods, experiences long recessions and base flows with a frequent discharge of 100–200 l min^{-1} owing to contribution from spring water. The lower site has a typical ephemeral regime owing to large transmission losses between the two sections. At the upper, deprived of bedload, section differences between observed D_{50} and calculated D_{50} exceed 80%, indicating

Table 3 Hydraulic characteristic and entrainment potential - upper and lower sections.

Discharge 5 $m^3 s^{-1}$	Parameters	Upper section	Lower section *Pools*	Lower section *Steps* ($\tau^*_c = 0.03$)
Hydraulic radius	R (m)	0.7	0.5	0.4
Gradient m m^{-1}	S	0.022	0.008	0.046
observed D_{50}	D_{50} (mm)	53	39	90
Avg. shear stress	τ_0 N m^{-2}	151	39	180
Estim. D_{50} for avg. shear stress	$*D_{50}$ (mm)	313	81	358
Estim. shear stress for observed D_{50}	τ_c N m^{-2}	25.6	18.8	43.6
Textural fining	$D_{50}/*D_{50}$	0.17	0.48	0.25
Entrainment potential	$(\tau_0 - \tau_c)$ N m^{-2}	126.4	20.2	128.4

high entrainment potential. Each flow produces sufficient competence of 17–58% to entrain sediment sizes up to the D_{50} (~50 mm). During a flood with the peak discharge of 12 $m^3 s^{-1}$, the "starved" flow caused extensive bank erosion which increased the cross-sectional area by about 15% and contributed a mixture of grain sizes to the channel. Short-term response to dam construction typically appears in the form of channel widening following bank erosion (Church, 1995). The eroded bank material was transported downstream and selectively deposited in the step-pool sequence of the lower section. Occurrence of cluster bedforms is indicative of channel stability as pebble clusters influence sediment entrainment, transport and deposition (Reid *et al.*, 1992; Best, 1996). Shear stresses required for the entrainment of clustered material might be twice that necessary to entrain open bed sediments (Lawless & Robert, 2001b). Clusters in the upper section occupy ~40% of the bed material. However, following the December 2001 flood (12 $m^3 s^{-1}$) the occurrence of open bed particles increased to 66% owing to the new input from the banks. This, coupled with the extensive bank erosion, emphasizes the relative instability of the upper section. Approximately 500 m downstream, at the lower section, channel morphology shows indications for relative stability, as defined by the following parameters: (a) step-pool morphology as generally associated with steeper channel gradients between 0.015 and 0.065 (Grant *et al.*, 1990; Chartrand & Whiting, 2000); (b) minor changes in the channel cross-sections and longitudinal profile indicating minimal deposition/erosion following the two large floods and also during the period 1999–2004; (c) about 60% of the step sediments are arranged in clusters, compared to ~40% at the upper site. The difference between the observed and predicted D_{50} is only 50% in the pools and 33–75% in the steps (depending on the value of τ^*_c). Step pool reaches were identified as fundamental and adjustable aspects of channel morphology in steep mountain streams (Chin, 2002). Although step-pools are generally considered to be stable forms with low sediment transport and supply rates (Grant *et al.*, 1990), their stability depends on particle size (Chin, 2002). Compared to the upper contributing section, sediment transport in step pool channels is usually supply-limited and appears to be regulated by the step-pool morphology (Grant *et al.*, 1990). The ephemeral regime of the lower section is well represented by the better sorting of the pool sediments compared to the steps (4.2 and 8.9, respectively). At the upper section, longer recessions and base flows which may exceed several weeks, winnow and transport finer sediment downstream causing poor sorting. The impact of the Pond on the upper section as indicated by channel instability is reduced 500 m downstream, where the site demonstrates relative minor changes based on the step pool morphology, the cluster frequencies and the minor changes in the longitudinal and cross-sectional profiles.

CONCLUSIONS

The construction of a large pond has little effect on the hydrological regime downstream but has a major impact on sediment supply. Sedimentological and morphological comparisons between the two sections, –50 and 500 m downstream of the Pond enables the assessment of channel adjustment to the imposed interference. Our results indicate: (a) both sections experience the same floods, however, large losses of water between the upper and the lower sections forms longer recessions and base flows with frequent discharge of 100–200 l min^{-1} at the upper site. The lower section is typically ephemeral with relative rapid recession; (b) floodwater in the upper section is deprived of bedload, consequently, entrainment potential is higher and the differences between measured D_{50} and calculated D_{50} exceed 80%; (c) each flow produces effective flow ($Q > 0.125$ m^3 s^{-1}) of between 17 and 58% of the flood duration, sufficient to entrain sediment of the D_{50} size (~50 mm); (d) during large flows, significant bank erosion at the upper reach contributed a mixture of loosely deposit material to the channel and a consequent increase in the relative occurrence of open bed sediment; (e) enlargement of up to 15% in the cross-sectional area due to bank erosion was documented at the upper reach *vs* minor channel degradation of <0.4 m at the lower section; and (f) base flows in the upper section winnow and transport poorly-sorted open-bed sediment downstream. The transported particles are selectively deposited at the lower section: the finer material in the pools and the coarser sediment at the steps and (7) the impact of the reservoir is no longer evident 500 m downstream where the channel shows indications of relative stability. Based on our limited dataset, it is evident that the Nahal Oren channel is highly resilient and dynamic equilibrium is rapidly re-achieved. The actual effect of the pond is limited to the immediate section downstream. In spite of the ephemeral flow regime of the Mediterranean climate of Mt Carmel channels, the intensity, duration and mainly the long recession of the flows seems to be sufficient to maintain the stability of the channels.

REFERENCES

Best, J. (1996) The fluvial dynamics of small-scale alluvial bedforms. In: *Advances in Fluvial Dynamics and Stratigraphy* (ed. by P. A. Carling & M. R. Dawson), 67–125. John Wiley, Chichester, UK.

Buffington, J. M. & Montgomery, D. R. (1999) Effects of hydraulic roughness on surface textures of gravel-bed rivers. *Water Resour. Res.* **35**(11), 3507–3521.

Chartrand, S. M. & Whiting, P. J. (2000) Alluvial architecture in headwater streams with special emphasis on step-pool topography. *Earth Surf. Processes Landf.* **25**, 583–600.

Chin, A. (2002) The periodic nature of nature of step-pool mountain streams. *American J. Sci.* **302**, 144–167.

Church, M. (1995) Geomorphic response to river flow regulation: case studies and time-scale. *Reg. Rivers Manage.* **11**, 3–22.

Grant, G. E., Swanson, E. F. & Wolman, M. G. (1990) Pattern and origin of stepped bed morphology in high gradient streams, western Cascades, Oregon. *Geol. Soc. Am. Bull.* **102**, 340–352.

Hassan, M. A. & Church, M. (2000) Experiments on surface structure and partial sediment transport. *Water Resour. Res.* **36**, 1885–1895.

Kondolf, G. M. (1997) Hungry water: effects of dams and gravel mining on river channels. *Environ. Manage.* **21**(4), 533–551.

Laronne, J. B. & Carson, M. A. (1976) Interrelationship between bed morphology and bed-material transport for a small, gravel-bed channel. *Sedimentology* **23**, 67–85.

Lawless, M. & Robert, A. (2001a) Three-dimensional flow structure around small-scale bedform in a simulated gravel-bed environment. *Earth Surf. Processes Landf.* **26**, 507–522.

Lawless, M. & Robert, A. (2001b) Scales of boundary resistance in coarse-grained channels: turbulent velocity profiles and implications. *Geomorphology* **39**(3–4), 221–238.

Marion, D. A. & Weirich, F. (2003) Equal-mobility bed load transport in a small, step-pool channel in the Ouachita Mountains. *Geomorphology* **55**, 139–154.

Milan, D. J., Heritage, G. L., Large, A. R. G. & Charlton, M. E. (2001) Stage dependant variability in tractive force distribution through a riffle-pool sequence. *Catena* **44**, 85–109.

Montgomery, D. R. & Buffington, J. M. (1997) Channel-reach morphology in mountain drainage basins. *Geol. Soc. Am. Bull.* **109**(5), 596–611.

Petts, G. S. (1980) Morphological changes of river channel consequent upon headwater impoundment. *J. Water Engineers Science* **34**, 374–382.

Powell, D. M (1998) Patterns and processes of sediment sorting in gravel-bed rivers. *Progress Physical Geography* **22**(1), 1–32.

Reid, I., Frostick, L. E. & Brayshaw, A. C. (1992) Microform roughness elements and the selective entrainment and entrapment of particles in gravel-bed rivers. In: *Dynamics of Gravel Bed Rivers* (ed. by P. Billi, R. D. Hey, C. R. Thorne & P. Tacconi), 253–276. John Wiley, Chichester, UK.

Wittenberg, L. (2002) Structural patterns in coarse gravel river beds: typology, survey and assessment of the role of grain size and river regime. *Geog. Ann.* **84** A (1), 25–37.

Zimmermann, A. & Church, M. (2001) Channel morphology, gradient profiles and bed stresses during flood in a step-pool channel. *Geomorphology* **40**, 311–327.

Key word index

Author index

SEDIMENT BUDGETS 1 & 2

IAHS AISH

www.iahs.info

*A pair of edited proceedings were produced for an IAHS ICCE symposium co-sponsored by UNESCO, held in April 2005 at Foz do Iguaçu, Brazil, a contribution to the International Sedimentation Initiative (ISI) of the UNESCO International Hydrological Programme (IHP VI). The preface and abstracts of the papers in each volume can be seen at **www.iahs.info**.*

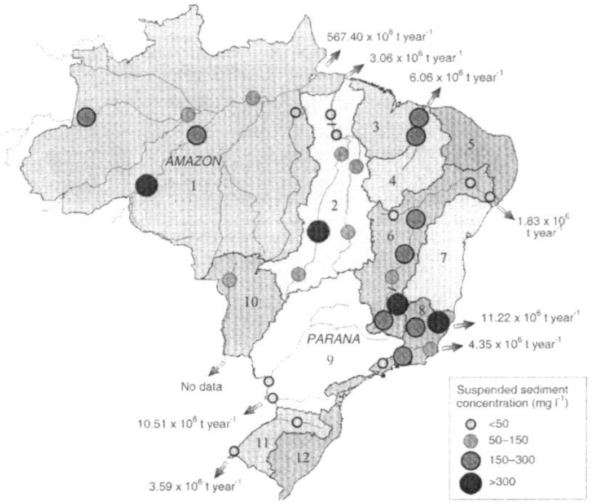

A sediment budget provides an integrated view of the sediment sources, transfers, sinks and outputs of a drainage basin, and in so doing draws together the many different aspects of erosion, and sediment mobilization, transport, storage and yield. Several key developments have influenced recent sediment budget investigations:

– increasing awareness of the environmental implications of increased sediment loads in rivers and of the need for effective sediment management and control strategies.
– advances in the techniques and models available for establishing and predicting catchment sediment budgets including use of ^{137}Cs, ^{210}Pb and ^{7}Be as sediment tracers; continuous recording of suspended sediment flux; application of distributed modelling, remote sensing, and GIS
– recognition of the need to quantify sediment-associated chemical fluxes.

The papers reviewed and edited for these volumes exemplify these developments internationally.

Sediment Budgets 1 *Edited by Des E. Walling & Arthur J. Horowitz*

IAHS Publ. 291 (*April 2005*)
ISBN 1-901502-87-2; + 372 + xii pp.; **Price £63.00**

Volume 1 focuses on the main components of the sediment budget:

– *Sediment mobilization and sediment sources*
– *Sediment transport and transfer*
– *Sediment storage*
– *Sediment yield*

Sediment Budgets 2 *Edited by Arthur J. Horowitz & Des E. Walling*

IAHS Publ. 292 (*April 2005*)
ISBN 1-901502-92-9; + 358 + xii pp.; **Price £63.00**

Volume 2 emphasizes the value of sediment budgets in providing an integrating framework, for both scientific investigations and environmental management:

– *Sediment budgets*
– *Modelling sediment budgets and their components*
– *Human impact on sediment budgets*
– *Sediment problems and sediment management strategies*

Please send book orders and enquiries to:

Mrs Jill Gash
IAHS Press, Centre for Ecology and Hydrology
Wallingford, Oxfordshire OX10 8BB, UK

jilly@iahs.demon.co.uk
tel.: + 44 1491 692442
fax: + 44 1491 692448/692424